Entangled Geographies

Inside Technology
edited by Wiebe E. Bijker, W. Bernard Carlson, and Trevor Pinch

Janet Abbate, *Inventing the Internet*

Atsushi Akera, *Calculating a Natural World: Scientists, Engineers and Computers during the Rise of U.S. Cold War Research*

Charles Bazerman, *The Languages of Edison's Light*

Marc Berg, *Rationalizing Medical Work: Decision-Support Techniques and Medical Practices*

Wiebe E. Bijker, *Of Bicycles, Bakelites, and Bulbs: Toward a Theory of Sociotechnical Change*

Wiebe E. Bijker and John Law, editors, *Shaping Technology/Building Society: Studies in Sociotechnical Change*

Wiebe E. Bijker, Roland Bal, and Ruud Hendricks, *The Paradox of Scientific Authority: The Role of Scientific Advice in Democracies*

Karin Bijsterveld, *Mechanical Sound: Technology, Culture, and Public Problems of Noise in the Twentieth Century*

Stuart S. Blume, *Insight and Industry: On the Dynamics of Technological Change in Medicine*

Pablo J. Boczkowski, *Digitizing the News: Innovation in Online Newspapers*

Geoffrey C. Bowker, *Memory Practices in the Sciences*

Geoffrey C. Bowker, *Science on the Run: Information Management and Industrial Geophysics at Schlumberger, 1920–1940*

Geoffrey C. Bowker and Susan Leigh Star, *Sorting Things Out: Classification and Its Consequences*

Louis L. Bucciarelli, *Designing Engineers*

Michel Callon, Pierre Lascoumes, and Yannick Barthe, *Acting in an Uncertain World: An Essay on Technical Democracy*

H. M. Collins, *Artificial Experts: Social Knowledge and Intelligent Machines*

Park Doing, *Velvet Revolution at the Synchrotron: Biology, Physics, and Change in Science*

Paul N. Edwards, *The Closed World: Computers and the Politics of Discourse in Cold War America*

Andrew Feenberg, *Between Reason and Experience: Essays in Technology and Modernity*

Michael E. Gorman, editor, *Trading Zones and Interactional Expertise: Creating New Kinds of Collaboration*

Herbert Gottweis, *Governing Molecules: The Discursive Politics of Genetic Engineering in Europe and the United States*

Joshua M. Greenberg, *From Betamax to Blockbuster: Video Stores and the Invention of Movies on Video*

Kristen Haring, *Ham Radio's Technical Culture*

Gabrielle Hecht, editor, *Entangled Geographies: Empire and Technopolitics in the Global Cold War*

Gabrielle Hecht, *The Radiance of France: Nuclear Power and National Identity after World War II*

Kathryn Henderson, *On Line and On Paper: Visual Representations, Visual Culture, and Computer Graphics in Design Engineering*

Christopher R. Henke, *Cultivating Science, Harvesting Power: Science and Industrial Agriculture in California*

Christine Hine, *Systematics as Cyberscience: Computers, Change, and Continuity in Science*

Anique Hommels, *Unbuilding Cities: Obduracy in Urban Sociotechnical Change*

Deborah G. Johnson and Jameson W. Wetmore, editors, *Technology and Society: Building Our Sociotechnical Future*

David Kaiser, editor, *Pedagogy and the Practice of Science: Historical and Contemporary Perspectives*

Peter Keating and Alberto Cambrosio, *Biomedical Platforms: Reproducing the Normal and the Pathological in Late-Twentieth-Century Medicine*

Eda Kranakis, *Constructing a Bridge: An Exploration of Engineering Culture, Design, and Research in Nineteenth-Century France and America*

Christophe Lécuyer, *Making Silicon Valley: Innovation and the Growth of High Tech, 1930–1970*

Pamela E. Mack, *Viewing the Earth: The Social Construction of the Landsat Satellite System*

Donald MacKenzie, *Inventing Accuracy: A Historical Sociology of Nuclear Missile Guidance*

Donald MacKenzie, *Knowing Machines: Essays on Technical Change*

Donald MacKenzie, *Mechanizing Proof: Computing, Risk, and Trust*

Donald MacKenzie, *An Engine, Not a Camera: How Financial Models Shape Markets*

Maggie Mort, *Building the Trident Network: A Study of the Enrollment of People, Knowledge, and Machines*

Peter D. Norton, *Fighting Traffic: The Dawn of the Motor Age in the American City*

Helga Nowotny, *Insatiable Curiosity: Innovation in a Fragile Future*

Ruth Oldenziel and Karin Zachmann, editors, *Cold War Kitchen: Americanization, Technology, and European Users*

Nelly Oudshoorn and Trevor Pinch, editors, *How Users Matter: The Co-Construction of Users and Technology*

Shobita Parthasarathy, *Building Genetic Medicine: Breast Cancer, Technology, and the Comparative Politics of Health Care*

Trevor Pinch and Richard Swedberg, editors, *Living in a Material World: Economic Sociology Meets Science and Technology Studies*

Paul Rosen, *Framing Production: Technology, Culture, and Change in the British Bicycle Industry*

Richard Rottenburg, *Far-Fetched Facts: A Parable of Development Aid*

Susanne K. Schmidt and Raymund Werle, *Coordinating Technology: Studies in the International Standardization of Telecommunications*

Wesley Shrum, Joel Genuth, and Ivan Chompalov, *Structures of Scientific Collaboration*

Charis Thompson, *Making Parents: The Ontological Choreography of Reproductive Technology*

Dominique Vinck, editor, *Everyday Engineering: An Ethnography of Design and Innovation*

Entangled Geographies

Empire and Technopolitics in the Global Cold War

edited by Gabrielle Hecht

The MIT Press
Cambridge, Massachusetts
London, England

© 2011 Massachusetts Institute of Technology

All rights reserved. No part of this book may be reproduced in any form by any electronic or mechanical means (including photocopying, recording, or information storage and retrieval) without permission in writing from the publisher.

For information about quantity discounts, email special_sales@mitpress.mit.edu.

Set in Stone Sans and Stone Serif by the MIT Press. Printed and bound in the United States of America.

Library of Congress Cataloging-in-Publication Data

Entangled geographies : empire and technologies in the global Cold War / edited by Gabrielle Hecht.
    p. cm.—(Inside Technology)
Includes bibliographical references and index.
ISBN 978-0-262-51578-8 (pbk. : alk. paper)
1. Cold War 2. Decolonization. 3. Geopolitics. 4. Technology—Political aspects. 5. Technology and state I. Hecht, Gabrielle.
D843.E58   2011
325'.309045—dc22
                                                                                                  2010030321

10 9 8 7 6 5 4 3 2 1

Contents

Acknowledgments   ix

1   Introduction   1
Gabrielle Hecht

2   Islands: The United States as a Networked Empire   13
Ruth Oldenziel

3   The Uses of Portability: Circulating Experts in the Technopolitics of Cold War and Decolonization   43
Donna Mehos and Suzanne Moon

4   On the Fallacies of Cold War Nostalgia: Capitalism, Colonialism, and South African Nuclear Geographies   75
Gabrielle Hecht

5   Rare Earths: The Cold War in the Annals of Travancore   101
Itty Abraham

6   Nuclear Colonization?: Soviet Technopolitics in the Second World   125
Sonja D. Schmid

7   The Technopolitical Lineage of State Planning in Hungary, 1930–1956   155
Martha Lampland

8   Fifty Years' Progress in Five: Brasilia—Modernization, Globalism, and the Geopolitics of Flight   185
Lars Denicke

9   Crude Ecology: Technology and the Politics of Dissent in Saudi Arabia   209
Toby C. Jones

10 A Plundering Tiger with Its Deadly Cubs? The USSR and China as Weapons in the Engineering of a "Zimbabwean Nation," 1945–2009   231
Clapperton Chakanetsa Mavhunga

11 Cleaning Up the Cold War: Global Humanitarianism and the Infrastructure of Crisis Response   267
Peter Redfield

Bibliography   293
About the Authors   329
Index   331

Acknowledgments

The contributions for this volume took shape at a pair of workshops entitled Bodies, Networks, Geographies: Colonialism, Development, and Cold War Technopolitics, held at the University of Michigan (2005) and the Technische Universiteit Eindhoven (2007). But the idea first formed during meetings of the Colonialism, Decolonization and Development group, a branch of the Tensions of Europe network, in the early 2000s. Funding from the National Science Foundation (NSF award SES-0129823), the Nederlandse Organisatie voor Wetenschappelijk Onderzoek (NWO dossier 360-53-020), the Foundation for the History of Technology, and the host universities enabled these various meetings to convene. Harry Lintsen and Tom Misa served as principal investigators for the grants. Harro Maat and Dianne van Oosterhout worked on the NWO project and presented at both workshops. Ruth Oldenziel and Johan Schot offered leadership and insight, and—along with their families—truly remarkable hospitality in the Netherlands. Very special thanks are due to Donna Mehos, who conceived and drafted the NWO grant, organized the Eindhoven workshop, and did much else besides; without her, this project may well not have come to fruition.

In addition to these scholars and other volume authors, discussions at both workshops benefited from the valuable contributions of many participants: Håkon With Anderson, Alec Badenoch, Chandra Bhimull, Stuart Bloom, Peter Boomgaard (also a leader of the NWO project), Charles Bright, John Carson, David William Cohen, Matthew Connelly, Frederick Cooper, Mamadou Diouf, Matthew Hull, Marcia Inhorn, John Krige, Pauline Kusiak, Shobita Parthasarathy, Adriana Petryna, Alexandra Minna Stern, Hans Weinberger, and Michael Wintle. Marguerite Avery and three anonymous reviewers expended considerable time and thought to help us improve the volume. The preparation of the final manuscript relied on the steady encouragement and editorial talent of Paul Edwards, Nina Lerman, Douglas Northrop, and Jay Slagle.

1 Introduction

Gabrielle Hecht

From its earliest days, the Cold War proceeded in uneasy tension with empire. Tensions ran through global disputes over politics, economics, society, and culture. They were also enacted in struggles over technology. Technological systems and expertise offered less visible—but sometimes more powerful—means of shaping or reshaping political rule, economic arrangements, social relationships, and cultural forms. This volume explores how Cold War politics, imperialism, and disputes over decolonization became entangled in technologies, and considers the legacies of those entanglements for today's global (dis)order.

Our project began with an effort to see what insights three domains of scholarship might offer each other:

1. The anthropology and history of development. Scholars have explored the Cold War roots of modernization theory, development economics, and related social science knowledge, revealed the construction of development, poverty, and illness within international agencies, and conducted deep ethnographic and historical inquiries into development projects.[1]
2. Diplomatic history. Moving beyond the minutiae of superpower relations, this field has morphed into "the new international history." Scholars now pursue a global understanding of the Cold War, search for local perspectives on "proxy" wars, examine the dialectics between foreign policy and domestic racial practices in Western nations, and delve into connections between decolonization and Cold War imperatives.[2]
3. Science and technology studies (STS). This multidisciplinary field has largely separated analysis of the Cold War from the study of colonialism. STS scholars have examined how Western and Soviet technoscientific projects shaped Cold War politics and culture, Meanwhile, STS explorations of colonialism—focusing mainly on the period prior to 1940—have analyzed the roles of science, technology, and medicine in colonial practices and structures.[3]

For the most part, these three domains have flourished independently of each other. "Development," "the Cold War," and "technology" have been examined—separately—as entities whose meanings and practices are contested, negotiated, and historical. Scholars who unpack "development" have tended to treat technology as an exogenous force whose ideology might be critiqued but whose material form remains largely unexamined. They historicize "the Third World," rightly locating its invention in a longer historical trajectory but implicitly treating the Cold War as a stable referent—as simply the latest ideological motivation for modernization, rather than as an entity whose specificities are worthy of dissection. In the process, "the Second World" has dropped out of their analyses (or, more to the point, was rarely there to begin with). Scholars of international history, meanwhile, have engaged with the Cold War inflections of modernization theory and development projects, but they tend to focus on moments and places of conception rather than on zones of application. In both literatures, the result is that technology—when it appears at all—looks flat.[4] Put another way, technology seems merely a tool of politics, rather than a mode of politics.

STS, meanwhile, has devoted a great deal of effort to exactly this issue, exploring technology and science as multi-dimensional forms of politics (and culture, and other social forms). Yet to date, when it comes to topics specific to the Cold War, STS scholars have focused on flashy flagships: nuclear weapons, space exploration, computerized command and control. They have shown, for example, how the design of missile guidance systems, computers, and satellites—far from following an internal technological logic—expressed and shaped superpower grand strategy and Cold War culture. Taking such technologies as starting points, however, has made it difficult to stretch geographically—not because these systems don't extend elsewhere, but because the richness of metropolitan archives, the fascination with hegemonies, and the seduction of revealing the hidden politics lurking in large systems all make it seem as though the most important stories remain grounded in the superpowers and in Europe.[5]

All research proceeds by unpacking some processes while holding others constant. Today, the very depth made possible by these various exclusions enables scholars to connect technology, empire, and the Cold War in new ways. Consider, among other examples, Timothy Mitchell's probe into the regional, national, and global technopolitics that shaped modern Egypt; or Michael Adas's exploration of the technological dimensions of American imperialism before, during, and after the Cold War; or Nick Cullather's analysis of the spatial arrangements and narratives enacted by Green

Revolution technologies. By mapping the projection and the practice of power in new ways, these scholars and others force us to reconsider the legacies of the Cold War and of decolonization and their interactions.[6]

This essay collection builds on such efforts to "deprovincialize" the Cold War. How, we ask, did Cold War and postcolonial imaginaries—each with claims to global purview—shape material assemblages? How did such assemblages fuse technology and politics? What strategic—and what unexpected—forms of power did they enact?

Underlying much of our analysis is the notion of technopolitics, a concept that captures the hybrid forms of power embedded in technological artifacts, systems, and practices. In my book on French nuclear power, I used the term to describe the strategic practice of designing or using technology to enact political goals. Such practices, I argued, were not simply politics by another name; they produced systems whose design features mattered fundamentally to their success and shaped the ways in which those systems acted upon the world.[7] Similarly, in describing the rule of experts in twentieth-century Egypt, Timothy Mitchell uses the term "techno-politics" to emphasize the unpredictable power effects of technical assemblages—that is, the unintentional effects of the (re)distribution of agency that they enacted. These two usages are compatible, and this volume embraces both in order to explore a range of ways in which technologies become peculiar forms of politics. Intentions matter, but they are not determinative. The material qualities of technopolitical systems shape the texture and the effects of their power. Technologies can also, however, exceed or escape the intentions of system designers. Material things can be more flexible—and more unpredictable—than their builders realize. The allure of technopolitical strategies is the displacement of power onto technical things, a displacement that designers and politicians sometimes hope to make permanent. But the very material properties of technopolitical assemblages—the way they reshape landscapes, for example, or their capacity to give or take life—sometimes offers other actors an unforeseen purchase on power by providing unexpected means for them to act.[8] Some of us in this volume focus directly on technopolitical practices, while others treat technopolitics more as a heuristic backdrop. Either way, we want to draw attention to how the material properties of technologies shaped the exercise of political power in the second half of the twentieth century.

In most accounts, atomic bombs are the defining technology of the Cold War. The Swedish writer Sven Lindqvist observes, however, that Cold War nuclear imaginaries descended directly from the colonial warfare of earlier eras. For centuries, Europeans had maintained that different moral structures underlay the rules of war for battles between "civilized" nations and conflicts with "savages." In twentieth-century empires, aerial bombs joined machine guns as tools of extermination. Even as ecstatic prophets proclaimed the airplane's ability to ensure world peace, the British experimented with strategic bombing in Baghdad and the French bombarded Damascus. So perhaps it was inevitable that atomic energy (and other late-twentieth-century tools of war) should follow suit. At one stage, Lindqvist invites us into the creepy prescience of a 1920s German science-fiction novel:

> Should atomic power remain in the hands of whites? Or should we share our secret with the peoples of the world? . . . A world conference is convened to settle the question. . . . Licenses should be issued only to dependable people, and only for economic purposes. But immediately voices are raised, accusing Europe of wanting to use atomic power for imperialistic purposes. The conflicts seem endless.
> "They will never stop," says Professor Isenbrandt [an atomic physicist]. "The gulf between the races is too great. No bridge can cross it." . . .
> Quite right: one day some black miners in South Africa gang up on a smaller group of whites and drive them away "for a trifling reason." . . . In Algeria, in Tunisia, wherever blacks are working for European companies, the flag of revolt is raised. The whites are defeated by overwhelming black masses. Then the message arrives that the Chinese are on the move. All the colored races unite under the leadership of the Chinese against the whites.
> Then Isenbrandt explodes his superweapon over the Mongolian masses. "He watched the magnificent spectacle, his work, with the joy of the master. He was the one who had freed the element and bent it to his will. Even now he was filled entirely with the great task of acting as the protector and savior of the threatened colonies."
> "It was wrong," he says sharply, "when our prophets of the past promised the same rights to everyone in the world. Now everywhere on earth the black, brown, and yellow races are calling for freedom. . . . Woe betide us if we grant it! Our power and even our existence would soon be at an end.
> The superweapon will be the white race's, and thus humanity's, salvation. For "only the pure white race can fulfill the task it has been given."[9]

Two decades later, in a Pacific war fraught with racial overtones,[10] several hundred thousand Japanese became the first victims of the "white superweapon." While the Atomic Bomb Casualty Commission industriously erected colonial scientific structures to study the aftermath,[11] the United States, Britain, and France scoured colonies in Africa and elsewhere in a desperate bid to monopolize the magic stuff new stuff of geopolitical power: uranium. "Black miners in South Africa" would be among those who dug it up.[12] Once the weapons were built, the imperial cycle began anew, with atomic bombing—more palatably referred to as "nuclear testing"—of the Marshall Islands, the Sahara, the Navajo Nation, Maralinga, Moruroa, and other colonized spaces.

Nuclear weapons were not the only threads that entangled the Cold War with empire, however. In 1949 US President Harry Truman famously presented his Point Four Program, articulating a vision of how technological progress would help poor "peace-loving peoples" transcend colonialism via capitalist democratization rather than socialist revolution. Yet in practice, much of US foreign policy through the 1950s explicitly supported the maintenance of European empires, delaying decolonization.[13] Meanwhile, the rapidly declining European empires feared that US technological dominance constituted a new form of imperialism, of which they would soon find themselves subjects rather than masters.[14]

At the heart of these (post)colonial Cold War entanglements lay a refiguring of global technopolitical geographies. The "new imperialism" of the nineteenth and twentieth centuries had found legitimation in ideologies that measured human advancement by achievements in industrial technologies and Western scientific and medical practices.[15] Cold War thought and practice turned such justifications into a futurist vision. Prominent Western intellectuals and strategists argued that democracy and technology could work together to offer a fundamentally non-ideological mode of action. Capitalist modernization theory posited that, with the right sort of assistance, any human society could climb the ladder of progress: on each successive rung, industrialization and democratization would proceed hand in hand.[16] The Soviet vision offered a development path that led to socialism through (often large-scale) industrialization. Apart from its rejection of the "free market," however, the Soviet model of progress differed little from the Western one.[17] Through their claims to modernity, both capitalism and communism proclaimed the power to provide rational means of explaining and transcending global inequalities. In both cases, the very claim to rationality depended on an imperial objectification that lumped emerging nations together under the rubric of "underdevelopment."[18]

Both flavors of developmentalism often escaped the boundaries imagined by their promoters. Both would prove seductive for nationalist leaders elsewhere in the world, particularly when accompanied by promises of material and military assistance. Elites in decolonizing nations understood the power of technopolitics, not just in the global pecking order, but also within their new nations. For example, Indian leaders challenged "First World" ownership of nuclear things by proclaiming nuclear development to be a fundamental building block of India's postcolonial national identity.[19] Indonesian officials seeking technical aid for agricultural development resisted the economic models inspired by US Cold War imperatives in favor of their own national and nationalist economic agendas.[20] In Senegal, the state sought to break with a colonial geography of production and export, refiguring national space into development zones, and thus was able to engage in totalizing infrastructural, educational, and production projects.[21]

Even as the "darker nations" affirmed independent historical trajectories—even as their leaders formulated the Third World project[22]—development schemes formed the infrastructure of global entanglements. The establishment of the Non-Aligned Movement did not obviate the Cold War in the South, though it certainly shaped its meanings and power. Technological exchanges between those fully committed to the superpower struggle and those who sought to combat its hegemony made the Cold War inescapable. Sometimes, though, the blind, blundering logic of the Cold War could be subverted or inverted via the very technologies to which it laid claim. Development—in all its multiple meanings and practices—offered post-colonial leaders routes to power not foreseen by Cold Warriors.

Our essays sketch how Cold War ideological struggles, decolonization, postcolonial nation building, and new (or refashioned) imperial projections became entangled in technopolitical projects and practices. The volume as a whole thus contributes to the historiography of what Odd Arne Westad and others have called "the global Cold War." This phrase gestures toward the many relationships among the superpower struggle, decolonization, global inequalities, and imperial difference. The research sites we have chosen—India, Brazil, Saudi Arabia, and South Africa, among others—implicitly align us with Westad's conclusion that "the most important aspects of the Cold War were neither military nor strategic, nor Europe-centered, but connected to political and social development in the Third World."[23] Critics have noted that this formulation forces an unnecessary choice, and that the "Third World's" importance in the Cold War does not obviate that of the superpowers. Clearly, the Cold War's technopolitical legacy remains strong in the North's military-industrial complex, in the structures of its universities and their scientific research, in the enduring environmental and social impacts of weapons production, and so on. While we agree with these critics, we also appreciate Westad's insistence on distributing the political history of the Cold War more widely. His stand serves as a stage for our discussion of its technopolitical history.

We had debates and disagreements along the way, most notably around the notion of (global) Cold War itself. Do we take this as merely a temporal label, and lump everything that happened between the end of World War II and the fall of the Berlin Wall into this slot? Clapperton Mavhunga argues that this forced alignment would deny historical agency to much of the colonial and postcolonial world. Itty Abraham similarly notes that regional histories follow cadences of their own, whose contingencies would be lost if we surrendered to a periodization dictated by the superpowers' hegemonic fantasies. We have thus tried, in a variety of ways, to hold different historical temporalities in tension. The related move of taking the Cold War merely as an etic category (of historical analysis) poses parallel problems. Ruth Oldenziel, Martha Lampland, and Peter Redfield show how this can create a false sense of rupture: an implicit argument that everything changed when the Cold War began, and changed again when it ended. Cold War technopolitics were not created from scratch in response to superpower tensions and the division of Europe. Perhaps paradoxically, understanding their longer histories helps to explain their power, and helps to deprovincialize the Cold War both temporally and spatially. In the end, many of us attend to the Cold War as an emic category, seeking to make visible how historical actors understood, invoked, or deployed it: as legitimation, resource, rupture-talk, organizational logic, or object of contestation.

Our collection trains a technopolitical lens on the Cold War while simultaneously attending to multiple spatial, temporal, and political scales: global, transnational, international, imperial, colonial, postcolonial, national, regional, local. Each essay traces different entanglements among scales. Some of the geographies we outline are centered in places typically considered peripheral to the Cold War. Others suggest alternative maps of polities and technologies typically considered central. Reaching back to nineteenth-century US territorial practices (and forward to those of Bush-era war-making), Ruth Oldenziel rethinks American geography in technopolitical terms. Cold War America, she insists, extended well beyond the continental mainland and European bases and allies: it was technologically distributed in—and dependent upon—islands scattered over the globe. Cold War logic, nuclear and otherwise, imagined these islands as empty. Attending to the forced evacuations that enabled this illusion and its attendant fantasy of a non-imperial US, Oldenziel invites us to contemplate the labor geographies and technological systems that underwrote America's global Cold War thrust.

Nationalisms powered or reinvigorated by nuclear weapons obscured the colonial relationships necessary to their existence. Nuclear states mined their fuel in colonized territories and tested their bombs in imperial waters. My essay suggests how agencies and treaties that sought to define the global nuclear order, such as the International Atomic Energy Agency and the Nuclear Non-Proliferation Treaty, claimed to temper Cold War moral injunctions with postcolonial ones. Yet the specter of planetary destruction conveyed a certain temporal and material urgency that could serve as a powerful trump. The nuclear imperatives that drove the permanent removal of Kwajalein residents, discussed in Oldenziel's essay, also supported South Africa's efforts to build international legitimacy while remaining the West's last colonial power. Portraying nuclear development, "the market," and their relationship as apolitical terrains unsuited for anti-colonial claims-making, the apartheid state crafted commercial circuits that entangled its uranium with American, European, and Japanese nuclear systems. I argue that such entanglements reverberate into the present.

Even during the Cold War, nuclear technopolitics sometimes took unexpected turns. Itty Abraham suggests that in the struggle over the decolonization of the Indian subcontinent, the presence of radioactive thorium in the sands of Travancore ended up subverting—rather than supporting—the possibility of that state's independence from India. Much as Frederick Cooper has argued that Africa's current nation-states were not the only option for the political organization of the continent at the moment of decolonization, Abraham shows that the possibility of Travancore's statehood was very real, thanks to thorium. Only later did the Indian atomic leadership's appropriation of the rare earth cut off that possibility. Sonja Schmid turns her gaze on relationships between the Soviet Union and East Germany and Czechoslovakia, analyzing the Soviet transfer of nuclear power and expertise to these countries as uneven colonial practice. The Soviets were reluctant to share their technology, fearing that their technopolitical leverage would be undermined if secrets and materials leaked out. When they eventually did engage in technology transfer, the Soviets sought tight control over the terms of the exchange. India's leaders, it turns out, were not alone in seeing nuclear technology as a motor for nation-building; so too did Eastern European leaders. They deployed socialist ideologies of progress to argue for ever-greater access while also developing indigenous expertise. Ultimately, Schmid shows, the strategy to refashion Soviet-supplied power plants into instruments of nuclear nationalism—and political and economic power—met with very different fates in East Germany and Czechoslovakia. Both Abraham and Schmid demonstrate how the material dimensions of nuclear things shaped larger political possibilities (without fully determining them).

It was not just in nuclear matters that Cold War technopolitics embodied settlements between the ideals of universality and the inescapability of locally specific conditions. Donna Mehos and Suzanne Moon tackle this theme head-on in their exploration of portable knowledge. The logic of colonial rule, they argue, privileged place-based knowledge. Decolonization—the formal transfer of sovereignty—destabilized the economic value of such expertise. When political upheaval sent Western corporations packing, agricultural specialists and others who had made their careers in colonial companies had to find ways of delocalizing their knowledge, of making it portable. One result was the emergence of globally oriented consulting companies. But making knowledge portable, Mehos and Moon argue, was more than simply a strategy for economic survival: in the corridors of UN agencies, portable knowledge became central to the technopolitics of internationalism. A central mission of the UN, technical assistance was never just about the aid itself; it was also, always, a gesture toward transcending Cold War politics. This dynamic played out in the composition and the practices of aid teams. In a bizarre twist, some construed subalternity itself as a universalizing epistemology, so that an expert from Haiti was thought to have privileged insight into a project in Afghanistan simply by virtue of having witnessed poverty first-hand. Martha Lampland echoes this theme in her analysis of state planning and development economics. She treats these bodies of expertise as scientific instruments and forms of technopolitics, which slipped across the grand divide between capitalism and communism both before and during the Cold War. The planners who managed Hungary's transition to socialism had acquired their expertise by developing scientific management and economic theory in the interwar period. Their technopolitics camouflaged the capitalist antecedents of the practices they used to build a socialist state. Attending to state planning as technopolitics, Lampland concludes, makes visible the parallels between Second World and Third World state-building.

As instruments of rule that distributed agency across material assemblages, technopolitical strategies certainly proved seductive to many who aspired to state-building. We see this time and again in these essays, from Sir C. P. Ramaswamy Aiyar in Travancore to Haile Selassie in Ethiopia; from apartheid leaders in South Africa to Robert Mugabe in Zimbabwe; from the ministries of the Saudi kingdom to the air-borne fantasies of Brasilia. Inverting the perspective offered by Oldenziel, Lars Denicke considers globalized Cold War geographies from the standpoint of Brazil, a potential node in the planetary networks formed by air travel. Brazilian leaders hoped the technologies of flight (not just airplanes, but also airports) would allow them to claim stakes in Cold War geopolitics. This possibility provided the inspiration for re-mapping the nation-space to make a new capital and a new modernity. Reshaping territory, hope for a new place (both spatial and political) in a new world order: these certainly contributed to the seduction of technopolitical strategies. But, as other authors show, so did the displacement and redistribution of power. By turning the social and political difficulties of rule into problems of resource and labor management, statemakers hoped to neutralize dissent.

Reshaping technopolitical geographies could have unpredictable effects—consequences not necessarily inscribed in the design of systems or readily apparent in the territorial rearrangements they enacted. Toby Jones argues that the technopolitical gesture itself—the attempt to sublimate social difference via the al-Hasa irrigation project—helped to catalyze Shi'i dissent and led to the 1979 rebellion whose story frames his essay. The Saudi state may have found resources for rule lurking in Cold War dynamics, but the subjects of rule understood this gesture all too well. Their dissent drew vigor not just from the material devastation of the irrigation system, but also from the Cold War alliances on display in the project. Freedom fighters in late colonial Rhodesia, meanwhile, were no mere pawns of Sino-Soviet power. Adopting an expansive understanding of "weaponry," Clapperton Mavhunga argues that nationalists inverted the technologies and rhetorics provided in the name of the Cold War to their own ends, to engineer Zimbabwe out of the infrastructure and detritus of colonialism. Nationhood, he emphasizes, was just the beginning: as president, Robert Mugabe continues to reconfigure these tools and rhetorics to violently consolidate his power.

Finally, lest we feel overly tempted to scoff at the hubris of experts, Peter Redfield reminds us of the virtues of displaced, materialized, portable expertise—while not losing sight of its limits. His discussion of Médecins sans Frontières signals the significance of portable medical kits in coping with the violence engendered by, during, and after the Cold War. The kits distributed expertise, experience, and an ethics of global intervention into technical assemblages, ready to expedite at a moment's notice. Such portability, however, constituted merely a technopolitics of emergency: a momentary intervention, rather than a permanent solution to political violence.

For many Americans, the final death knell of the Cold War was sounded not by glasnost, or the fall of the Berlin Wall, or even the formal dismantling of the Soviet Union, but rather by the explosions of September 11, 2001. US Secretary of State Colin Powell called the moment the post-post-Cold War: rupture squared. In the moment of tragedy, calling attention to the ironies and continuities represented by the attacks (that the attackers exploited weaknesses inherent in American technological systems, that they had learned to fly in American flight schools, that they had learned terror tactics in camps instituted by the United States during the Soviet-driven Afghan war) seemed tantamount to treason. The authors in this volume, however, suggest that if we are indeed in a post- or a post-post-Cold War world, then it's the same kind of "post-ness" that we find in the "post-colonial." In other words, the infrastructures and discourses of global Cold War technopolitics continue to shape the possibilities and limits of power, just as the infrastructures and discourses of empire do. Technopolitical assemblages are not static. By enacting historical and geographical entanglements, they continually generate new effects and new meanings. This book excavates the roots of these technopolitical entanglements in the hope of better understanding their power and their peril.

Notes

1. Cooper and Packard 1997; Cooper and Stoler 1997; Escobar 1994; Ferguson 1990; Gupta 1998; Scott 1998.

2. Among many examples: Borstelmann 2001; Connelly 2002; Westad 2005; Von Eschen 1997.

3. For the first mode, see Collins 2002; Edwards 1996; Krige 2006; Leslie 1993; Lowen 1997; Wang 1999. For historiographic overviews of works on science, technology, and empire, see Arnold 2005; McLeod 2000.

4. Some of the contributions in Ong and Collier 2005 offer exceptions to this observation, but they are concerned more with present-day "globalization" than with the longer historical moment surrounding the Cold War.

5. Though not impossible: see Hecht and Edwards 2008.

6. Abraham 1995; Adas 2005; Connelly 2008; Cullather 2004; Hecht 2006a; Leslie and Kargon 2006; Mitchell 2002; Moon 2007; van Oosterhout 2008.

7. Hecht 1998/2009.

8. The relationship between technology and political power is an old and important theme in historical and social analysis, dating back at least as far as Karl Marx's discussions of means of production as a controlling factor in political economy and continuing on into the twentieth century with public intellectuals such as Lewis Mumford and Jacques Ellul. The modern field of Science and Technology Studies has vastly expanded and deepened such reflections. Landmark studies include (among many others) Bijker et al. 1987, Hughes 1983, MacKenzie 1990, and Winner 1986.

9. Plot summary of Hans Dominik's *The Trail of Genghis Khan, A Novel of the 21st Century* in Lindqvist 2001, section 128.

10. Dower 1987.

11. Lindee 1994.

12. Hecht 2009.

13. Borstelmann 2001; McNay 2001.

14. Kuisel 1993; de Grazia 2005; Krige 2006.

15. Adas 1989.

16. Latham 2000; Gilman 2003; Engerman et al. 2003; Engel 2007.

17. Adas 2005; Westad 2005.

18. Cooper 2005; Cooper and Packard 1997; Escobar 1995; Mitchell 2002.

19. Abraham 1998.

20. Moon 1998.

21. Diouf 1997.

22. Prashad 2007.

23. Westad 2005: 396. A similar argument, albeit with different empirical emphasis, lies at the core of Mamdani 2004.

2   Islands: The United States as a Networked Empire

Ruth Oldenziel

In the spring of 2003, US President Bush, British Prime Minister Blair, Spanish Prime Minister Aznar, and their host, Portuguese President Barroso, landed on the island of Terceira to hold a press conference and present Iraq with an ultimatum for war. The location of their press conference in the Portuguese Azores—a constellation of nine Atlantic islands far from mainland Europe—puzzled commentators. Reporters speculated that this far-flung setting, best known as an exotic holiday destination rather than a convincing projection of US power, symbolized the marginal European support for an invasion of Iraq.

In ways that commentators did not realize, the choice of the Azores brought into focus a projection of American power rooted in networks and islands. Washington's geographical selection was not a remote launch pad for war, but a manifestation of power that often and purposefully remains hidden from view.[1] The Azores transformed from a remote Portuguese outpost into a US hub of information, communication, and military systems. The process of transformation reconfigured the Azores and several other colonial island chains into nodes in the American projection of power. That process fundamentally remapped the globe during the Cold War.

The character of American power has been widely discussed. The historian Arthur Schlesinger Jr., articulating the Cold War consensus, once argued that the United States, though "richly equipped with imperial paraphernalia [such as] troops, ships, planes, bases, proconsuls, local collaborators, all spread around the luckless planet," should be understood as an "'informal' empire, not colonial in polity."[2] That argument became a dominant narrative frame during the Cold War as the US faced ideological competition from the USSR.[3] Since then, others have refined the idea to claim the US is a reluctant empire, an empire by invitation, or the world's indispensable nation.[4] All of these arguments turn on the notion that the US wields a strikingly different kind of power because it lacks overseas possessions. Indeed the US does not occupy vast tracts of land outside the American continent like the Roman, British, and Russian empires of yore. But the US does rule over extensive—but to its citizens, invisible—island possessions.[5] The US territories include thousands of islands in the Commonwealth of Puerto Rico, Guam, American Samoa, Johnston Atoll, Navassa Island, Micronesia, Marshall Islands, the Commonwealth of the Northern Marianas, Palau, and the US Virgin Islands of St. Thomas, St. John, and St. Croix. Most of these possessions have been in US hands for more than a century. These US territories are the largest of the post-colonial era, exceeding the combined population of the overseas territories of Britain and France.

The residents of these islands are second-class citizens lacking the full protection of the law as well as federal voting rights and representation in the US House, Senate, and Electoral College. In their political limbo, garment workers in the Northern Marianas sew "Made in America" labels for American clothing companies such as The Gap, Wal-Mart, Liz Claiborne, and Calvin Klein while receiving 60 percent of the US minimum wage. Puerto Ricans are subject to the death penalty under US federal law, even though their commonwealth law forbids it. Islanders may serve in the US military yet compete in the Olympic Games and beauty pageants as nationals from countries distinct from the US.[6] The islands and their residents belong to, but are not part of, the United States.

America's territories are modest in size, their 4,000 square miles barely larger than the state of Connecticut. But the small size of islands like the Azores and Marianas masks their political, economic, legal, and technical weight. The islands in US domain have been critical nodes in multiple global networks. Home to capital-intensive, low-labor-intensive technologies, islands have helped to nurture America's self-image as a post-colonial, post-imperial power in the era of decolonization and globalization. They also have bridged exceptionalist American history and European colonial history.

This essay casts the United States' islands possessions as a narrative anchor in an alternative cartography of Cold War paradigms by looking at the configurations of large global systems. Skirting along the edges of empire, it seeks to understand the function of islands during a time when both the US and the USSR disavowed territorial expansion as a matter of ideological principle. By looking through the lens of technology, this essay offers an alternative view to the characterization of the US as an informal, deterritorialized global power. It thus anchors the Cold War in the technopolitical geographies of islands to understand how archipelagic areas like the Portuguese Azores have become the central nodes of US global power.

Where in the World Is America?

America has an uncanny ability to be both everywhere and nowhere, omnipresent yet deterritorialized. The historians Charles Bright and Michael Geyer pose the intriguing question "Where in the world is America?"[7] Indeed, it is the geographical and ideological location of America's exercise of global power that I explore here, using an iconic episode: the Spanish-American-Cuban War of 1898. After US President William McKinley received news from Admiral George Dewey of the naval victory in Manila Bay, he admitted he "could not have told where those darned islands were within 2,000 miles."[8] Probably apocryphal, this tale of a president's inability to locate islands that his military was poised to conquer nonetheless brings to mind the geographical illiteracy endemic to Americans. The anthropologist and geographer Neil Smith notes that this malady stands in stark contrast to the extraordinary resources that the US government has spent gathering geographic intelligence. The Department of State, the Central Intelligence Agency, the Department of Defense, and the National Security Agency all have maintained departments staffed with geographers. Collaborating with these agencies, the National Imagery and Mapping Agency represents the geographic nervous system for US global strategy, Smith argues.[9] It is no coincidence that the citizens of today's superpower have difficulty locating lands over which their country exerts power. It has been a matter of policy.

To understand this history we need to go back to the British, who invested in islands to lay the foundation for their globally networked power. The British Empire, combining the principles of the old landed empire and the new networked empire, ruled by way of land masses like India and oceanic nodes. Its policy makers pioneered two closely intertwined technopolitical foundations of naval strategy anchored in islands: a global network of underwater communication cables and a comprehensive chain of coaling stations. British politicians and engineers learned that oceanic cables making landfall at small islands offered a far more efficient communications system than short lines strung across hostile countries.[10] Submerged cable infrastructures foiled wire-cutting insurrectionists and circumvented unreliable regimes like the Ottomans and the Egyptians.[11] The British obsession with implementing nationally controlled lines through island possessions eliminated interference from other countries.[12] The construction of a network of coaling stations for refueling, repair, and trade route protection laid the second technopolitical foundation rooted in islands.

This British technopolitical model became an article of faith for American expansionists. The naval theorist and historian Alfred Thayer Mahan

(1840–1914) argued that the nation with the strongest navy would dominate seas and markets.[13] The US should therefore construct both a powerful navy and a chain of stations to provide for coaling, supplies, and repairs. The political will of Senator Henry Cabot Lodge and his friend President Theodore Roosevelt transformed Mahan's theory into a grand strategy. The US already possessed facilities at Midway (1867), Samoa (1878), and Pearl Harbor on Hawaii's O'ahu Island (1887). Within two decades of the Spanish-American-Cuban War, the US built a navy second only to Britain's, constructing Pacific and Caribbean naval nodes for control of ocean spaces.[14] As Mahan had envisioned, the war enabled the US to acquire sites for coaling stations and underwater cable nodes through strategically placed islands that remain (with the exception of the Philippines) in US possession: Cuba's Guantanamo Bay, Puerto Rico (with its strategically important islets, Culebra, Vieques, and Mona), Guam, and American Samoa. In 1917 the US bought the Virgin Islands from Denmark to complete the chain of coaling stations. Expansion through control of the ocean—politicizing and militarizing oceanic space—thus created a global system of international relations in which islands, peninsulas, and littoral spaces played a key geopolitical role.[15]

In what was not a foregone conclusion, Americans went on to perfect this form of global power building. The US was, at the outset, a commercial and coastal territory that fully participated in the oceanic world. In the nineteenth century, the US became preoccupied with territorial conquest, building its nation around a vast continental homeland and incorporating territories like Oklahoma (1907) and New Mexico (1912). Large infrastructures of canals, roads, railways, post, and telegraphy helped forge an internal cohesion as land-starved European nations looked to overseas expansion.[16] During this nation-building phase, representing the American West as virgin and uninhabited became crucial for America's national identity as a republic. When the US reversed its geopolitical orientation in the 1890s, a cognitive gap developed. The myth of a domesticated American West now vied against a reinvigorated projection of American power overseas.[17]

The US Supreme Court's legal frame in the Insular Cases (1902–1922) marks the first indication that the technopolitics of islands helped bridge the cognitive gap between America's self-representation as a republic and its projection of imperial ambitions. The representation of islands as liminal yet crucial sites occupied center stage in the court's legal discussion. At the core of the debate was the status of the 5.2 square miles of Navassa Island in the Caribbean. In 1889 the Supreme Court heard the case of African-American laborers contracted by a Baltimore company to scrape guano (fertile deposits of bird feces used in commercial agriculture) off the island's rocks. Venting rage against inhumane conditions and lethal abuse, the workers killed five white supervisors. The court readily accepted the death penalties sought by the prosecutor. The question that occupied the justices was whether the US had jurisdiction over this outlying guano island. The court developed the argument that islands like Navassa "belonged to" but were "not part of" the United States. This ruling exerted a lasting influence following the 1898 annexation of Puerto Rico, the Philippines, Cuba, and Guam. Its legal precedent undergirded the Insular Cases, which established that an "unincorporated territory" was to be "foreign to the United States in a domestic sense," its inhabitants neither aliens nor citizens.[18] This doctrine ended the automatic incorporation of territories and enabled further expansion by introducing separate-but-equal status for overseas territories, exempting them from full legal rights.

After World War II, the ambiguous language of the Insular Cases set the stage for absorbing many other archipelagos into US possession. America took as bounties from the defeated Japanese and German empires the Marshall Islands, Palau, and the Northern Marianas. Despite America's professed distaste for colonialism, most of these islands remained in its possession and resembled old-fashioned protectorates in a time of decolonization.[19]

Anchoring Islands for War during the Era of Decolonization, 1931–1945

Island technopolitics further helped the United States build a deterritorialized empire that was based on global communications along with air, nuclear, space, and other technical systems. World War II provided the underpinnings for Cold War expansion when the US enacted a technopolitical regime and an ideological discourse of anti-imperialism and democracy that was more systematic than in previous decades. This discourse resulted in political preferences, military strategies, and design choices of technological systems that shaped each other profoundly.[20] Franklin D. Roosevelt first steered America's old naval and air interests clear of the ideological strong winds of decolonization. He used the island territories to bridge the gap between imperial expansion and the new demands of decolonization. As a teenager, he had avidly studied his uncle Theodore's work and Alfred Thayer Mahan's naval theory. As Assistant Secretary of the Navy (1913–1920) under Woodrow Wilson, Theodore Roosevelt put into practice the new philosophy of the US as an island power. He realized Mahan's vision of a chain of stations by supervising the construction of bases, shipyards, and other facilities. When he became president, the geographically astute

Roosevelt helped to revise navalism for an era of air power. He also updated the imperial thrust for an era of decolonization.[21]

Roosevelt's administration devised a legal basis for turning islands into stepping stones of a combined naval and air strategy, revising the geographic logic of large technological systems well before World War II.[22] The Department of State dusted off the Guano Island Act to justify claims for islands that by the 1930s had become commercially useless.[23] This 1856 law, which resulted from farm interests' lobbying Congress to address the British monopoly on the wonder fertilizer of Peruvian guano, gave full government support to any American entrepreneur who found and claimed an uninhabited island worth mining.[24] To neutralize critics, the law promised that the US would relinquish jurisdiction over these islands when the guano was exhausted or the claims were abandoned.[25] By the 1930s, not only had artificial fertilizer supplanted guano; the Department of State had discovered that the islands' legal statuses were in disarray despite the provision. The US government quickly placed more than twenty private claims on various islands (including Howland, Jarvis, Baker, and Johnston) under US federal (but unincorporated) jurisdiction. The US thus laid the legal foundation for building civil and military aviation systems almost a decade before the Japanese attack on Pearl Harbor.

When war broke out, the islands were ready for further incorporation into the American orbit. Military engineers transformed Hawaii, Midway, Wake, Johnston, Palmyra, and other obscure Pacific islands into stepping stones. With similar lightning speed, the barren Aleutian Islands were also pressed into strategic service.[26] Using what it called an "island-hopping" strategy, the US military leapfrogged from one Pacific or Aleutian island to the next, bypassing Japanese strongholds, cutting Japanese supply lines, and starving out the stranded Japanese troops.[27] The term "leapfrogging" sounded innocent, but the strategy left damaging footprints. The Navy's construction battalions of civilian contractors and engineers, known as the Seabees, built naval and air bases. For airstrips the engineers preferred flat, cleared, and cultivated spaces that often coincided with the best farmlands. Appropriating farmland displaced agricultural laborers, whose skills were then mobilized for the construction projects. Blowing up coconut palms to level the ground and dredging up coral reefs to build runways, the construction battalions transformed landscapes within weeks. They left as quickly as they had come. This military-industrial machine operated like an assembly line. From Bechtel's Calships wharfs at Terminal Island in Los Angeles Harbor to Hawaii and on to Midway, an average of 112 base facilities were built per month. The effort dwarfed all earlier ones.[28]

The technopolitics of islands allowed the US to straddle two ideological roles when the global expansion justified by war became politically problematic during peacetime. One role was the US as a self-contained, anti-colonial homeland. The other was America as a fully engaged superpower that claimed no interest in overseas territories. Already during the war, American policy makers, intellectuals, and historians shunned words of conquest such as "expansion," "colonies," "dependencies," and "protectorates," preferring instead the terms "territories," "commonwealths," "insular areas," and "outlying areas."[29] One advisor urged the British colonial office to avoid colonial terminology and speak about US overseas possessions in terms of union, self-government, and federation.[30]

The Roosevelt administration fully exploited the islands' ideological value in service to America's self-representation as a non-colonial—even anti-colonial—power. While the Department of State put intense pressure on the British to dismantle their territories, the US Navy and the Joint Chiefs of Staff saw many strategic advantages in occupying the many Pacific islands under British control. The Navy nevertheless understood the radically changed political climate and Roosevelt's anti-colonial stance. "We cannot allow ourselves to be charged with imperialism," said Admiral Richard Byrd. Yet, Byrd noted, the 130 Pacific islands presented an unique opportunity that "may never come again for a comprehensive far-flung chain of bases." He argued that the US occupation of the islands should not be construed as territorial expansion: "None of the islands in question possesses natural features of value from other than the military standpoint . . . [and therefore] cannot constitute territorial aggrandizement." Although the islands possessed no economic value and were, more importantly, "empty," they were crucial for air routes, landing fields, and combined "commerce and political and military strategy."[31] In 1945, Secretary of War Henry Stimson and President Harry Truman repeated the semantically self-serving position that islands should be considered prospective military bases, not annexed or colonized territories. The annexed islands, Stimson insisted, "are not colonies; they are outposts, and their acquisition are appropriate under the general doctrine of self-defense."[32]

The war and the anti-colonial pressures also established a US preference for leasing instead of annexing territory. The Department of Defense believed that the strategy helped to solve the explosive issue of colonialism, enhancing "our reputation for integrity of international agreement and traditional lack of imperialistic ambition."[33] On the basis of the Destroyers for Bases and Lend-Lease agreements in 1940 and 1941, for example, the US had assimilated an imperial infrastructure by taking over British ports in the North Atlantic and the Caribbean on a 99-year rent-free lease.[34] The Americans also forced the British to give up their monopoly on global communications lines. In effect, the US used the wartime cover of "partnership" to take over the networked part of the British Empire—its chain of bases and communication systems—but left the territorial pieces alone.[35] This deal gave the British enough nominal sovereignty to maintain their fantasy of empire while promulgating the fiction of an anti-imperial US power. After the war, the US military faced political pressure to bring troops home and shut down half of its bases. Peace forced the Navy to abandon its ambition of keeping an "Offshore Island Perimeter." Instead, the Department of Defense began to make separate deals, negotiating with Denmark for base rights on Greenland, with Portugal for rights on the Azores, and with Iceland and Britain for rights on additional territories. George Kennan articulated a foreign policy of containment that helped justify the Navy's "forward strategy" encircling the Soviet Union and China.

However, by the mid 1950s, with the Cold War in full swing, US military planners believed the situation had become acute when the old colonial powers were no longer capable of holding down the fort against independence movements. As David Vine has shown in his fine study of Diego Garcia, the newly established Long-Range Objectives Group at the Pentagon articulated a comprehensive "Strategic Island Concept." In view of "anti-colonist feelings or Soviet pressures," the Department of Defense systematically looked for "strategically located, lightly populated, isolated islands still controlled by friendly Western powers." Planners believed that "remote colonial islands with small [colonial] populations would be the easiest to acquire, and would entail the least political headaches."[36] Stu Baker, the author of this strategy, urged the US to stockpile base rights before these islands became independent nations. A race ensued to rack up as many islands as possible before independence movements could take the helm of the local political machine.

Thus in the mid 1950s the US embarked on a systematic policy to lease hundreds of islands, peninsulas, and littoral spaces from declining empires and emerging nations. These agreements—supported by a legal framework of extraterritoriality that asserted the right to apply laws beyond a nation's territory—henceforth set the format for America's global arrangements with other nations.[37] They bolstered the legal basis for nonterritorial forms of American expansion, representing, in the words of one scholar, "a floating island of American sovereignty."[38] Constructing islands as demographically empty, geographically "thin," and economically worthless—but strategically vital and legally "thick"—helped the US fill the cognitive gap between its self-representation as a republic and its ambition for empire in a new era. The Cold War reified the strategic claims into a discourse of anti-imperialism, anti-colonialism, democracy, and capitalism.[39] As we will see, large global technical systems anchored in putatively empty islands provided the connective tissue between these opposing goals.

Erasing Space, Filling Technology

The legal and geopolitical mobilization of islands into the American orbit took a specific technological shape that would dominate the postwar era: emptying out space by filling the islands with technologically intense systems that obscured the political imprints of the United States. A closer look at three territories demonstrates how the island chains physically anchored the technological systems that wired the US into a networked empire. In the Azores, in Kwajalein, and in Diego Garcia, the US displaced local peoples, supplanting their presence with layers of technological systems—nodes in a global network of power.

"The First Cold War": The Azores and Kwajalein

The transformation of the Portuguese Azores symbolizes the beginning of what historians have called "the First Cold War." A strategic nineteenth-century node for communication and coal, the Azores became a temporary base during World War II, remaining in US orbit henceforth as a logistical centerpiece of the First Cold War.[40] After weaning the Portuguese dictator Antonio Oliveira Salazar from the Axis powers, the US built naval bases and airports on the Azores that were not completed until one week after Germany signed the peace accord. Nevertheless, the bases proved immediately useful. The airport on Maria Island was used to shuttle more than 50,000 troops home from Europe, and Lajes airport on Terceira Island was used to secretly divert aircraft from Europe to the Pacific during the final months of the war.[41] Then in 1948, just as they were about to close down, the Azores bases were pressed into service as a logistical link in the Berlin Airlift.[42] Two years later, the US pressured the Portuguese government for a permanent US military presence never to leave the American domain again, as the 2003 press conference pressuring Iraq into war testifies so well.

Kwajalein, one of many Pacific islands mobilized for the nuclear age, served as a Cold War proxy on the other side of the globe. Seeking to counterbalance the Soviet Union's supremacy in manpower, the United States saw its nuclear monopoly as a capital-intensive and knowledge-intensive investment that could replace troop power with a technological system.[43]

This was not the only way in which labor was displaced. Colonized, recently decolonized, or tribal lands had become the Western power's favored testing grounds for nuclear weapons and other controversial technologies.[44] Americans favored testing outside their borders, using distant colonies where populations were sparse and the political costs minimal. Meanwhile, the Soviet Union tested bombs within its borders.[45] Because of the atomic bomb, the remote Pacific islands emerged as a nuclear laboratory founded on colonial relationships. As the Cold War accelerated, many of the old guano islands (e.g., Johnston Atoll) and the newly annexed archipelagos (Bikini, Enewetak, Kwajalein) came to operate as offshore labs and testing sites for chemical, biological, and nuclear technologies.[46] Of these, Kwajalein best represents the fate of the islands in the transition to the nuclear age.

First, Kwajalein had to become a lab. It was scientifically and politically critical to "empty out" islands for nuclear experimentation by forcibly removing the population. Appropriating islands for military purposes found its most tragic precedent in the Bikinis in the late 1940s. The US Navy relocated the residents of Bikini and Rongelap to the atoll of Kwajalein before using the islands in nuclear testing programs that continued until 1958. The center of US nuclear and ballistic testing then shifted to the 93 islands of Kwajalein. Residents of the Marshall Islands, Bikini, Enewetak, Rongelap, Rongerik, and Utirik sought compensation from the US government. The politics of "emptying out" spaces to fill them with "pristine," high-tech, prestigious, but geographically "thin" technologies were repeated around the world. The peoples of the Aleutian Islands in Alaska, Vieques and Culebra in Puerto Rico, Thule in Greenland, and Okinawa in Japan were removed from their islands. Once emptied out, the islands were filled with technological systems for similar geopolitical purposes in the postwar era.[47] These are well-known stories. Here, I focus on the less familiar, but perhaps more significant labor geographies of the technical systems that sustained the Cold War struggle. The geographically "thin" technologies had their photo negative in the "thick" labor-filled sites that sustained them but remained hidden.[48] This geographic division of labor came to characterize the particular US exercise of global power during the Cold War, as the examples of Kwajalein and Ebeye show.

The crowded Pacific isle of Ebeye metaphorically orbits Kwajalein at the center of the US nuclear and missile program in the Marshall Islands. Kwajalein is an atoll officially listed as uninhabited except for the residing military personnel. After serving as a naval base during World War II, the island became a temporary settlement for Bikini and Enewetak islanders displaced by nuclear testing.[49] More than a decade later, Kwajalein—along with Wake Island—was transformed into a missile testing ground vital to President Reagan's missile defense initiative. It also hosts the ground station for the US Navy's NAVSTAR, which spun off into the commercial and hugely successful Global Positioning System (GPS), making it a technologically thick-layered place. In preparation for tests of the Nike Zeus and Nike X weapons systems, Kwajalein islanders and the already displaced Bikini and Rongelap residents were moved to neighboring Ebeye Island, whose small population grew to several thousand amid slum conditions. To create an extra "mid-corridor" for the missile-testing site, the US military also relocated the people of Roi-Namur, Lib, Meck, Lagan, and Ningi to Ebeye. Then the US government cordoned off Kwajalein for military personnel only.

Ebeye is the gritty mirror image to Kwajalein's pristine technology. With 12,000 people, the 80-acre Ebeye serves as the missile site's overcrowded camp for non-US workers. The island's population included the original Kwajalein residents, the expelled residents from the Bikini archipelago, and skilled Micronesian workers. The workers in US employ were ferried to Kwajalein and back to Ebeye each workday, but were prohibited from shopping, eating, swimming, and using the library at Kwajalein's facilities. This commuter workforce was also forbidden from taking highly valued consumer goods off the well-stocked island. A journalist testifying before Congress in 1984 compared the living conditions of the two neighboring islands thus: "Kwajalein is like . . . one of our Miami Resort areas, with palm-tree-lined beaches, swimming pool, a golf course, people bicycling everywhere, a first-class hospital and a school; and Ebeye, on the other hand, is an island slum, overpopulated, treeless filthy lagoon, littered beaches, a dilapidated hospital, and contaminated water supply, and so forth."[50] The acting director of the Department of the Interior's Office of Territorial Affairs observed a broader pattern. He equated the island of Ebeye with the many other labor ghettos that had sprung up around American military installations throughout the world.

Indeed geographically "thin" and technologically sophisticated islands like Kwajalein cannot be understood without the correlates that "track" them, the labor-intensive and "thick" sites like Ebeye. A chain of seedy camp towns runs through South Korea, Guam, Okinawa, Palau, Ologapo, and beyond. Here, poor women and war orphans eke out a living as prostitutes, entrepreneurs, and criminals looking to earn dollars. Sanctioned by the US government, these are places of "rest and relaxation," in the official parlance of the American military. For the women who offer their sexual favors in neon-lit bars, massage parlors, and discos, they are places of work.[51] In short, places like Ebeye are the photo negatives of America's technologically thick military-industrial complex.

"The Second Cold War" and Beyond: Diego Garcia

The technological layering that symbolizes the transition to "the second Cold War" is even better illustrated in the waters of the Indian Ocean, midway between Africa and Asia, on the British atoll of Diego Garcia. Situated along oil-shipping lanes in the Chagos Archipelago, this 66-square-mile paradise lies around a large lagoon that is shaped like a footprint on a beach. In November 1968, a contingent of four American geographers, five Filipino technicians, and their cook arrived on Diego Garcia to install a tracking station as a part of the global Satellite Triangulation Program sponsored by the US Department of Defense. The military geographers found a vibrant Creole plantation economy living without electricity, telephones, or postal service, but also an abundance of lobsters that the crew caught in the island's water and preserved in a self-powered freezer to sell to passing ships.[52] Until then off the electrical grid, the island of Diego Garcia was fully wired within a decade. The island came to host a plethora of ground stations for global systems. It has served as a launch pad for Special Forces for many of their military actions, including the failed 1980 mission to rescue hostages in Iran. It was a base for American B-52 and B-2 bombers in the Gulf War of 1991 and the Iraq War of 2003. Recently, British officials have admitted that Diego Garcia has been one of the CIA's infamous "black holes" where suspected terrorists have disappeared without a legal trace. At least one US ship has been "used as a floating prison for high-profile prisoners while it was in the vicinity of Diego Garcia."[53]

Like the Bikinis and Kwajalein, Diego Garcia is listed as uninhabited except for its 2,000 military personnel. It has been deliberately constructed as politically empty so that it could be loaded with technologically complex systems. In the early 1960s, in search of suitable sites for its Strategic Island Concept, the US demanded that the British "sweep" and "sanitize" the Chagos Islands. Only then could America turn Diego Garcia into a node in its global power network. The British readily obliged by eliminating any local opposition and expelling over 2,000 residents. To provide the basis for the US-UK agreement, the British government created the legal fiction of the British Indian Ocean Territory (BIOT) in 1965. This arrangement transformed the residents of Diego Garcia—French-speaking British subjects who had worked on the island's coconut plantations for five generations—into temporary contract workers originating from Mauritius and Seychelles. The BIOT scheme, which nullified the Mauritian claim on the Chagos

Figure 2.1
A Diego Garcia coconut worker photographed by a US geographer on the eve of population's removal from the island in 1968. Source: NOAA Geodesy Collection. Courtesy of NOAA, Washington.

Archipelago in exchange for Mauritian independence, paved the way for a US base at Diego Garcia a year later. British policy makers designated the Diego Garcians as "a floating population" of migrant workers before expelling them through a policy of harassment, starvation, and deportation. Residents who left the island for medical reasons or to visit family elsewhere in the archipelago were not allowed to return, for example. The remainder were forcibly removed. Using the exhaust fumes from military vehicles, a manager in the employ of the British killed all the island's dogs and donkeys. The policy was intended to prevent the residents from claiming that they were an "indigenous population" while keeping them useful as a workforce.[54] On the eve of their deportation, the last islanders helped incoming American geographers and Filipino technicians to unload kits filled with parts for the global satellite system.

Having removed the island's native population, the US rapidly began to fill Diego Garcia with technical systems. The Department of Defense, in collaboration with the Coast and Geodetic Survey, established satellite camera observatories as part of a global network to compete with the Soviet Union. In choosing locations for satellite ground stations, government engineers

Figure 2.2
The last Diego Garcians help US engineers to unload the material for the base camp for the Triangulation Satellite Program, 1968. Courtesy of NOAA, Washington.

hewed to political rather than scientific and technological demands. The worldwide satellite triangulation program mapped the shape of the earth using sets of two island stations that photographed satellites against a fixed background of stars. Marketed as a science project, it was strategic from the beginning. The program promised to cover the earth's surface in mathematically perfect triangles. But in execution, the project neatly reflected Cold War political geography by adjusting the global net to the political map. This geographic distribution did not match the scientific maps the agency proudly presented in its public relations campaign. In reality, the points of the triangles faithfully followed the islands under American control, including Maui, Puerto Rico, Guam, Samoa, Tinian, Wake, and the Aleutians. Also mapped were the strategically important territories fading from the grasp of America's reliable allies: British Diego Garcia, St. Helena, Ascension, and Tristan da Cunha; Bermuda; French Seychelles; and Australian Christmas Island (now Kiritimati). Touted as a truly global system, the project in fact represented a Cold War geography. For example, the engineers failed to cover the Soviet Union and China because they lacked access to stations needed to complete the triangles.[55] Moreover, the construction

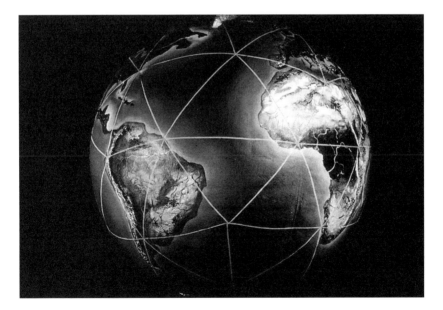

Figure 2.3
Artist's impression of the US Triangulation Satellite Program suggesting the project covered the earth's surface in mathematically perfect triangles, 1968. Source: NOAA Geodesy Collection. Courtesy of NOAA, Washington.

BC-4 WORLD PRIMARY NETWORK

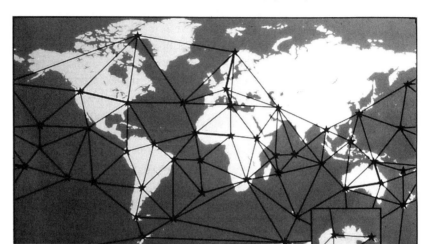

Figure 2.4
Grounded in islands, a map of the Satellite Triangulation Network as a Cold War reality without perfect triangles and excluding China and the Soviet Union, 1968. Source: NOAA Geodesy Collection. Courtesy of NOAA, Washington.

of the global satellite triangulation system during the 1960s used many of the islands under US jurisdiction as earth stations even when their locations did not make scientific sense or failed to generate the perfect triangles presented publicly.[56]

The satellite triangulation program that brought America to Diego Garcia laid the groundwork for subsequent ground stations of the Echelon spy network, NASA's Mercury Project, and the Global Positioning System, for example. As a node of several global networks, Diego Garcia linked myriad technical systems. For example, the global surveillance system Echelon brought together the British and American systems, personnel, and stations under a secret 1947 agreement. The British Commonwealth countries of Canada, Australia, and New Zealand joined in the American-British network and were followed by Norway, Denmark, Germany, and Turkey.[57] So, too, Echelon's spy network anchored its ground stations on the islands of Guam, Kunia, Hawaii, Diego Garcia, and the Japanese-controlled Iwakuni. In the late 1950s, NASA's Mercury Project sought to put a man into orbit around the Earth. For its earth stations, the Mercury Project relied on Diego Garcia, Cyprus, Canton, and Enderbury. During the 1970s, the earth links for the Global Positioning System (GPS) were located on Ascension Island in the Atlantic, Diego Garcia in the Indian Ocean, and Kwajalein and Hawaii in the Pacific. In each instance, the islands chosen fell in the extraterritorial domain of US jurisdiction and power.

The satellite triangulation program and its many successors were part of American-based espionage, space exploration, and satellite systems thus anchored in an island empire that had come into being over the course of a century.[58] Again and again, these large Cold War technical networks were grounded in colonized islands in an era of decolonization. In all these projects, satellite systems were linked closely to submarine warfare. Technological systems included ocean acoustics, deep sea bathymetry, and satellite altimetry of sea surfaces.[59] These complex technopolitical nodes (and their commercial spin-offs) did more than that. They integrated oceans, airways, and outer space into a single system under US global command.

The global networks were not just part of the struggle between the superpowers. The meaning of "the global" varied according to political context of these technological systems. During the 1950s, satellite ground stations outside Western Europe often opened with great fanfare. The stations served both as symbols of American success in the superpower struggle against the Soviet Union and as arguments against British colonialism. Commonwealth nations and decolonizing countries believed that American domination of global networks promised to circumvent the British colonial stranglehold on communication systems and to provide a symbol of national independence.[60] For their part, the British sought to prolong their empire by subscribing instead to "the global." In the case of Diego Garcia, for example, the British invested heavily in symbols to claim their sovereignty, issuing commemorative island stamps, flying the Union Jack, arranging a visit by the Duke of York, and preserving old plantation buildings at East Point.[61] The investment in the imperial symbols of British sovereignty could not mask the *de facto* status of Diego Garcia as a US territory, however.

Diego Garcia well represents the Cold War; it also prefigures the post-Cold War world. After its initial buildup in the guise of a communication and geodesic tracking center (1966 and 1973), the tiny island of Diego Garcia became the strategic answer to President Nixon's search for bases free of political headaches like those associated with Vietnam. By then Diego Garcia served as the US hub for an updated Cold War strategy, the so-called Second Cold War.[62] The island was the pivot point in the Carter administration's plan to protect America's access to Persian Gulf oil after the 1973 oil crisis, the Soviet Union's invasion of Afghanistan, and the hostage crisis in Iran.[63] The Reagan administration, which likewise declared Persian Gulf oil a vital American interest, used the island as a springboard to project American power. In Pentagon doctrine, that projection of power hardened in the design of Diego Garcia as an "empty" island, eventually filled with complex and layered technological systems that spanned the globe.

Diego Garcia, which the Reagan administration called "the footprint of freedom," has also become the model for future bases. The island transformed into a mobile invasion kit to alleviate the military's dependence on vast German- and Korean-style US bases or politically instable regimes. The naval kit consisted of the Marine Amphibious Brigade: seventeen fully loaded vessels, including cargo ships that were "packed with all the supplies needed for a Middle East invasion, already loaded into trucks. Everything right down to water tankers for thirsty troops."[64] The principle of the kit—a mobile self-sustaining system for a limited time—was to provide enough supplies and spare parts to allow the integrated naval and air unit to operate for 90 days without external support. Flexible, integrated naval and air kits were designed to roll out a complete war machine within days. Military planners designed these war kits to eliminate the dependence on local politics and geography altogether. Diego Garcia became the model. Although these mobile kits, in combination with long-range flight, airborne refueling, and massive aircraft carriers, seemed to signal the end of the usefulness of the geographical positions of islands like Diego Garcia as anchor points for US power, nothing is further from the truth.

Islands as Boundary Objects of the Networked Empire
Islands went through careers of sorts. Once technologically useful, they lay dormant at times before being pressed into use for novel exploits.[65] For example, the nineteenth-century geographic logic that demanded a chain of island coaling stations became obsolete when the US Navy turned to oil for power. Samoa became a backwater.[66] Midway and Guam, used as landfalls for underwater cables during World War I, lost out to radio soon after.[67] The advent of air power changed the geographic logic once again. Guam and Midway were re-enlisted as stepping stones for civilian and military air travel during World War II. The coming of long-range flight, airborne refueling, and massive aircraft carriers threatened to render the Azores obsolete as a transatlantic stopover. The logical conclusion came when the Reagan administration launched a shipbuilding program to free the US of military bases tied to territories by developing units that could roll out as an invasion kit. Even though in each instance the technical and geographic logics changed, the political rationales for keeping islands within the US orbit remained remarkably stable over the course of

a century or so. Technical obsolescence rarely resulted in abandonment or restoration of sovereignty.

Instead it was the extraterritorial status replete with legal vagaries that made islands so politically desirable. On the map of decolonization, islands were not simply specks on the globe or solitary dry surfaces in a vast ocean, but alluring, brightly colored colonial thumbtacks. Far removed from centers of power on the most peripheral of peripheries, most archipelagos have been located at the center of major twentieth-century historical events.[68] These dots on the map have allowed America to continually renounce territorial ambitions while expanding to become the sole global power after the Cold War. One could well argue, as Chalmers Johnson has, that the US is not an empire of islands but one of bases connected through a military chain of command lacking civilian oversight. It is an empire in the business of maintaining absolute power, controlling communication through eavesdropping stations, preserving economic control of petroleum flows, and reproducing an institutional income system for the military-industrial complex. Its network of bases maintains an extraterritorial comfort zone with social and medical benefits that include clubs, apartments, gyms, golf courses, swimming pools, and shopping malls—amenities often inaccessible to the ethnic minorities and lower classes in the continental US.[69]

Yet islands are not just privileged sites for employees of the military-industrial complex. As the legal scholar Christine Duffy Burnett has argued, the guano islands operated as imperial boundary objects that could expand and contract as needed.[70] The same flexibility applies to the thousands of islands now in US possession and to the many bases and littoral spaces not formally under US dominion. The legal and technopolitical moorings of islands have helped US power to expand, contract, and change as cultural movements and political administrations have waxed and waned. Many islands transformed into novel extraterritorial spaces, some even turning into engines of globalization that seemed to have little to do with military bases. At the height of the Cold War, for example, only two Export Processing Zones (EPZ) and Free Trade Zones (FTZ) existed; by the year 2000 that number had exploded to 800.[71] Samoa, once a coaling station, currently serves as America's exclusive tuna-processing zone. The American Virgin Islands and Saipan in the Northern Marianas, no longer just military sites, are free-trade zones for the garment industry and transnational Internet companies. American companies, paying their workers wages well below US standards, maintain sweatshop conditions in these territories that are exempted from American wage and immigration laws and from US tariffs.[72] Other Exclusive Economic Zones (EZZ) are eagerly explored for their mineral resources. These extraterritorial zones are neither hollowed out nor adjusted to the nation-state, as some critics of globalization have feared. The zones have permitted the US to exercise sovereignty with its techno-military apparatus while supporting the demands of the free market. In fact, these liminal spaces have helped sustain the military's need for strengthening the American nation-state while also meeting the demands of corporate America.[73]

Other islands have become precious—if precarious—havens of biodiversity. Given the protracted clean-up of nuclear and military sites, the irony is that many of these islands have turned into environmental showpieces at once pristine and polluted. The Pacific's former guano islands Jarvis and Midway, as well as the Caribbean island of Navassa, now host the NASA earth-mapping projects advertised as balm for the planet's ecological woes. The listening stations of the Sound Surveillance System (SOSUS), which once tracked Soviet submarines, now eavesdrop on migrating whales.[74] Ecological concerns have been mobilized to justify closure of public lands and to restore a semblance of sovereignty in some locales. Vandenberg Air Force Base in California, from which many space shuttles and satellites were launched, boasts many endangered species. The British government recently established a society dedicated to the protection of Diego Garcia's environment and history. In the words of a high-ranking British official, the US military and the island's remoteness had luckily "spared the impact of mass tourism and factory fishing, the environmental banes which are despoiling more and more of the rest of the Indian Ocean."[75]

Islands, in short, went through many careers but nevertheless firmly remained within the US domain to become the anchor points of a worldwide, interconnected, and integrated system of nodes. This network has not only aided in sustaining US democracy but fostered a global gulag kept off the official map. The legal distinction made by the US Supreme Court in the death penalty case of the Navassa workers (1898) served as the basis for the status of the Philippines, Puerto Rico, and Guam a decade later. It also made possible the imprisonment of "enemy combatants" at Guantanamo in 2002.[76] Guantanamo Bay, as Amy Kaplan correctly argues, rather than the exception, is the rule for how the US has exercised its power. Thousands of other archipelagic spaces under America's domain become visible only when events rupture the powerful narrative of America's deterritorialized power. Anchored in "empty" islands, the reach, power, and prestige of large technical systems—from the first telegraph communications to current outer space systems—have come to replace territorialized empire as an indicator of geopolitical power.[77]

Anchoring the Cold War: Historiography Revisited

A more complete understanding of the geopolitical locations of large technical systems also has consequences for our scholarship, however. No doubt for political reasons, many of the earth stations and technology systems for outer space are named to obfuscate their geographic locations. For instance, Anderson Military Base often stands in for the island of Guam. This erasure replicates military protocol of secrecy and army habits of community building. It follows the long-standing colonial practice of appropriation as an act of power as demonstrated in the renaming of the Hawaiian Island of Kalama as "Johnston Island." The habit has also percolated into scholarship. In the history of technology, technical systems like the Satellite Triangulation Project, the Mercury Project, the Geographical Positioning System, the Strategic Defense Initiative, and OAO-2 are often analyzed as geographically neutral systems difficult to pinpoint on a map. Even if, as we have seen, high-tech systems map faithfully to island possessions in America's domain, historians have generally ignored the colonial contexts. Similarly, theories of networked societies often portray these arrangements as disembodied entities in a new transnational arena.[78]

A focus on islands offers a fresh reading in the wide-ranging public debates on America's specific exercise of global power. For one thing, it renders visible the geographical moorings of the technopolitics that displaces workers through capital-intensive, labor-poor technologies, launches proxy wars, supplies ground stations for space systems, and provides corporations with havens of cheap labor. Many of the island groups have transformed into critical nodes in exclusionary, globe-spanning systems that make up America's networked empire. Geographically, this island empire is indeed "thin" and "invisible" but technologically "thick." Moreover, focusing on the edges of the American empire helps explain why Cold War historiography exhibits both a persistent difficulty of grasping the territorial basis of America's global power and a recurrent self-definition of the US as an exceptional world power.[79] The discourse of deterritorialization of America's global position has been a powerful narrative indeed.

In the latest incarnation of territorial blindness, Joseph Nye, a scholar of international relations and a Clinton administration official for foreign policy and global markets, introduced the term "soft power" to characterize America's disembodied, deterritorialized exercise of global power, insisting that this form of power is far more important than the hard power of military might.[80] Even Michael Negri and Antonio Hardt present the US as the alternative to the classic territorial model, calling America a networked—not a landed—empire across an unbounded terrain.[81] The historian Charles

Maier correctly criticizes Negri and Hardt's metaphorical scripting of the networked empire as one without actors or institutions. Yet even Maier concludes that US power should be considered a post-territorial empire of production and consumption.[82] This weak understanding of US power vis-à-vis its global geography also inflects the discourse of American Studies.[83] It casts America as a topos without geography, an engine of capitalism that is everywhere and nowhere, a cultural production that is ubiquitous and elusive. The lack of geographic precision is thus deeply ingrained in the representation of America's global position. Perhaps it is not so ironic that Americans have trouble pinning down the US even though most of the world-spanning projects of the Cold War involved intense global mapping. As we have seen, the contradiction has been a matter of policy and design.

We need to understand that the US is an empire grounded in networks stretching across the globe and masked by islands. Technical nodes of global networks have been purposefully anchored in politically weak regimes on islands that are strategically constructed as empty to support the notion of a deterritorialized American power.[84]

The successful construction of an island's emptiness devoid of political headaches has nevertheless come back to haunt. In 2000, after years of litigation, the expelled people of Diego Garcia unexpectedly won their right in Britain's highest court to return to their home. In response, the Blair government annulled the court's judgment in 2004, arguing that the US military's occupation made the ruling a *de facto* impossibility. In March 2006, the British tried to escape the awkward situation by allowing more than a hundred Chagossians to pay respects to their forebears at the graves on the island's East Point Plantation. The expelled Creole inhabitants, accompanied by two priests, a stonemason, a doctor, a nurse, and a British official, held a mass and left behind a memorial marker. Oliver Bancoult, who had been forced to leave the islands as a boy, led the Chagos Refugees Group. "This is not the end of the matter," he vowed. "We maintain our objective of returning to live in our birthplace." To avoid further embarrassment, the British prohibited journalists from reporting on the occasion. Two weeks later, the US Supreme Court denied the defendants their right to return to Diego Garcia. Still, the story did not die. In May, the British High Court dismissed the Blair government's annulment. In 2008, the UK government overturned the court's judgment on appeal. The Chagossians now hope the expiration of the lease in 2016 will offer them the opportunity to return.

Once in a while, other displaced peoples of technologically thick spaces—e.g., the Inuit of Thule or the original people of Okinawa—rupture the narratives of Western newspapers and demand their right of return to

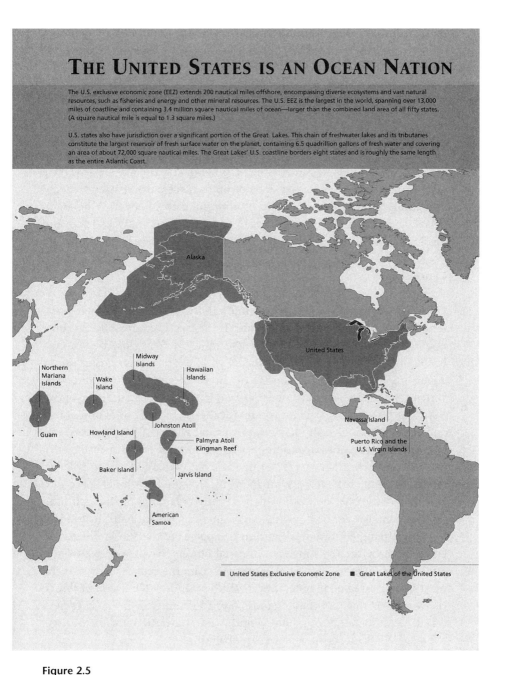

Figure 2.5
The Exclusive Economic Zone (EEZ) of the United States in the Pacific comprises 3.4 million square nautical miles, an area larger the land area of the fifty states. Source: http://aquaculture.noaa.gov.

the places that barely register with the rest of the world. We may dismiss these struggles as insignificant hiccups in the global scheme of things. But what islands lack in land mass they make up in political and strategic significance. The prime value of these liminal spaces lies in their insular and extraterritorial status—a status reinforced over the last decades with the expansion of the 200-mile island radiuses that have become US Exclusive Economic Zones. These zones encompass 3.4 million square nautical miles, an area about 20 percent greater than the entire land area of the US.[85] Even the square miles in land and oceanic mass fail to map the real expanse if we take into account these dimensions of outer space encompassed by technological systems. Strung together in powerful global networks, the archipelagic areas offer the American nation-state extraordinary political and ideological flexibility for an era of decolonization recast in Cold War terms.

Conclusion

To the American nation-state, the prime value of islands lies in their extraterritoriality and their offshore status. These island groups have become nodes in exclusionary, globe-spanning technical systems closely connected to the military hardware of a networked empire. The insular prison wards, guano mines, and coaling stations of the nineteenth century were converted to twentieth-century nodes for underwater cables, military operations, nuclear testing, satellite communications, and off-shore processing, and even into biodiversity havens. With their weak political systems, low environmental standards, and improvised labor laws, islands became critical nodes in technical systems during the Cold War. The Satellite Triangulation, Echelon, and Mercury programs reconfigured territorial space into an integrated global network. This alternative mapping of the technical, political, and economic topography of America's islands helps us to identify the striking continuities but also to appreciate the subtle differences between European and American imperial powers throughout the twentieth century. Spread over vast expanses of ocean, the islands are cast as the most peripheral of peripheries. Together, they render US power invisible to the world. Yet many of these islands played a central role in the twentieth century's most important events: World War II, the birth of the nuclear age, the Cold War, and the next wave of globalization.

It is striking how effective America's island empire has been. Island infrastructures and networks have connected ocean floors, littoral areas, and outer space. Most of all, they have remained off the political map. The phenomenon has buttressed the national myth of preserving a continental nation or even a federal republic while projecting a global vision of democratic principles. No matter what fundamental incompatibility exists between its insistence on its identity as an exceptional nation and its desire to spread universal values, the US has used its island empire to resolve the problem. The paradox has become a technopolitical reality.

Acknowledgments

I would like to thank Stuart Blume, Charles Bright, John Cloud, Lars Denicke, Park Doing, John Krige, Maria Paula Diogo, Peter Redfield, Damon Salasa, Ana Paula Silva, Erik van der Vleuten, and the members of the Cornell University STS seminar and the TU Eindhoven for their comments and suggestions. In particular, my gratitude goes to Gabrielle Hecht and Nil Disco for their extensive comments on various drafts, to Jay Slagle for his fine editorial hand, and to John Cloud for locating the rich photographic collection at NOAA in Washington.

Notes

1. Ed McCullough, "Azores: A Safe, Convenient Place to Decide War and Peace," Associated Press, March 16, 2003.

2. Schlesinger 1986: 141.

3. Williams 1959. See also Williams 1969; Williams 1980; Perkins 1962.

4. Lundestad 1986.

5. For studies comparing the United States with the Roman and British empires, see Ferguson 2004; Johnson 2006; Mayer 2006.

6. Sparrow 2006: 212–214; Clarren 2006; Stuart 1999: 7.

7. Bright and Geyer 2002.

8. LaFeber 1963: 361; Smith 2003: 1.

9. Smith 2003: 3.

10. Kennedy 1971; Headrick 1991: 24. See also Hugill 1999.

11. Cited in Headrick 1991: 98; Maurer 2001: 474.

12. Dewey to Secretary of the Navy, May 7, 1898, Department of the Navy, Navy Historical Center; Kennedy 1971: 732, 738, 740–743; Lieutenant Cameron McR. Winslow, "Cable-Cutting at Cientguego, May 11, 1898," *The Century* 57, 5 (March 1899); Headrick 1981; Headrick 1999.

13. Mahan 1890; LaFeber 1963: 90–95; Kelly 2003: 357–359.

14. Seager 1953: 508–510; Buhl 1974.

15. Mancke 1999; Bender 2006, chapter 1.

16. LaFeber 1963.

17. See Bright and Geyer 2002; Hietala 2003[1985].

18. Kaplan 2005: 841; Burnett 2005: 791; Sparrow 2006; Ferguson 2004: 48; Fairchild 1941: 109.

19. Sparrow 2006, chapters 8 and 9.

20. The Soviet Union remained a land-based empire and relied on manpower with no particular interest in American-style global communications or maritime strategies.

21. See also Bacevich 2002, chapter 2. On Franklin Roosevelt, see Louis 1978, passim.

22. Smith 2003: 359.

23. Nichols 1933.

24. Nichols 1933: 506; Orent and Reinsch 1941: 458.

25. Cited in Burnett 2005: 785.

26. Perras 2003. Kaho'olawe in Hawaii served as the training ground for Iwo Jima and Vietnam; Puerto Rico's Isla de Culebra (1939–1975) and Vieques (1975–2004) sustained troops for almost seven decades.

27. Firth 1997: 312–313; MacLeod 2000: Introduction.

28. Firth 1997: 312–313; Bechtel 1998: 40–41.

29. The British had a Colonial Office; the Americans called theirs the Office of Territories. Stuart 1999: 15; Howe et al. 1994: 229.

30. Louis 1978: 111; Smith 2003: 359, 409.

31. Cited in Louis 1978: 265–269.

32. Hanlon 1998: 44–45; Howe et al. 1994: 229; Gaddis 2005: 41.

33. Louis 1978: 265–269; see also Smith 2003: 359, 409.

34. Gerson 1991: 11.

35. Headrick 1991, chapter 14. See also Sanders 2000.

36. Vine 2009: 41–43. Vine reconstructs the emergence of the concept in his finely researched and just published study of Diego Garcia. In chapter 3, he comes to the same conclusion as this essay, which was written before its publication.

37. The British government repaid the loans only in 2006 ("Britain to Make Its Final Payment on World War II Loan from US," *International Herald Tribune*, December 28, 2006). On other extraterritorial arrangements in Germany and South Korea, see Gerson 1991: 16.

38. Ruskola 2005; Silva and Diogo 2006. For similar agreements and loss of sovereignty, see Johnson 2006, chapters 4 and 5; LaFeber 2003: 30; Headrick 1991; Bello 1991; Kent 1991; Kennedy 2004.

39. See Williams 1980: vii, ix.

40. Silva and Diogo 2006. Before the introduction of weather satellites, they served as gathering and transmission stations for meteorological data for European weather forecasting. Lepgold 1997; Lajes Field, "US. Air Force Fact Sheet. Lajes Field History—Santa Maria and Beyond" (available at http://www.lajes.af.mil); Headrick 1991.

41. "US Air Force Fact Sheet."

42. In 1973, when all European countries denied the United States access to their bases it used the Azores to support Israel in the fourth Arab-Israeli War.

43. Maier 2006: 196–197; Gaddis 2005: 35–36, 66–67.

44. Hecht and Edwards 2008. On reproductive technologies testing on Puerto Rico's poor and ill-educated women for the pharmaceutical industry's in clinical trials because of less restricting laws and researchers' colonial attitudes, see Oudshoorn 2003.

45. Firth and von Strokirch 2003: 324.

46. Nero 1997.

47. Vine 2009: 65–68.

48. Hecht and Edwards 2008: 25.

49. The following is based on chapter 7 of Hanlon 1998.

50. Cited in Hanlon 1998: 201.

51. Enloe 2000[1989], chapter 4; Moon 1997. Johnson 2004: 6–7.

52. Kirby Crawford, "The Very First Americans," at http://www.zianet.com.

53. Jamie Doward, "U.K. Island 'used by US for rendition,'" *The Observer*, March 2, 2008.

54. Pilger 2004a,b; Winchester 2001; Edis 1993; Vine 2009, chapters 6 and 7. On Diego Garcia's strategic position, see Walker 1991: 41; Doyon 1991.

55. US Department of Commerce 1966: 37–44. See also photographic archives chronicling the project's construction at www.photolib.noaa.gov; Schmid 1974;

Reilly et al. 1973; Satellite triangulation in the Coast and Geodetic Survey, Technical Bulletin 24 (Government Printing Office, 1965); Berkers et al. 2004: 32.

56. US Department of Commerce, *The Coast and Geodetic Survey*: 38–44.

57. Cloud 2002; Duncan Campbell, "Inside Echelon: The History, Structure and Function of the Global Surveillance System Known as Echelon," at http://echelononline.free.fr.

58. The G. W. Bush administration's policy toward extending it to outer space is outside the scope of this essay, but see chapter 3 of Johnson 2004.

59. Johnson 2004, chapter 3; personal communication with John Cloud, September 15, 2003.

60. Headrick 1991, chapter 14; Slotten 2002: 349.

61. Edis 1993: 87.

62. Halliday 1986.

63. Odom 2006.

64. Walker 1991: 41. For an elaboration on the kit, see Peter Redfield, this volume.

65. I thank Park Doing, Nil Disco, and Lars Denicke for helping to articulate these techno-geographical logics.

66. DeNovo 1955; Howe et al. 1994: 245.

67. Headrick 1997: 3–7; Headrick 1991, chapters 7–10; Redfield 2000: 120–122.

68. Hanlon 1998: 1. Hau'ofa, "Our Sea of Islands" cited in Nero 2003: 441.

69. Johnson 2000; Johnson 2004: 7; Johnson 2006. See also Evinger 1998.

70. Burnett 2005 passim; Kaplan 2005.

71. Palan 1998.

72. Clarren 2006; Maurer 2001.

73. For the archipelagic nations the Exclusive Economic Zones (EEZ) regulating sea zones and exploitation rights of marine sources (fishing and oil drilling) is crucial. The US pushed hard to protect its own natural resources in 1945 when President Truman extended US control over the continental shelf. Between 1946 and 1950, several Latin American countries also extended their sovereign rights to a distance of 200 nautical miles to cover their Humboldt Current fishing grounds. Most nations honor 12 nautical miles.

74. Whitman 2005. Environmental groups claim the Navy's use of sonar tracking systems confuses whale and dolphin communication systems, threatening their habitat.

75. Edis 1993: 87.

76. Burnett 2005: 794–796; Kaplan 2005: 841–842.

77. Hecht and Edwards 2008: 22.

78. Castells 1996. For a critique of his thesis, see Smith 1996.

79. See e.g., Bender 2000; Bender 2006.

80. Nye 2002; Nye 1990.

81. Hardt and Negri 2000: 160–182.

82. Maier 2006, chapter 2 and pp. 280–283.

83. Even non-US-based Americanists subscribe to a discourse of a deterritorialized global US power, best summarized in Pells 1997 and critically analyzed in Van Elteren 2006 and Fishkin 2005.

84. Johnson 2006.

85. US Geology Service, Western Coastal and Marine Geology, "Research Project: Pacific EEZ Minerals," at http://walrus.wr.usgs.gov.

3   The Uses of Portability: Circulating Experts in the Technopolitics of Cold War and Decolonization

Donna C. Mehos and Suzanne M. Moon

Technical experts saw their professional geographies change, sometimes dramatically, as a consequence of the Cold War and decolonization in the years after World War II. Stunned businesspeople watched the cozy relationships they enjoyed with imperial powers disappear under decolonization, and the stable and profitable colonial business environment with it. Entrenched companies sometimes pulled up stakes and moved in the face of revolution, nationalized industries, or collapsed systems of supply. Technical experts who had spent careers amassing knowledge suited to the specific social, political, and ecological environments in one stable colonial territory suddenly found themselves trying to operationalize that knowledge somewhere entirely new. United Nations technical aid programs, which were designed to diminish the global tensions of the Cold War, offered new opportunities for experts to put their knowledge to use around the world. Experts who had developed careers in one country or colony could now choose globetrotting careers as technical advisors, putting their knowledge to use in disparate political and social environments under the umbrella of the UN. Changing institutions, political boundaries, and degrees of political stability gave technical experts reason to circulate far beyond the well-trodden paths laid down by their predecessors.

This changing mobility of experts was not merely an unintended artifact of political change. Increasing the portability of experts and expert knowledge became a common strategy of Cold War technopolitics. Following Hecht, we define technopolitics as "the strategic practice of designing or using technology to enact political goals."[1] While the UN hoped to ease political tensions through the use of technical aid, private businesses found it necessary to collaborate with the changeable technopolitical modernization agendas of host governments to ensure their own survival. Experts, seen as a scarce resource in some parts of the world, became valuable mediators of technopolitics. Both UN administrators and businesses operating in the decolonizing world found it increasingly important to make sure their experts could function successfully in a variety of ecological, social, and political settings. In this essay we explore both how and why such portable experts and expert knowledge, particularly in the area of agriculture, became notable actors in the technopolitics of Cold War and decolonization. We contrast portable knowledge that can find some application in many areas with place-based knowledge that might be effective only in one geographical region. Before and after the Cold War, both kinds of knowledge mattered; indeed, they were interdependent. We argue that the increased emphasis on portable knowledge changed the character of the interrelationship between place-based and portable knowledge during the Cold War. While some kinds of expert mobility certainly pre-dated the Cold War, long-term colonial relationships had made it possible and desirable for foreign businesses or governments to invest heavily in the acquisition of place-based knowledge. Experts could move if they chose, but many could lead successful careers based on their specialized knowledge of one particular area. In this essay we show how the technopolitics engendered by decolonization made experts whose knowledge could be operationalized widely far more valuable and made portable knowledge far less risky than the place-based knowledge that had been generated in colonial contexts.

It is important to draw a distinction between what scholars call "indigenous" or "local" knowledge and what we are describing as place-based knowledge. Indigenous knowledge is usually understood to be limited to a particular community or region and to be created outside the paradigms of Western scientific tradition.[2] The place-based knowledge we highlight in this essay is not necessarily indigenous knowledge, although the scientists and technical experts we discuss may have obtained this knowledge by interacting with indigenous communities. Rather, it is the kind of scientific or technical knowledge that is most useful for a particular area but of limited interest or applicability elsewhere in the world. The details of soil composition of a particular valley would be an example of place-based knowledge; the techniques of analyzing soil would be an example of highly portable knowledge. Why and under what circumstances did the balance between attention to place-based knowledge and attention to portable knowledge shift in the Cold War?

In this essay we explore two organizations whose histories offer complementary perspectives on the uses of portability in Cold War technopolitics. The first is a long-established colonial plantation enterprise, the Handelsvereeniging "Amsterdam" (HVA), which was ousted from the decolonized Republic of Indonesia in the late 1950s. In a match brokered by the UN, the

HVA dispatched technical staff from Java to build a sugar-producing operation in Haile Selassie's Ethiopia only to watch history repeat itself as their business was nationalized in Ethiopia's revolution of the 1970s. Confronted with the uncertainties of decolonization and Cold War, the HVA gradually began to specialize in increasingly portable consultant work to capitalize on its in-house expertise and to avoid risking capital investments in unstable regions. Without the good offices of the United Nations, however, the HVA might not have had the opportunity to move to Ethiopia. To understand better the changing global circulation of experts, the second institution we examine is the United Nations itself. Acting on the conviction that shared prosperity would reduce the chance of cataclysmic warfare, leaders at the UN played matchmaker between displaced enterprises and decolonizing nations and created their own technical aid program, the Expanded Program of Technical Assistance (EPTA).[3] Technical experts from the academy, private business, and government service circulated around the developing world at the UN's behest, making portable experts and portable knowledge both ideologically and pragmatically essential.

It is no accident that these two very different kinds of institutions came to value portability in technical knowledge and personnel. The common denominators were the technopolitics of Cold War and decolonization and the changing political order in which old ties and collaborations became unstable or disintegrated. Portable knowledge and experts became indispensable technopolitical tools for creating institutional stability in this dynamic and unpredictable environment. Indeed, they may have been every bit as important for facilitating collaboration as the technological projects each organization proffered. Investigating the character of the portability embraced by these organizations and the uses to which it was put offers insight into how the politics of Cold War and decolonization reshaped the circulation of technical experts and the nature of the knowledge they produced and used.

Colonialism and the Decolonization of Agricultural Production: The Handelsvereeniging "Amsterdam"

To continue doing business in the decolonizing world, the HVA, like all international knowledge- and technology-intensive industries, adapted to new political environments and new ways to collaborate in the technopolitical goals of host nations. As it learned in both Indonesia and Ethiopia, a company might change from desirable collaborator to unwanted foreigner virtually overnight because of shifts in government or ideologies of modernization. A new business model that operationalized place-based knowledge as portable knowledge gave it the flexibility to accommodate a range of technopolitical circumstances. Unable to mold political environments to their needs, businesses survived only when they shaped their practices to a destabilized world. In this section we explore how the Dutch company HVA (founded in 1879), which owned multiple plantations on colonial Java and Sumatra, steered a new course during the crises of World War II and subsequent Indonesian decolonization. We elucidate how the Cold War influenced both the company's business model and the circulation of its technical employees.

Thriving colonial plantation industries required and produced much technical expertise in tropical agriculture and in processing. In Europe, colonial powers promoted colonial agriculture by investing in technical education to train their nationals as engineers and agronomists for colonial service. In the colonies, companies financed both in-house research and collective research, most notably in large-scale experiment research stations that produced detailed knowledge of local situations. However, as world wars and decolonization destabilized colonial holdings, plantation industries collapsed. Companies, often losing significant fixed capital investments, sought new avenues for tropical agricultural enterprise. Forced to flee their countries' former colonies, technical staffers who had gained expertise of place in these specific environments faced new challenges, particularly in redefining and using their specialized skills in new ecological, geographical, and political contexts. Whereas in the colonial period experts had circulated primarily between their European home countries and the colonies they controlled, decolonization created new conditions and paths upon which they could circulate professionally.

In the first half of the twentieth century, the HVA operated plantations exclusively in the colonial Dutch East Indies. Founded as a trade company that exported products to the colony, it took over coffee and sugar plantations on Java that had suffered losses during an economic crisis in the 1880s. Thereafter, the HVA turned to agricultural production and research by gaining access to land and by planting not only more coffee and sugar cane but also cassava and agave. Furthermore, it expanded to the island of Sumatra, where it cultivated palm oil, rubber, agave, and tea. By the outbreak of World War II, the HVA controlled 100,000 hectares (250,000 acres) on Java and Sumatra and employed 800 staff members (most of them Dutch) and 170,000 indigenous laborers.[4] Political instability forced this company down new entrepreneurial paths.

A significant interruption of the HVA's colonial business occurred in 1940 when Germany's occupation of the Netherlands commenced. With the HVA's headquarters and its board of directors in Amsterdam paralyzed, the HVA's office in Surabaya took over the task of selling products that no longer could be shipped to the war-torn European continent. Japan's invasion of the Dutch East Indies in 1942 proved more disruptive. Dutch experts not evacuated to their homeland found themselves interred in camps that many did not survive. The lucky ones were recalled from the camps to sugar cane fields and factories to maintain some sugar production.[5] However, most fields and mills emerged from the war neglected and damaged.[6]

Two days after Japan's capitulation in 1945, revolutionary leaders declared the independence of the Republic of Indonesia. Republican forces occupied Java and Sumatra and dashed the HVA's hopes for immediate reconstruction. In 1947 the Dutch army attacked and repossessed some areas. Following closely were Dutch businessmen with their technical staffs who had returned from the Netherlands prepared to resume agricultural production. Some were protected by military forces. During the guerilla war of independence in the following years, Dutch businesses reinvested tens of millions of guilders to rebuild their destroyed plantations. Despite decreased production capacity and significant loss of experienced personnel—both educated (Dutch) technical staff and local laborers who had suffered enormously under the Japanese forces and later in the terror of the war—they profited greatly from worldwide shortages and subsequent high market prices.[7] Amid the violence of war, the HVA, like other firms, demonstrated complete confidence that the Dutch would repossess their colony and enterprise would flourish. The Dutch government only suspended military efforts of re-colonization when the United States threatened to withdraw all Marshall Plan aid to the Netherlands.[8] Expecting the end of unrest and violence, industrialists ultimately welcomed decolonization in 1949 when the Dutch government recognized Indonesian sovereignty. Moreover, the ex-colonial and post-colonial governments reached formal agreements that granted the Netherlands most-favored-nation status in Indonesia.[9]

Although peace did not return immediately, the HVA continued to invest in Indonesia and in their place-based technical expertise. Significantly, even the revolutionary Indonesian government claimed to value the Dutch place-based knowledge in its initial Foreign Policy Manifesto of November 1945 stating "we will need foreign, technical, intellectual, and capital assistance in building up our country" and that the Dutch were the preferred foreigners "because they are already here and familiar with conditions here."[10] However, in the new context Dutch experts would not circulate as freely between the Netherlands and its former colony. The new republic, committed to improving the prosperity of Indonesians with industrial and economic development, accepted alien businesses because there were too few Indonesians trained to manage industrial endeavors or to build up national capital.[11] While Indonesia needed foreign investments, the foreigners had to accept the government's condition that companies "in Indonesia should be primarily places for Indonesian citizens to work."[12] The HVA, though welcome, would have to rely less on its imported Dutch technical workforce and would have to promote Indonesians. By 1950, the HVA was training indigenous men for its technical staff.[13]

Even as Indonesian politics moved increasingly to the left throughout the 1950s, the HVA felt secure and expected continued support from President Sukarno. In fact, assuming Indonesia's economic dependence on the Netherlands, the HVA remained confident that the new state could not manage without Dutch industry. It thus came as a shock when, in 1957 and 1958, Indonesia expelled Dutch nationals and nationalized all foreign industry, and foreign firms lost their assets. This move would ultimately be disastrous for the Indonesian economy, yet the appearance of complete independence from the Dutch had become central to Sukarno's political strategizing.[14] Many HVA experts lost their jobs, and faced unemployment upon return to the Netherlands or explored opportunities in tropical agricultural industries across the world's continents. The HVA—still unable to digest its expulsion—maintained a skeleton technical crew in Amsterdam prepared for possible return to Indonesia and sought ways to re-deploy its in-house experts and expertise.[15]

Early Diversification: Java Displaced to Ethiopia

In the first half of the twentieth century, the HVA proved reluctant to invest outside of the East Indies. Some scholars have portrayed the company as a model for corporate diversification because it moved from imports to plantation production, from coffee and sugar to many other tropical crops, and from Java to Sumatra, all the while developing new expertise and techniques.[16] However, the HVA concentrated investments in Dutch Southeast Asian colonies. This is not to say that the HVA did not explore possibilities elsewhere. In fact, in the 1920s and the 1930s small teams of HVA technical experts traveled to various areas in Africa and Asia, studied the environmental conditions, and assessed the possibilities for beginning new agro-business ventures. Only the study of the Congo resulted in a favorable recommendation, yet the HVA chose to invest in Java (probably as a result of prodding from the Dutch Minister of Finances).[17] After World War II, concerned about political instability in Southeast Asia, the HVA's board discussed geographical expansion and sent a group of experts, headed by the agronomist A. Kortleve, on exploratory trips to South America and Africa, where they conducted extensive scientific studies. Only one country—Angola—was deemed suitable. The reason? Its climate, soils, and environmental conditions resembled Java more than any other African, and some South American, countries.[18] Thus, having excelled at developing agriculture in one specific climatic region, the HVA analyzed the potential to expand to new areas in relation to its in-house, place-based knowledge of Java and Sumatra. Despite employing hundreds of knowledgeable experts on various crops, processing techniques, and factory technologies, who in principle could be redeployed, and despite its extensive experience in crops that were being cultivated across the tropical world, the HVA's highest-level agricultural experts and its decision-making board members searched for, and in fact required, environmental conditions similar to the Dutch East Indies. They assessed their company's expertise, in general, as place-based rather than as portable subject-based knowledge applicable across the tropical world. As long as the political context appeared secure, the HVA maintained a geographically conservative strategy of expansion.

Ironically, the singular exception to the HVA's unwillingness to invest outside of Indonesia before 1958 proved crucial to its survival. In 1950–51, after being approached by a representative of a United Nations committee, the HVA agreed to cultivate sugar cane in Ethiopia, the one country that the HVA had absolutely excluded from consideration for development in its studies just a few years earlier.[19] Apparently, it could not resist the deal proposed by Emperor Haile Selassie I (reigning 1930–1974), who offered them vast tracts of land. The manner in which the HVA cultivated and manufactured sugar in Ethiopia was to displace, part and parcel, its Java operation.[20] Viewing Ethiopia as politically safe, in part, because it had little history as a colonized state, the HVA could not envision this empire's vulnerability to revolution.[21]

This imperial invitation to the HVA must be understood as a form of technopolitics—indeed Ethiopia's history presents a tangle of international players with technopolitical ambitions. Throughout World War II and the Cold War, Haile Selassie masterfully manipulated international tensions to realize his imperial industrialization and modernization goals. Occupied only briefly by Italy (1936–1941), Ethiopia gained new geopolitical strategic value in 1940 when Italy declared war on the Allies. Britain, in an effort to secure the Suez region, granted Ethiopia military and civilian aid.[22]

In 1942, it recognized Ethiopia as a sovereign state. Selassie, however, felt London held too much power over his empire, and sought to balance it with a US alliance.[23] Similarly, after World War II, the US courted Ethiopia to balance British and French colonial legacies and made the Ethiopian empire its most important trading partner.[24] The US facilitated Ethiopia's entry to the UN in 1942, and in 1943 a US treaty granted Ethiopia arms and sent a technical mission to report on its needs.[25] American experts drew up blueprints for Ethiopian economic development that included millions of dollars in aid and loans for industrial development.[26] The US interest in Ethiopia included control of a radio broadcasting facility in Eritrea that had been built by the occupying Italians and taken over by the Americans in 1942. It became an important technopolitical locus, as the Americans used the facility to broadcast global intelligence to Washington.[27] While most US aid went to military and industrial rather than agricultural development, Ethiopian profits from the coffee it exported (primarily to the US) bought many forms of technical assistance from other countries.[28] Furthermore, Selassie maintained relations with Eastern Bloc leaders that gave him bargaining power in the West.[29] It was in response to this context of technopolitical alliance building, competitive international aid, and Selassie's ambitions for modernization that the HVA relocated to Ethiopia, creating the private sugar monopoly that has been dubbed Ethiopia's "big success story of post-1941 industrialization."[30]

When the UN dispatched technical consultants upon Selassie's request, the emperor had hoped that they—former high officials in colonial Dutch East Indian affairs—would locate firms interested in investing in agricultural production.[31] After meeting with one of the UN advisors, the HVA directors sent Kortleve, the expert responsible for the African and South American reports in 1946, to Ethiopia twice. On the first exploratory trip, he investigated possibilities for sugar production and was accompanied by a coffee planter from Java. On the follow-up trip, he was accompanied by a surveyor, an infrastructure engineer, and a former colonial administrator. Ultimately, Kortleve recommended that the HVA cultivate sugar cane in Ethiopia. The HVA commenced negotiations with the Ethiopian government, which granted it a concession for the use of 1,500 hectares (3,750 acres) of land in the Awash Valley, with options for more. By the mid 1960s, it had cultivated 12,000 hectares (30,000 acres) to grow sugar cane, which was processed in the Wonji, Shoa, and Metahara factories.

The move to Ethiopia added a third point to the two-point circulation pattern between the Netherlands and its colonies. While one could argue that this was the start of HVA's strategy of portability, the move to Ethiopia was neither as easy nor as swift as the term "portable" suggests. The HVA clearly saw the Ethiopian project as a long-term enterprise, and therefore invested greatly in the region. To transform the dry Awash Valley, the HVA relocated a technical team from Java in 1951–52. Because the valley lacked infrastructure, a crew of HVA civil engineers constructed bridges, roads, transport, and water and irrigation facilities.[32] Agronomists, expert in cane cultivation, soil chemistry, and research, organized and equipped a research station where they carried out laboratory experiments and coordinated research in the fields. When the HVA purchased new sugar-processing equipment from the Dutch manufacturer Stork, HVA requested that Stork dispatch 25 technicians to install the factory. Moreover, the HVA had to provide enough experienced personnel to teach the local workers every aspect of sugar work, from digging ditches to running experiments. In addition to the reconstruction efforts in Indonesia, the Ethiopian initiative gave the HVA opportunity to employ its experts and operationalize their expertise.

Whereas in Indonesia indigenous workers had skills and experience in fields, factories, and laboratories, few indigenous Ethiopians in the Awash Valley had any experience—and perhaps few had any interest—in such work. In fact, at the start of HVA's Wonji project, several cultural groups of peasants and nomads inhabited the Awash Valley. The Dutch, furthering one of Haile Selassie's political goals, transformed their territories and lifestyles. Every worker had to be trained both for the specific, often grueling, work and for the discipline of wage labor.[33] The Stork personnel built the first factory and trained the first generation of Ethiopian factory technicians.[34] For other training, the HVA brought in 200 staff members—experienced primarily, if not exclusively, in the Dutch East Indies—to demonstrate, teach, and supervise the laborers to produce profitable sugar. The Dutch staff gradually diminished as Ethiopians participated in HVA training programs, gained experience, and graduated from Ethiopia's new technical schools.[35]

During the early years of its work in Africa, the HVA gradually shifted from a Java-centric view of cane production to one that incorporated knowledge of Africa. In its first years of operation in Ethiopia, the HVA assessed most factors—such as production statistics, soil quality, and cane growth—relative to Java. The first harvest in 1954 provided yields competitive with other African sugar-producing regions but was disappointing in comparison with the average yield on Java. Soon recognizing that sugar cane cultivation in Africa varied vastly from that on Java, the HVA experts began to compare Wonji results with other African yields and to seek new cane varieties and planting methods.[36] The HVA Java sugar experts traveled to explore sugar industries in African climates and to acquire skills and knowledge effective on the African continent. Though one can argue that they acquired new place-based knowledge, the process signaled a new flexibility to increase knowledge of various environmental conditions for cultivation and to build expertise that gradually became more portable, or applicable in numerous regions. Inadvertently, the Dutch UN advisors to Selassie hung a safety net that saved the Dutch firm from bankruptcy as a consequence of Indonesian decolonization. The HVA's only geographic diversification outside of the former Dutch colonies—a product of the UN—both turned a profit and provided the company with a new region in which to expand and to recover from Indonesian decolonization. It also led the company to increase the portability of its experts, who would later operationalize their knowledge in new geographic regions, thereby safeguarding the firm during political upheaval.

In view of its good relations with the emperor, the HVA felt secure in Ethiopia during the global turmoil of decolonization and Cold War geopolitical struggles. It continued to rely on revised place-based plantation knowledge, first developed in Java and Sumatra and later re-established in Ethiopia. Significantly, many HVA technical employees were transferred to Ethiopia from Indonesia and, in fact, perceived employment at Wonji as professionally more promising than in Java.[37] After 1958, the HVA expanded sugar production and investigated new commercial crops, thereby repeating its previous diversification strategy. In 1962, the HVA built both a second factory in the Wonji complex, called Shoa, and a candy factory. Later, it negotiated another contract with Selassie's regime and expanded sugar production to the Metahara region, where its third factory opened in 1969. In the early 1970s, the firm launched tea, vegetable, and strawberry cultivation.[38] These investments suggest that the HVA perceived little risk in the rising social and political unrest.

Selassie's modernization plans benefited primarily the Ethiopian urban elite.[39] Little foreign aid received by Ethiopia went to rural agriculture. Lack of attention to local food production and the problems of widespread poverty, inflation, and unemployment proved fatal to Selassie's hold on power.[40] In the early 1970s, northern Ethiopia was ravaged by drought and subsequent famine; government forces faced armed conflict with Eritrean secessionists (Eritrea had been federated with Ethiopia via a UN resolution of 1950), and power struggles ensued between Selassie and the military as well as within the military. In 1974, junior military officers backed by the Soviet Union deposed the emperor. Government officials were imprisoned and executed. Haile Selassie died in prison in 1975. That year, virtually all industries, including the HVA agro-businesses, were nationalized. Ethiopia, under the leadership of Mengistu Haile Mariam, enjoyed significant Soviet military assistance and Cuban military advice. The American government and President Jimmy Carter no longer tolerated Mengistu's Red Terror.[41] In 1977, it sent a technopolitical message to the regime by closing the US communications base Kagnew—the radio broadcasting facility it had controlled and rented from Ethiopia since 1942.[42] Later that year, Ethiopia lost all US military aid for violations of human rights.[43] The HVA lost significant assets in the only country where it held considerable fixed capital.

Expansion of Portable Knowledge

After being almost bankrupted in Indonesia and again in Ethiopia, the HVA turned to the marketing of increasingly portable knowledge and experts. Its expulsion from Indonesia in 1958 led the HVA to reorganize into three distinct subsidiaries: the Amsterdam headquarters, focused on trade activities, the Ethiopian agro-business, and the new diversification efforts to explore agro-industry throughout the tropical world.[44] Ultimately the third subsidiary saved the concern. The HVA developed into a consultancy firm that sold technical expertise not only in agricultural production and processing but also in the management of tropical agro-businesses. Its employees who acquired specific place-based knowledge in factories and plantations in Indonesia, and who explored African sugar production to improve the operation in Ethiopia, saw characteristics of their expertise gradually transformed. The company's new direction into consultancy required subject-based portable expertise that could be operationalized when it contracted with governments and private industry in new regions. This corporate policy reduced the risks to fixed capital and characterizes the HVA's survival strategy in the Cold War context of political instability.

While focused on Ethiopian sugar production, the HVA took its first step into consultancy when it was approached for advice by the Kilombero Sugar Company (KSC) in Tanganyika after it suffered losses from its first harvest in 1963.[45] Significantly, the internationally financed KSC was largely supported by the Dutch state via the Netherlands Overseas Financing Association (Nederlandse Overseese Financieringsmaatschappij), an organization founded to finance new endeavors of Dutch industrialists who lost their assets and livelihoods in the process of Indonesian decolonization.[46] The HVA sent Kortleve with a team to study regional conditions and to offer technical suggestions. Thereafter, the HVA contracted with Kilombero to provide technical assistance. In 1965 it signed a 5-year contract to manage the concern. It thereby capitalized on in-house management and technical expertise without making vulnerable investments.

In contrast to Ethiopia and Indonesia, at Kilombero the HVA continued profitable business activities despite major political ruptures. In 1964, Tanganyika and Zanzibar joined to form Tanzania. Its first president, Julius Kambarage Nyerere, aimed to create an uncorrupt, economically independent socialist state. After Kilombero was nationalized in 1967, and throughout Nyerere's tenure (which ended in 1985), the HVA maintained its management contract. Nyerere supported the sugar industry in general, and more specifically the Dutch management at Kilombero. Furthermore, Nyerere appealed to the HVA to manage the sugar-producing area Mtiba, which was nationalized in 1975. That same year, Kilombero II opened—a complete new sugar concern funded by diverse development agencies including the Dutch and Danish governments and the World Bank. Built under the supervision and planning of the HVA, the factory was fitted with equipment supplied by the Dutch company Stork. Collaborations of corporate, national, and international aid organizations enabled by influential bureaucrats friendly with HVA directors and managers were not unique.[47] They have elicited both harsh criticism[48] and defensive reactions.[49]

Through the 1960s and the 1970s, the HVA sold increasingly portable knowledge as it took on consultancy and management contracts across the tropical world and diversified its expertise to encompass cattle, tea, oil, vegetables, and more.[50] In the mid 1960s, for example, government representatives of the Dutch colony Suriname asked the HVA for technical and financial cooperation in palm oil production. The HVA entered this project and profited from extensive exploratory research performed in 1960–61 in Brazil, a country bordering Suriname. Although in the 1960s the directors chose not to produce palm oil in Brazil, in the early 1970s they agreed to manage a new palm oil endeavor initiated by the Brazilian government. The HVA invested in this project, which was co-financed by the Dutch government via the Netherlands Development Finance Company (Financierings-Maatschappij voor Ontwikkelingslanden, the successor to the organization that financed much of the Kilombero Sugar Company) and promoted by the Dutch ambassador to Brazil. The HVA benefited from the portable knowledge produced by its experienced staff members who circulated between Europe, Southeast Asia, Africa, and South American as it cooperated with the Dutch government and international development agencies, and contracted with governments and companies across the globe. After 1970 when the company actively pursued this new strategy, its circulating experts generated much income running agro-businesses in at least 30 countries, spanning Africa, South and Central America, the Middle East, Southeast Asia, and the Caribbean.[51] In 1958 it employed 15–20 expert consultants in Amsterdam.[52] By 1975, 80 advisors worked in Amsterdam and 120 abroad.[53] Income from technical advice increased from 1 million guilders in 1970 to 8 million guilders in 1975.[54] The HVA's board of directors, reporting to its shareholders in 1976, did not exaggerate when it stated "the major strength of the company is its specific knowledge, unique in the Netherlands, in the area of (sub)tropical large-scale agriculture as well as in working in the tropics."[55] As a result of decolonization and Cold War technopolitics, the HVA's directors and agriculture experts transformed place-based knowledge of colonial Java and Sumatra into portable expertise of the tropical world. Although successful agro-business certainly required place-based knowledge, portable expertise became more widely marketable in tropical agriculture.

By taking the new direction into consultancy and agro-business management, the HVA maneuvered internationally within various political and economic contexts and in countries with both capitalist and socialist economies. The HVA profited when it mobilized experts no longer employable in Indonesia and Ethiopia, and when it marketed the company's scientific and corporate expertise rather than agricultural products. By moving away from ownership into consultancy and management contracts, the HVA simultaneously gained from its in-house knowledge and reduced the risk of fixed capital. Significantly, political instability was not necessarily a threat to its contractual work. In contrast to colonial contexts where the HVA depended on Dutch political power, or in Ethiopia where its presence was predicated on its relationship with Haile Selassie and his capitalist industrializing vision, the HVA's new direction was largely independent of political change because it shared a critical goal with its contractors—capitalist or socialist governments and companies—to build or maintain successful agro-industries.

The case of the HVA demonstrates clearly that the instability of the Cold War and decolonization created not only financial dangers but also opportunities for former colonial capitalist industries. The HVA's portable experts became important assets and facilitators of a new kind of technopolitical collaboration that depended less on investments in fixed infrastructure and long-term local research establishments and more on trade in expertise. That is not to say that place-based knowledge disappeared. Technical experts working in these corporations had greater career stability if they possessed (or could claim to possess) knowledge germane to many geographical settings.

Technical experts circulated more widely during the Cold War than in earlier colonial periods. While developing nations negotiated with private companies, they also courted aid from national development organizations and from the UN and other international agencies. Moving back and forth between the corporate world and the world of publicly funded aid programs was in fact quite common as both kinds of organizations saw opportunities for political and financial profits in the developing world, opportunities mediated by technical experts. Adriaan Goedhart, for example, employed for decades by the HVA and ultimately the chairman of its board of directors, also served the UN as a member of its Industrial Cooperative Program of the Food and Agricultural Organization (FAO) and of its Industrial Development Organization (UNIDO). He worked with the African Development Bank, with the Asian Development Bank, and with the International Bank for Reconstruction and Development, among other organizations.[56] The UN, pursuing technical collaborations with developing countries, created new paths of circulation for technical experts during the Cold War. Portable experts became central to its technopolitics of internationalism.

The UN's Program in Technical Assistance

Whether by linking nations with technological businesses or by running their own technical aid programs, UN officials used technology as a tool to produce greater political stability in a polarized and fractious world. In the aid programs established in the early 1950s, portable experts and knowledge were more than just a defense against changing political environments as was arguably the case with HVA. They were also essential to making technical aid achieve its political ends. Circulating experts would become agents of technical change and the UN's internationalist idealism, which would make them important to the UN's political agendas. Further, not only did the UN provide paths for expert circulation; it also defined the characteristics that made experts and expert knowledge sufficiently portable, going to great administrative lengths to compensate for the consequent loss of place-based knowledge. The UN created in the process a substructure for expert circulation allowing technical experts to achieve UN goals more efficiently.

To understand why the UN needed portable experts to make technical aid a successful exercise in technopolitics, it is helpful to understand why the UN engaged in technical aid at all. The earliest technical assistance sponsored by the UN operated under the auspices of the UN's autonomous specialized agencies, including the Food and Agriculture Organization

(FAO) and the World Health Organization (WHO).[57] The first formal technical assistance mission took place in 1946 when Greece requested that the FAO provide experts to consult on agriculture, fisheries, and "related industries."[58] The Greek mission contrasted with the work of the United Nations Rehabilitation and Relief Administration (UNRRA), which helped countries repair damage from the war. While the UNRRA provided short-term help with rebuilding infrastructure, the FAO mission took a longer-term view of economic development, recommending new ways to utilize water, and new technologies to improve agricultural, forestry, and fishery yields. Though initially imagined as an agricultural project, the aid team noted that their recommendations occasionally strayed outside of agriculture, because they found that they could not divorce issues affecting agriculture from other aspects of industry and economic life.[59] The aid team recommended that the UN send an advisory mission to Greece that would include representatives from other specialized agencies, including the International Bank for Reconstruction and Development.

By 1947, the FAO, the WHO, and the International Labor Organization (ILO) had created joint committees of experts to help deal with problems that required expertise not found within a single agency, for example on the subjects of agricultural labor and agricultural health.[60] This growing interest in integrated development plans culminated in a UN mission to Haiti in July 1948. This new kind of mission, organized by the UN's Economic and Social Council rather than by the specialized agencies, brought together experts from the United Nations Secretariat (experts in public administration), from the FAO, from the International Monetary Fund, from the United Nations Scientific and Cultural Organization (UNESCO), and from the WHO.[61] On December 4, 1948, a resolution of the General Assembly authorized the Secretary-General to "arrange for the organization of international teams consisting of experts provided by or through the United Nations and the specialized agencies for the purpose of advising those Governments in connection with their economic development programmes."[62] In addition to sending teams of experts at the request of member governments, the resolution gave the Secretary-General power to provide fellowships for foreign study, to arrange training programs for local technicians in "under-developed" countries, and to provide some facilities that would assist governments in obtaining technical personnel. The funds provided in the first year were meager (roughly $300,000) but demonstrated a willingness to embrace technical assistance as an appropriate task for the UN itself and not just for the specialized agencies.[63]

Thus, in January 1949, when President Harry Truman gave his inaugural address, introducing what became known as the Point Four Program, the United Nations had already formulated a program for technical assistance. Truman's initiative gave technical assistance programs a higher profile and greater hope for more substantial funding. By mid 1949, the United Nations and its specialized agencies jointly administered and operated the Expanded Program of Technical Assistance for the Economic Development of Less-Developed Countries.[64] This special program would be funded by voluntary contributions from member nations, coordinated by a Technical Assistance Board made up of the heads of the specialized agencies and chaired by the Assistant Secretary-General in charge of Economic Affairs. The EPTA's mission (like that of the Point Four Program) was to provide expertise rather than capital, which made the process of circulating experts central to its political and technical ends.

"No Nation Has a Monopoly on the Best Techniques"

Because of the prominence of the Point Four Program, the UN found itself defending the special value their experts brought to technical assistance, particularly in the face of the UN's more modest budgets. Although Truman suggested that technical aid ought to be funneled through the United Nations, the United States put a considerable amount of funding into its own bilateral Point Four initiatives. Consequently, the UN and the US competed both for experts and for the opportunity to provide aid. The Point Four Program (and such successor programs as the Technical Cooperation Administration and the US Agency for International Development) benefited from the deep pockets of the Americans and from what one analyst called "the high-pressure salesmanship" of American officials, who hoped to nudge countries to request technical aid that could then be tied to political cooperation.[65]

The fundamental justification for UN technical aid differed little from that for Truman's Point Four Program, but UN officials and other advocates had strong ideological reasons for wanting the UN to be the primary source of aid.[66] In an article written in 1950, David Owen, the first chairman of the Technical Assistance Board, reiterated the basic reasons to provide aid:

> The overriding duty of the United Nations is to promote peace and security throughout the world. Clearly it would be a shortsighted attitude to assume that all this requires is for the United Nations to act as a kind of world policeman. If we are to have real and stable peace, genuine security, we must attempt to eradicate the conditions which lead to international unrest and friction.[67]

Yet in the political environment of the Cold War, it wasn't just a matter of supplying aid to ameliorate poor conditions and then waiting for peace to break out. Had it been, the UN could have left the work to the eager and prepared United States. Who provided aid and how they did so mattered just as much as the aid itself. The UN specified that its assistance should "not be a means of foreign economic and political interference in the internal affairs of the country concerned, and shall not be accompanied by any considerations of a political nature."[68] To diminish the perception of political interference, Owen emphasized that all UN teams would be multinational in character—no single nationality would be overrepresented.[69] Further, the EPTA required that assistance be given to governments as much as possible in the form requested by the governments themselves. Bilateral aid such as that offered by the US carried too much potential for political interference and offered a tempting opportunity to use technical assistance to manipulate the domestic and foreign politics of needy nations. In contrast, the UN's policy made a particular point of respecting the autonomy of requesting governments and prioritizing an ethic of self-determination even in matters of technological change. For supporters of the UN, keeping manipulative politics out of technical cooperation programs created a form of international cooperation that neutralized the polarized power plays of the Cold War through technical practices. Only multilateral technical cooperation could become a tool for building the international relationships that could prevent the escalation of Cold War to a hot war: "The peoples of the world have to think and live and work together in a community of international co-operation if they are to be spared the final catastrophe."[70]

In 1950, David Owen made a point-by-point case for multilateral aid in preference to bilateral aid without naming names—a necessity since the US provided much of the funding for the EPTA. He pointed out that many "underdeveloped"[71] countries were sensitive about any appearance of interference in their internal affairs, and that aid from a single industrialized power sometimes looked like an effort to create new forms of foreign domination.[72] Walter Sharp, a political scientist from Yale who had no need to exercise Owen's diplomatic tact, bluntly argued that American bilateral aid often pushed aid in directions of special interest to the US, offering as an example Thailand, where the US had offered military cooperation and technical aid in tandem or not at all. Some nations, striving to maintain a neutral position in the Cold War, consequently refused American demands and did without it, as was the case with both Burma and Syria in 1952.[73]

But Owen didn't limit his critique to politics. He also argued that multinational teams of UN experts were technologically more desirable than single-nation teams. He argued that when all experts came from a single nation, even with good intentions and free of a secondary political agenda, they were likely to view the problems of the recipient country in a one-sided way, and to create solutions that adhered closely to the technical practices of the team's home country, which might not be appropriate to the recipient country's social and political culture.[74] Multinational teams would, by definition, provide multiple points of view and therefore a better chance of finding appropriate technical solutions. Thus the virtues of international organization in political affairs and the declarations of political equality for the world's nations found their technical doubles in the rhetoric defending international technical aid: "They—the United Nations and the specialized agencies—constitute a clearinghouse for the technical knowledge of all member states. No nation has a monopoly of the best techniques."[75]

Owen's rhetoric of international participation linked diversity of national origin to technological diversity. It left unstated the possibility that some technical experts might be educated within traditions that would limit the diversity of technical opinions even within an international team. Indeed, the recognition of expert status in some fields required narrow training.[76] Though it was plainly advantageous for the UN to conflate national diversity with technological diversity, even critics of the UN seemed not to be concerned with this issue. The UN's publications made much of the national diversity of its teams, frequently highlighting the national backgrounds of otherwise unnamed team members.[77] As important as the multinational character of teams was the contribution of "underdeveloped" nations to the larger project of multinational cooperation. The UN tried to avoid the divisive perception that technical assistance operated between mutually exclusive groups of givers and receivers, and argued instead that its programs represented a "unique blending of the experience of almost every country in the world."[78] Just as in the United Nations at large, where any nation, regardless of size, wealth, or age, had an equal voice (at least in theory), so too in the project of technical assistance would any expert, from any nation, find an outlet for his or her skills in service to the world.

"Cross-Fertilization," Circulating Experts, and Portable Knowledge

This construction of diversity affected how experts circulated through UN programs. The right balance of appropriate technical expertise and variety of national backgrounds mattered in determining the ideal makeup of a mission. By the late 1950s, David Owen boasted that widespread recruiting had enlisted more than 8,000 experts from over 70 countries (although most came from Western Europe and the United States).[79] Because it offered a slate of candidates from which the requesting country could choose, the UN put special emphasis on having a wide array of candidates, allowing many experts to circulate. Geopolitical factors could restrict the circulation of some experts, however. The UN was sometimes reluctant to assign former colonial administrators to newly decolonized countries.[80] In 1956, Egypt rejected any experts who were Jewish or had spent time in Israel. Some nations made a point of requesting only internationally well-known experts, eliminating those who did not have a global reputation before the war and thus leaving out many experts from outside Europe and the United States.[81] The brevity of assignments factored into the ways that experts circulated as well. Aid missions, by definition, were meant to be short-term projects, with a focus on teaching and advising. The longest projects rarely lasted longer than two years; the shortest might last no more than a few months. Though many of these 8,000 experts took on no more than one assignment, those who agreed to multiple assignments could expect to move frequently, and across a wide geographical area. Both the demand for particular technical expertise at any moment and the desirability of an individual's national background from the point of view of providing a new or fresh perspective would affect that person's opportunities.

What seemed remarkably unimportant to the UN was the amount of place-based knowledge an expert possessed about his or her assignment. Experts who had little knowledge of a particular country were not necessarily less desirable than those who had more experience. Even language skills played a relatively small role in decision making about where to send experts. This isn't to say that the UN never appointed experts who had some knowledge of an area or a language. Rather, those skills, or the lack of them, became secondary to the need for national diversity on teams and the political requirements of supplicant nations.[82] Indeed, some experts on public administration thought too much local experience might create problems, as experts were in danger of becoming too embroiled in local politics. In the mid 1960s, the EPTA experimented with so-called regional or inter-regional experts who participated in a series of related missions in countries within a single region, justifying the practice on the ground that experience in one area brought added value to other regional projects. The primary objection to this approach was that it prevented an essential "cross-fertilization between regions."[83] A UN pamphlet titled World Against Want similarly emphasized the promise of diversity and, implicitly, the dangers of narrow worldviews:

Such a cross-fertilization of ideas and cultures, which is not following rigid cultural, political, or regional patterns, can give new impulse and response to the challenges of material problems which century after century in the long past brought great civilizations down into the dust.[84]

More prosaically, two public administration scholars cautioned that experts in US technical aid programs might "go native" when they were too familiar with the recipient country.[85] Ex-colonial experts with substantial place-based knowledge might become politically undesirable, which might trump the value of their experience. The short-term nature of most aid missions meant that even those experts who took multiple consecutive assignments probably would not return to the same country, which would make it difficult to build up or use extensive local knowledge and experience. Both political necessity and ideological preference drove the UN's need for portable experts and knowledge that was not place-based.

While the technopolitical ideal of "cross-fertilization" defended almost any kind of knowledge as potentially useful almost anywhere in the world, UN administrators did not see their portable experts as agents for simplistic, standardized solutions to diverse technical problems; quite the contrary. David Owens argued instead that nationally diverse teams would be better suited than bilateral aid groups to discern what solutions were most appropriate for local circumstances. In 1949, when the FAO resolved to join the EPTA, it requested that the Director-General consider "the need for ensuring that the approach to the organization and execution of technical assistance projects is through the culture of the local peoples, and accords with the accustomed ways and institutions of these peoples" when approving technical assistance activities.[86] In 1951, UNESCO passed a resolution "to bring together and to diffuse existing knowledge and to encourage studies of the methods of harmonizing the introduction of modern technology in countries in process of industrialization, with respect for their cultural values so as to ensure the social progress of the peoples."[87] In 1952, UNESCO resolved to study the effect of technological change on the "social and economic structures of communities."[88] From the earliest days of the program, EPTA officials emphasized the need for a non-trivial understanding of local society and culture in order for technical assistance to work in ways that were compatible with the broader goals of the United Nations.[89] However, there was also a need for experts to remain portable and circulate often. Area experts not only minimized cross-fertilization; they also ran the risk of becoming invested in local political concerns.[90] Experts needed to create solutions responsive to local conditions without having much place-based knowledge about those conditions. The UN's attitude toward experts stands in stark contrast to the attitude of colonial administrators, who greatly valued experts with extensive cultural, linguistic, and ecological knowledge.

UN administrators recognized that both portable and place-based knowledge were needed in aid projects. The question was how to make place-based knowledge available without compromising the portability of technical experts. At first, the logic of portability prevailed. The UN commissioned the anthropologist Margaret Mead to write a manual to help technical assistance experts understand how to deal with local circumstances, in effect framing place-based knowledge itself as something that could be universalized. Mead's manual emphasized the broad scope of social and cultural patterns affected by technological change, concluding with a list of recommendations that included the need to strip all technical artifacts of "cultural accretions" and seek out "congruent subgoals" among groups with conflicting expectations of technological change.[91] There is little evidence that the dense, jargon-laden manual got much practical use in technical assistance programs.

More successful were efforts that used new administrative infrastructures to mobilize place-based knowledge for the use of portable experts. The UN appointed Resident Representatives for multi-year terms as one reliable source of place-based knowledge, especially knowledge concerning local politics. In early technical aid projects, experts might arrive at their postings only to discover that their project had fallen to a lower priority or disappeared entirely as a result of local political shakeups, an interministerial squabble, or an inability to secure basic facilities for work, all of which happened on a UN mission to Afghanistan between 1950 and 1953.[92] In 1952, after experiments with UN representatives in field offices in Haiti, Pakistan, and Colombia, the Technical Assistance Board began to appoint professional administrators with relevant local experience to serve as resident representatives. Acting as liaisons between the UN Technical Assistance Board, foreign experts, and the local government, they helped experts locate appropriate local counterparts and provided briefings on both professional and personal matters that would help an expert and family adjust to the new environment.[93]

Pairing UN experts with local counterparts who could provide the experts with local insight into the problems at hand was another technopolitically meaningful practice. A local counterpart was (in theory) a professional with similar training as the expert. The counterpart acted both as an advisor to the expert and as a mediator to spread foreign expertise to other technical professionals within the country. Technical relationships with local counterparts allowed the UN to put into practice its political philosophy of respectful collaboration, with the people of the recipient country driving the technical agenda.[94] The counterpart system did not always work as planned, however. Experts often complained that local counterparts possessed neither the requisite training nor interest in the work, as was the case in the UN mission to Afghanistan mentioned earlier.[95] In other cases, the counterparts had the interest and training but lacked the political status to keep aid projects moving forward. Operating within a bureaucracy separate from the UN, local counterparts had to cope with pressures that could undermine their willingness to work on the terms requested by foreign experts. The right counterpart, therefore, could make or break an aid program. A change in counterpart assignment could bring a program to a halt. So could unhappiness over the usually large differences in pay between local and foreign experts.[96] On a more basic level, foreign experts with little international experience sometimes found it difficult to work through cultural misunderstandings with their counterparts, which made technical missions as heavily dependent on the mix of personalities as on technical expertise.[97]

As experience with aid programs accumulated, administrators began to seek portable personalities as much as portable expertise. Although not all administrators agreed with one author who claimed that an expert's personality was more important than his or her technical abilities, most agreed that people with particular personality traits were more likely to be successful.[98] Most prominent among these traits were those that presumably helped experts cope with their lack of familiarity with local cultures and conditions: sensitivity to local circumstances; adaptability with regard to working conditions, local habits, and cultural practices; open-mindedness; humility; a sense of humor; patience; and diplomatic skill.[99] One author argued that technical assistance required "a new type of man—one who is not only competent in a particular skill useful to the country concerned, but also sensitive to the customs and character of the people."[100] Analysts did not hesitate to ascribe failures both large and small to the lack of these necessary qualities among experts.[101] One lamented that these needed qualities did not always go hand in hand with other qualities demanded by recipient countries, namely age and international reputation, suggesting that older experts with "big names" did not always have the requisite flexibility and creativity that younger experts might possess.[102] Stories about technical aid experts emphasized not their know-how but their resourcefulness, stamina, and spirit of adventure. A report on a Bolivian mission described the "intermittent civil disorders, rumored and actual political plots, suppressive police action and emergency military measures" that the team faced, and underscored the resourcefulness of experts who on one occasion, trapped by an unexpected flood, pulled branches off trees to spell out "UN" in giant letters so that aircraft could drop supplies.[103] Experts who could cope with such difficulties were portable because they could operate successfully in a wide variety of environments, which put them in high demand.

Personal qualities could help a technical expert be more portable, but what about expert knowledge itself? As early as 1950, David Owen had suggested that technical development had to fit local society. Not every technical solution was fully portable from country to country, particularly, he emphasized, those solutions originating in the so-called developed countries. Owen and others argued that the experts the UN recruited from underdeveloped countries carried knowledge of particular value—the knowledge of "very similar" conditions as those in other underdeveloped countries.[104] Sometimes this meant merely that experts from poor countries were used to working with more limited resources than experts from developed countries. But more than one author also suggested that any expert from any underdeveloped country would bring special, locally appropriate knowledge to bear on the problems in other underdeveloped countries, knowledge that experts from developed countries were assumed to lack. Analysts were vague about the nature of this special knowledge. A UN pamphlet stated that the EPTA was "being asked for advisors who can give help at all stages of development, and according to various conditions of climate, of social welfare, and tradition." Advisors from poor countries were particularly helpful if "rubbing shoulders with such poverty has taught men the answer to a problem affecting peoples in some other part of the world."[105] One UN expert suggested that those from underdeveloped countries would be more likely to understand when it made more sense to adopt labor-intensive, rather than labor-saving, approaches.[106]

Whatever the nature of this knowledge, it was held to be particularly portable between underdeveloped countries. These experts and their expertise became, in the estimation of UN administrators, virtually local and able to stand in when there were few indigenous technical experts available in a community or a region. Analysts reduced the problem from one of cultural differences to one of "development" differences. By this logic, all poor nations suffered from an essentially similar "underdevelopment," which made it possible for one author to applaud the presence of Korean and Chilean rinderpest experts in Ethiopia as a move that "reduced cultural differences" and for another to argue that a Haitian sanitation expert brought special insights into the conditions of Afghan villages.[107] This virtually local knowledge was much in demand by UN administrators in the field. One resident representative working in Pakistan argued that the Asian experts employed in Pakistan were able to "understand Asians better and can often work more successfully with them than can Caucasian Americans." He extended the argument further to say that "American Negroes," by virtue of race and personality, also worked better with the local people, and that Americans should "make more use than at present of highly trained American talent among our citizens of the non-white races, especially the Oriental races in Hawaii."[108]

These techniques involved strategies of distribution, interpretation, and selection. By distributing the responsibility of place-based knowledge to local counterparts and resident representatives, they made foreign experts more mobile. By interpreting "underdevelopment" as a singular condition, they could define certain foreign experts as virtually local, combining the virtues of place-based and portable knowledge in a single person. And by recruiting experts of personal flexibility, they smoothed the difficulties of intercultural interaction. Taken together, these approaches helped the UN maintain the patterns of expert circulation central to the ideals of cross-fertilization that informed the EPTA.

Some critics pointed out the superficiality of these arrangements, drawing attention to more fundamental difficulties with the whole project of foreign technical assistance. One economist questioned whether development through top-down government action—a pre-requisite of UN intervention—was even possible: "It seems to me very doubtful whether a history of economic change, of innovation, or of economic growth in different societies supports this optimistic view of the role and capacities of governments."[109] A different form of critique called into question the basic assumptions of portability of knowledge. Two public administration experts blamed what they saw as a lackadaisical approach to training experts for foreign work on the blithe assumption of the Point Four Program that "know-how" was a simple commodity:

This is one consequence of the fact that the original Point Four program (and for that matter the UN technical assistance program) publicized the notion that the overseas job could be done by pushing the business of exporting "know-how" and assuming that, once the bearer of this magic phrase was on his way to the country of operations, he would miraculously catalyze the poor, benighted, unknowing, indigenous population.[110]

Nevertheless, for the UN the value of circulating technical experts and portable knowledge clearly outweighed the loss of local knowledge. The infrastructure of portable experts and portable knowledge allowed the UN to promote a technopolitical vision of global stability and equitable international relationships in the polarized world of Cold War international aid.

Conclusion

Although the UN and the HVA were profoundly different kinds of institutions—one public and one private, one driven by political agendas and the other by profits—seeing their stories in parallel highlights the importance of portable experts and expertise in the Cold War and helps to explain both the nature and the reasons for changing circulations of experts during that period. The UN and the HVA adopted strategies of portability to solve unique problems that their institutions faced as they reacted to the pressures and the opportunities of the Cold War and the process of decolonization. The HVA, facing the newly unpredictable politics of decolonized states, found that portable experts and expertise gave it superior flexibility in comparison with the entrenched capital investment that informed its colonial and early post-colonial business model. The HVA became a collaborator in the technopolitics of development as its increasingly portable expertise could be molded to suit many different requirements. And if perchance those collaborations went sour, it was easy to pull up stakes and pursue work in other countries. Furthermore, by redefining its expertise in terms of management skills, the HVA became a useful partner with the World Bank, which primarily offered loans for capital. In this way, the HVA became a player in the global technopolitics of aid.

The politics of the Cold War played a major role in the UN's adoption of portable experts. To diminish the polarizing effects of Cold War tensions, which many UN leaders feared would lead to a global hot war, the UN embraced technical aid carried out by teams who exemplified internationalist ideals. This was a highly technopolitical project intended both to produce greater prosperity and equity in the world and to prove that only international cooperation could produce the best results. This was "the world helping the world"; aid givers represented nationally diverse teams and did not make Cold War alliance the price of their help. The UN not only kept the teams diverse; it also kept them moving to reassure decolonizing nations that the organization had no desire or intent to be an agent of "recolonization." The significant efforts expended to compensate for the pragmatic difficulties introduced by portable experts show how important portability was to the UN's political agenda. This work in turn shaped the kinds of experts and expertise the UN could use and the career paths of countless aid experts who thrived only if they and their knowledge could easily criss-cross the developing world.

In the Cold War, portable knowledge, whether used for corporate survival or for global political relevance, became an invaluable tool. This did not mean that place-based knowledge vanished entirely, of course. Instead, what occurred was a shift in the balance between the two as investments in place-based knowledge became politically or economically risky. One significant consequence of this process was the increased mobility of certain experts, who found their horizons expanded as new paths of circulation and career development surfaced in the changing environment. Another important consequence was the effect on the process of knowledge production itself. In recent years, much work has criticized both the corporate world and international aid organizations for their lack of appreciation of local knowledge. Much of this criticism emphasizes the cultural arrogance of experts and the organizations that employ them, and their conviction that only their own (usually scientific) knowledge is necessary. Scholars have not given as much attention to the historical development of this disregard of local knowledge, especially in relation to changing political environments.[111] Yet political circumstances matter at least as much as cultural arrogance. The historical coincidence of the Cold War and decolonization produced dynamics that gave portable knowledge a higher priority than place-based knowledge (which might include local or indigenous knowledge) for many institutions. The circulating experts, whether in the public or private sector, had fewer motivations and opportunities to gain or use local knowledge in their peregrinations. As experts increasingly circulated, they expended greater efforts to produce knowledge that could move with them.

In our study of Cold War technopolitical experts and expertise, agrobusiness and UN internationalism have been fruitful areas for us to explore. We believe, however, that analyses of the changing balance between place-based and portable knowledge in other times, places, and political environments can shed light on the circulation of scientific and technological knowledge and the international character of scientific disciplines and institutions. By attending to the ways that knowledge and people are made to circulate—or stopped from circulating—we can gain new insights into the changing geographies of knowledge production and the technopolitical character of circulation itself.

Acknowledgments

We would like to thank Gabrielle Hecht, Henrika Kuklick, and participants in the two "Bodies, Networks, Geographies" workshops in Ann Arbor and

Eindhoven for their comments and suggestions. Donna C. Mehos is grateful for the support from the Nederlandse Organisatie voor Wetenschappelijk Onderzoek (Netherlands Organization for Scientific Research) under its Research Program in "Technology and the Civilizing Mission: Dutch Colonial Development in the European Context, 1870-1970" (dossier 360–53–020). Suzanne Moon would like to acknowledge Harvey Mudd College, and especially her faculty sponsor, Richard Olson, for hosting her as a visiting professor in 2007 and providing time to research and write the first draft of this article.

Notes

1. See Hecht and Edwards 2008: 4; Hecht 1998: 15.

2. For examples of studies that look at the issue of indigenous knowledge for studies of science and technology, see Scott 1996, Harding 1998, Chambers and Gillespie 2000, Turnbull 2000, and González 2001. Many studies of indigenous knowledge and science focus on the epistemological, legal, and political conflicts between the two, as well as giving attention to the status of indigenous knowledge relative to other forms of science. None of these issues are addressed in this essay.

3. The Expanded Program of Technical Assistance was also sometimes called the UN Expanded Technical Assistance Program (UNETAP). We have chosen to use the acronym EPTA because it seems to be somewhat more common in the literature.

4. Goedkoop 1990: 221.

5. Sutter 1959a: part 1, 151–155.

6. Sutter 1959a: part 1, 154.

7. Goedkoop 1990: 226–227.

8. Van der Eng 2003.

9. Sutter 1959b: 1292.

10. Sutter 1959b: 1279.

11. See, e.g., Indonesian plans to meet the need for technicians, teachers, and administrators with Dutch citizens and to educate Indonesians for the mining industry so their skills would be equal to foreigners (Sutter 1959a: 643, 486).

12. Sutter 1959b: 1304.

13. Brand 1979: 79. This is not to say that natives did not work as technical staff in the colonial period. Rather, the Indonesian government's policy intended to promote technical training and increased employment opportunities for natives. How-

ever, their promotion to high-level positions remained rare until the Dutch were expelled. See J. P. van de Kerkhof, "Defeatism Is Our Worst Enemy," at http://www.indie-indonesie.nl.

14. Legge 1972: 292, 293, 328–329.

15. Goedkoop 1990: 233.

16. Goedkoop 1990: 234, 239; Allen and Donnithorne 1957.

17. Goedkoop 1990: 235.

18. Goedkoop 1990: 236.

19. Ibid.

20. Goedkoop 1990: 237.

21. Goedhart 1999: 176.

22. Marcus 2002: 151.

23. Marcus 2002: 152, 161.

24. Ademjumobi 2007: 99.

25. Marcus 2002: 153–154.

26. Marcus 2002: 159.

27. Marcus 2002: 158.

28. Marcus 2002: 161.

29. Marcus 2002: 162–63. When Selassie was dissatisfied with the amount of US aid offered in 1959, for example, he negotiated a $100 million aid credit with East Bloc countries as a bargaining tool to increase US technical and financial aid. The US responded by increasing its rent payments for the broadcasting station, building roads, and increasing both military and financial aid.

30. Zewde 2001: 198.

31. Goedhart 1999: 175.

32. Brand 1979: 82–83.

33. Goedhart 1999: 201, 213.

34. Goedhart 1999: 201, 215.

35. Goedhart 1999: 217, 219.

36. Goedhart 1999: 184–185, 191, 193.

37. Van de Kerkhof 2007: 8.

38. Goedhart 1999: 197.

39. Westad 2007: 254.

40. Westad 2007: 250–287.

41. Marcus 2002: 156.

42. Westad 2007: 259–261. That Kagnew was a significant technopolitical point is clear from a message that Zbigniew Brzezinski, President Carter's National Security Advisor, received from the US embassy in Addis Ababa: "[T]here is a case to be made for taking advantage of the closing of Kagnew to convey the political message to the Ethiopian military regime that we are disengaging from them." (quoted on p. 260 of Westad 2007).

43. Westad 2007: 259. There were also rumors that Mengistu was preparing to expel all Americans from Ethiopia. Carter preempted him.

44. Goedhart 1999: 227–228; Van de Kerkhof 2007: 9.

45. Goedhart 1999: 236–243.

46. Hoebink 1988: 90.

47. Murray 1981: 36.

48. Hoebink 1988: 90–98; Murray 1981.

49. Goedhart 1999: 247–48.

50. Murray 1981: 22.

51. *Jaarverslag* 1976: 14.

52. Murray, 1981: 22.

53. *Jaarverslag* 1976: 14.

54. Goedhart 1999: 261. The discussion that follows is drawn from Goedhart 1999: 248–255.

55. *Jaarverslag* 1976: 10.

56. Goedhart 1999: 235, 343–344.

57. Other specialized agencies as of 1949 included the International Labor Organization, the United Nations Educational and Scientific Organization, the World Meteorological Organization, the International Civil Aviation Organization, and the International Monetary Fund.

58. FAO 1947: vii.

59. FAO 1947: xi.

60. Sharp 1948: 262.

61. United Nations 1949. All UN technical aid missions are requested by the country itself, not initiated by the United Nations.

62. United Nations General Assembly resolution 200(III).

63. Owen 1959: 27.

64. Ibid.

65. Sharp 1953: 343–347.

66. For information about US and UN budgets for technical aid c. 1952, see Sharp 1953: 343.

67. Owen 1950: 109–10.

68. UN General Assembly Resolution 200.

69. This clearly post-dates the formation of the UN team of Dutchmen sent to Ethiopia.

70. Broderman 1948: 7.

71. Terms like this one have been widely and justly criticized, having been used extensively from the 1950s through the 1970s. We have elected to use the term here not as an analyst's category but simply to highlight the worldview of the actors at the time and to avoid introducing any anachronistic terminology.

72. Owen 1950: 110. The issue of diversity of representation is not just visible in technical assistance, but in other areas of international administration. See, e.g., Swift 1957.

73. Sharp 1953: 346–347.

74. Owen 1950: 110.

75. Ibid.

76. For example, experts around the world had a more than trivial exposure to the work of German foresters. See, e.g., Rajan 2006. The debate among historians about the existence and extent of national styles in technological development is extensive; not surprisingly, it concludes that one cannot draw general conclusions. See, e.g., Cronin 2007; Kranakis 1997.

77. See, e.g., United Nations 1953; Gilchrist 1959; Sharp 1956.

78. Owen 1959: 25.

79. Owen 1959: 30.

80. Similarly, the World Bank seems to have been reluctant to grant contracts for industrial development in decolonized nations to companies owned or operated by nationals of the former colonial power. For example, the HVA was repeatedly denied World Bank contracts for agro-businesses in Indonesia. Goedhart (1999: 255–256) describes the practice of denying contracts on political grounds as a "public secret."

81. On Egypt, see Sharp 1956. For brief mentions of the problems of ex-colonial administrators in post-colonial countries, see Caustin 1967; Sharp 1953.

82. For a discussion of ways to deal with linguistic difficulties, see Alexander 1966: 85–90.

83. Alexander 1966: 73.

84. United Nations 1953: 15.

85. Trager and Trager 1962: 106.

86. FAO 1949.

87. UNESCO resolution 3.24, 1951.

88. UNESCO resolutions 3.231 and 3.232, 1952.

89. "Technical Assistance for International Development: Program of the United Nations and the Specialized Agencies," *International Conciliation* 457 (1950): 12–13.

90. Peter and Dorothea Franck were two analysts who recommended UN-sponsored area studies for experts, but noted that the UN could not support the cost of such a program. See Franck and Franck 1951: 69–71.

91. Mead 1953: 305–318.

92. Franck 1955.

93. See Schaaf 1960; Gilchrist 1959.

94. Some critics did point out how limited this view of collaboration was, assuming as it did a flat hierarchy of national interest that rarely reflected reality. See, e.g., Friedmann 1955.

95. Franck 1955.

96. See Sharp 1956: 253–254; Blelloch 1957: 41–43.

97. See, e.g., Sharp 1956; Willner 1953; Morgan 1953.

98. Willner 1953: 71.

99. The subject of personality requirements for experts gets a great deal of attention in the literature. See, e.g., Alexander 1966: 90–96. Alexander spends about three pages on professional technical requirements, and seven on personality requirements. American aid experts noted a similar set of desirable qualities in interviews

discussed in Sufrin 1966: 24–30. Further discussion appear in reports generated by particular missions. See, e.g.,: Lockwood 1956; Lepawsky 1952; Franck 1955.

100. Neal 1951: 116–117.

101. See, e.g., Franck and Franck 1951: 69–71; Sufrin 1966: 24–30.

102. See Sharp 1953: 363; Willner 1953: 71–80. The US Peace Corps was first proposed by Henry Reuss as a "Point Four Youth Corps."

103. Lepawsky 1952.

104. See Owen 1959: 30; United Nations 1953: 10–15; Alexander 1966: 163–166.

105. United Nations 1953: 11–12.

106. Morgan 1953: 29–32.

107. On the Haitian sanitation engineer, see United Nations 1953: 10–11. On the rinderpest project, see Sharp 1953: 361–362.

108. Gilchrist 1959: 517.

109. Frankel 1952: 303. Also see Friedmann's strongly worded critique (1955: 39–54).

110. Trager and Trager 1962: 93.

111. The best-known critique can be found in Scott 1996. For a fascinating look at the pressures on aid workers that discouraged the acquisition of greater local knowledge, see Hoag and Öhman 2008.

4 On the Fallacies of Cold War Nostalgia: Capitalism, Colonialism, and South African Nuclear Geographies

Gabrielle Hecht

As an old Cold Warrior, one of yesterday's speeches almost filled me with nostalgia for a less complex time. Almost.
—US Secretary of Defense Robert Gates, responding to Russian President Vladimir Putin's criticism of US foreign policy, February 2007

A peculiar nostalgia for the Cold War has pervaded American public discourse since September 11, 2001. Pundits and scholars alike invoke the Cold War as a time of clear, stark choices: capitalism vs. communism, good vs. evil, us vs. them. The oddly wistful tone of this false memory flows from the fiction that the Cold War remained cold, by which people usually mean that nuclear deterrence "worked": against all odds, the United States and the Soviet Union didn't annihilate the human race. Such nostalgia relegates proxy wars to near-irrelevance, not only because of the subalternity of their locations and victims, but also because no atomic bombs exploded on their battlefields (despite a few close calls). The implicit contrast is with the complexity of present-day geopolitics, especially the threats posed by non-state actors and their entanglements with "rogue" nuclear states.

In history and memory, the superpower arms race remains the identifying mark of Cold War technopolitics. The arms race raised the stakes of ideological struggle to the level of apocalypse, and perpetuated simplifications. Countries were either "nuclear states" or not. Bombs were either atomic or conventional, materials either radioactive or not. The ontology of the "nuclear" seemed beyond political dispute, apparently derived from the esoteric realm of physics and the technological achievements of brilliant engineers. Cold Warriors and their activist opponents might clash about the nature of the communist threat, but they agreed that nuclear systems formed the quintessential, exceptional terrain on which to wage (or oppose) the Cold War.

The Cold War filter made nuclear things appear primarily an affair of the North and of nation-states. Nuclear weapons would replace colonialism as a structure for situating nation-states in a global hierarchy. France and Britain saw them as technopolitical solutions to crises of national identity and security thrown up by their dwindling empires and the rising superpowers. Other nations later invoked nuclear systems (military, civilian, or mixed) as symbolic of their status as sovereign states. Nuclear systems became proof of post-colonial nationhood and offered solutions to newly configured problems of national and regional security. National nuclear expertise promised freedom from imperial dominance. International atomic organizations and treaties further codified nuclear nationalisms by differentiating "nuclear states" from all others.

In the first section of this essay I argue that multiple meanings and manifestations of the "nuclear" were produced and contested in the institutional accommodations of global Cold War. Nuclear ontologies—categories of things that did or did not count as "nuclear"—were ambiguous, not fixed. These ambiguities, I suggest, emerged from a set of dialectical tensions that played out in entities such as the International Energy Agency (IAEA) and the Nuclear Non-Proliferation Treaty (NPT): tensions between the geopolitics of colonialism and decolonization and those of Cold War East–West struggles, between managing the spread of nuclear weapons and encouraging the flow of other nuclear things, and between the moral high ground claimed by the prevention of planetary annihilation and the more mundane commercialization of nuclear power.

In its dual position as the West's most notorious colonial power (after 1960) and one of its primary uranium suppliers (and the two were linked), apartheid South Africa played a pivotal role in these tensions, and in managing the resulting ontological ambiguities. In the second section, I explore how South African uranium enacted the neocolonial accommodations and nuclear ambiguities embodied in the IAEA and the NPT. Producing and selling South African uranium involved continually negotiating relationships among colonialism (especially its manifestation as apartheid), nuclearity, Cold War, and markets. These negotiations played out in—and between—global, national, and transnational spaces. The resulting technopolitics shaped global nuclear (dis)order in ways that have lasted well beyond the Cold War. My conclusion takes up some of these legacies.

Colonialism and Capitalism in the Making of Global Nuclearity

In 1951, atomic bombs and colonial power induced twin anxieties in leaders of the dwindling empires of Britain and France. That year, Churchill's chief scientific advisor, Lord Cherwell, remarked: "If we have to rely entirely on the United States army for this vital weapon, we shall sink to the rank of a second-class nation, only permitted to supply auxiliary troops, like the native levies who were allowed small arms but no artillery." French parliamentary deputy Félix Gaillard echoed the sentiment: "Those nations which [do] not follow a clear path of atomic development [will] be, 25 years hence, as backward relative to the nuclear nations of that time as the primitive peoples of Africa [are] to the industrialized nations of today." Nuclear = colonizer. Non-nuclear = colonized.[1]

In 1951, nuclearity was still almost entirely about weapons: who had them, who could get them, how to use them, how not to use them. In 1953, Eisenhower's "Atoms for Peace" speech signaled the possibility of another state of nuclearity: atomic power plants. At one level, its rhetoric recycled the winged gospel into the atomic creed, whose first verse was world peace and boundless abundance thanks to an energy free-for-all. At another level, as John Krige has argued, "Atoms for Peace" was a panoptic Orwellian fantasy, doublespeak for "Atoms for War."[2]

As "Atoms for Peace" morphed into the International Atomic Energy Agency, the atomic creed left space for post-colonial leaders to challenge the technopolitical geography of nuclearity asserted by the West. India stepped in first. Nehru proclaimed nuclear development a fundamental building block of Indian national identity. During negotiations over the IAEA statute, Indian delegates raised a challenge. If representation on the IAEA's Board of Governors relied solely on technical achievement and a Cold War East/West balance, they charged, the agency would reproduce immoral global imbalances. Instead, qualification for Board membership should combine nuclear "advancement" with regional distribution.

The challenge worked. A complex formula allocated five permanent seats on the IAEA's Board of Governors to member states deemed globally "most advanced in the technology of atomic energy *including the production of source materials*," and another five according to geographical region.[3] The remaining seats were distributed on a rotating basis to uranium producers in the Eastern and Western blocs, "suppliers of technical assistance," and global regional representatives at large. The result tempered Cold War obsessions with technological rankings with an acknowledgment of the geopolitical importance of decolonization, and laid an ideological and structural foundation for a global nuclear order.

But meaning and practice never flow directly from ideology and structure. What made a nation count as "most advanced"? What were "source materials," and how significant a manifestation of nuclearity were they? For decades, the meanings of these phrases were negotiated and renegotiated in ways that reflected not only changing technologies, but also shifts in global Cold War geographies wrought by colonialism, decolonization, and nationalism.

Apartheid South Africa played a critical role in such negotiations early on. Its delegates to the IAEA statute talks, for example, had insisted on including "source materials" as an indicator of nuclear technological "advancement" in the agency's statute. South African contracts to provide the United States and Britain with uranium ore for their atomic weapons had made the production of that particular "source material" vital to South Africa's economy.[4] Anticipating that the IAEA would play a central role in shaping uranium markets, and knowing that its apartheid policies would impede its election to the Board, South Africa desperately sought a permanent statutory seat. The apartheid state represented the antithesis of the postcolonial settlement pursued by India, which wanted to demote South Africa to one of the rotating "producer" seats. At the time, South Africa's "nuclear" activities consisted only of uranium ore production underwritten by a very small research program. Prevailing on their British and American customers for support, however, South African delegates persuaded the others that "source materials" should count as an indicator of "advancement." In a technopolitical geography where the Cold War trumped racial inequality and other African producers remained under colonial control, South Africa's uranium production could serve as the pinnacle of *African* nuclearity.[5]

The question of whether "source materials" were sufficiently "nuclear" to warrant a governing seat on the IAEA begged the question of how to define "source materials" in the first place. Uranium ore had to undergo milling, refinement into yellowcake, conversion to tetrafluoride and/or hexafluoride, and enrichment before it could become fuel for nuclear reactors (or bombs). At exactly what point did uranium stop being "source material" and become "fissionable material"? The difference mattered enormously, because the two categories would be subject to different controls. In the words of one South African scientist, "the definitions would have to be essentially practical, rather than 'textbook' in nature, . . . legally watertight, and must take account of certain political implications." In the end, the IAEA abandoned the more ambiguous term "fissionable material" (preferred by Indian delegates) in favor of three other categories: "source materials," "special fissionable materials," and "uranium enriched in the isotope 235 or 233."[6]

These three categories—and the distinctions between them—mattered for safeguarding the nuclear order. How to control the flow of materials and technologies to ensure that instruments of planetary destruction didn't fall into the wrong hands? Who could be trusted with which systems? Which materials, knowledges, and systems were specifically nuclear? Exactly what would "safeguards" mean? Definitions alone didn't prescribe a control method.

Such questions had to be worked out in conjunction with the *raison d'être* of the IAEA, and on this score the West and the Rest were split. Rhetorically, everyone agreed that promoting the "peaceful uses of atomic energy" was the IAEA's central mission, one which endowed the agency with high moral purpose. But in practice, uranium producers and nuclear system builders had more mundane interests in mind: build reactors, make nuclear power commercially viable, and create a market for technologies and uranium that would sustain their own nuclear industries.[7] The West would sell to the Rest, while somehow (the details remained fuzzy) ensuring that ensuing programs would serve civilian rather than military ends. By providing a bigger market, the Rest would help the West commercialize its nuclear power systems. Spearheaded by South Africa's delegate, Donald Sole, representatives of the West argued that the IAEA should channel resources to countries that could develop nuclear infrastructures quickly, rather than function as yet another technical aid agency. But the Rest—led by India—saw things differently. The IAEA should ensure that emerging nuclear hierarchies not perpetuate global inequalities. They concurred that the agency should spread nuclear systems, but (for equally self-interested reasons) they argued for a broader distribution of resources.[8]

Also contentious were the conditions that would accompany whatever distribution of resources occurred. The United States wanted purchasers of nuclear technologies and materials to agree not to use their purchases toward military ends, and to accept international inspections in the name of averting apocalypse. Most other nations selling nuclear systems paid lip service to such measures. But buyers were underwhelmed by the prospect of controls, arguing (with India) that regulating access would perpetuate colonial inequalities and undermine national sovereignty.

Such high moral rhetoric obscured more mundane political and commercial issues. The US, the UK, and the USSR refused to accept inspections of their nuclear installations. Western European nations charged that IAEA safeguards inspectors could double as commercial spies, and accused the US and the UK of seeking competitive advantage. But it was their invocation of Cold War tensions (namely, that there could be no question of allowing Eastern Bloc IAEA inspectors into Western European nuclear facilities) that led to a 1961 agreement with the recently created Euratom, which granted its member states (France, Germany, Italy, and the Benelux countries) the right to inspect each other's facilities.[9] Third World nations with nuclear aspirations saw this Cold War-justified self-regulation as rewarmed imperialism.[10]

The Treaty on the Non-Proliferation of Nuclear Weapons (informally "the NPT"), meanwhile, expressed all the same tensions. Under the NPT, "nuclear weapons states" pledged not to transfer atomic weapons or explosive devices to "non-nuclear weapons states." The latter, in turn, renounced atomic weapons and agreed to accept IAEA safeguards and compliance measures. Strikingly, the NPT invoked human rights language and the rhetoric development:

1. Nothing in this Treaty shall be interpreted as affecting the *inalienable right* of all the Parties to the Treaty to develop research, production and use of nuclear energy for peaceful purposes. . . .
2. All the Parties to the Treaty undertake to facilitate, and *have the right to participate in*, the fullest possible exchange of equipment, materials and scientific and technological information for the peaceful uses of nuclear energy. Parties to the Treaty in a position to do so shall also cooperate in contributing alone or together with other States or international organizations to the further development of the applications of nuclear energy for peaceful purposes, especially in the territories of non-nuclear-weapon States Party to the Treaty, *with due consideration for the needs of the developing areas of the world.*[11]

In an effort to accommodate post-colonial morality into a Cold War paradigm, the NPT essentially declared that nuclearity—of the "peaceful" persuasion—was a fundamental right. It thus codified nuclear exceptionalism: no other international agreements referred to any scientific or technological activity as an *"inalienable* right" of special importance to "the developing areas of the world."[12]

The NPT made nuclearity a global "right," but left matters of safeguards and "technical assistance" to developing nations to the IAEA. Between 1961 and 1972, the IAEA produced five documents, each with a somewhat different solution to the ontological problem of which materials and technologies were sufficiently nuclear to demand safeguards and inspections. By then South Africa had cemented the technological justification for its Board seat with an extensive nuclear R&D program, and wanted to minimize external oversight of its uranium industry. It led other uranium producers in continually, and successfully, pushing to exclude mines and ore-processing plants from official definitions. The 1968 safeguards document, for example, defined a "principal nuclear facility" as "a reactor, a plant for processing nuclear material, irradiated in a reactor, a plant for separating the isotopes of a nuclear material, a plant for processing or fabricating nuclear material *(excepting a mine or ore-processing plant)*. . . ."[13] Uranium mines and mills were thus specifically excluded from the "nuclear"—even from the residual category of "other types" of "principal nuclear facilities." In 1972, uranium ore was specifically excluded, as well, from the category of nuclear "source material."[14] NPT signatories did have to inform the IAEA of yellowcake exports, but these were not subject to international tracking or inspections. Nor were uranium mines and mills. Restrictions on the end use of uranium ore were relegated to bilateral agreements or sales contracts; they thus depended on the two parties involved, and consisted only of unenforced, unverified pledges on the part of the recipient.

Together, the IAEA and the NPT thus enacted the technopolitics of global Cold War. Ideologically, they appealed to the highest possible moral purpose: the safekeeping of the human race and its planet. At the same time, they offered the possibility of reconciling this danger with the (also morally powerful) promise of unlimited modernity. They thus produced the "nuclear" as a mutual, but structurally unequal, engagement of the Cold War with the colonial and the postcolonial.

They also served as a forum to work out another dimension of the global Cold War, at least from the perspective of the West. It wasn't enough to keep the world safe from annihilation: the world also had to be made safe for capitalist markets. Nuclear things couldn't merely be instruments of state—they also needed to be instruments of capital. With the right rules in place (i.e., ones that favored nuclear suppliers), the IAEA could function as a rather effective trade organization. And no nation was more enthusiastic about this dimension of the agency than apartheid South Africa, struggling to hang on to its colonial state within an international order that increasingly condemned its practices.

Uranium Markets for Apartheid

In 1954, when negotiations over crafting the IAEA first began, the United States and Britain had active contracts for large quantities of South African uranium. In South Africa, the ruling National Party had begun elaborating the complex bureaucratic and spatial apparatus of apartheid. In the rest of the world decolonization had barely begun, and South Africa wasn't quite the international pariah that it would soon become.

In the late 1950s, things began to change. The US Atomic Energy Commission decided it had ordered enough uranium, and announced that it would stop buying foreign ore. "Stretch-out" agreements softened the economic blow to suppliers. These would expire in 1967, however, so South Africa had to find new customers if it wanted to keep that part of its industry going. This became increasingly challenging after 1960, which in South Africa marked the Sharpeville massacre and the country's withdrawal from the Commonwealth, and in the rest of the continent saw the formation of more than a dozen new postcolonial states.

As these newly independent states began entering the IAEA, they joined a push to evict South Africa. In 1963, some twenty nations signed a statement condemning the apartheid government and urging the IAEA to conduct "a review of South Africa's policy in the context of the work of the Agency."[15] Donald Sole, South Africa's representative on the agency's governing board, denounced the declaration as "purely political" and insisted that the IAEA was "not a proper forum" for such matters. With support from the US delegation and a few others, he prevented the motion from passing.[16] Clearly, however, South Africa would have to battle for a legitimate place near the top of the global nuclear hierarchy. Sole predicted that "the Afro-Asian upsurge, combined with the continued preoccupation with politics, will bring South Africa very much under the harrow." He kept pushing the IAEA to remain narrowly nuclear, not "just one more international organization for the provision of technical assistance which provides at the same time a platform for the propagation of varying ideologies, Western, Communist, Neutralist, anti-White, anti-colonialist and the rest."[17]

When he wasn't occupied with IAEA politics, Sole used the personal contacts he'd made at the agency to forge relationships with potential uranium customers. In 1959, he escorted two representatives of the South African Atomic Energy Board (AEB) all over Western Europe. This "sales survey team" sought to forecast supply and demand for the upcoming decade, guess at the probable price structure of commercial contracts, and assess how safeguards might constrain the sale of uranium.[18] The tour proved so fruitful that the AEB's sales committee repeated it regularly, expanding it to include Japan's burgeoning nuclear power program. Increasingly, these trips involved marketing South Africa at least as much as uranium. In the inimitably oblique prose generated by apartheid state bureaucrats, the 1961 mission cheerfully reported that it "had been very well received and in a number of cases it was able to correct glaring misconceptions which existed overseas where South Africa was concerned."[19]

Nevertheless, marketing South African uranium remained tricky. In 1962, the United Kingdom signed contracts to buy 12,000 tons of uranium from Canada. South Africans worried that this deal had been sealed by their nation's worsening international status—especially its recent withdrawal from the Commonwealth.[20] But they felt optimistic that the next two biggest consumers of uranium in the Cold War West, France and Japan, might care more about obtaining uranium free from bilateral safeguards—France because of its nuclear weapons program, Japan for reasons of national sovereignty.[21] Eliminating safeguards clauses from contracts—which in any event had appeared only in response to US and UK pressure—could significantly expand South Africa's markets.[22] The prospect of unrestricted uranium could trump squeamishness over apartheid. Furthermore, France was not just a potential customer for unsafeguarded uranium, but also a competitor. In the early 1960s, the French Commissariat à l'Énergie Atomique operated uranium mines in France, Gabon, and Madagascar, and had active prospects for large mines in Niger.[23] The French were interested in South African uranium to ensure continuity and diversity of supply, and also to free up some of their own reserves for sale. Contracts free of safeguards had clearly made French uranium attractive: South Africans suspected that Israel had "broke[n] off negotiations for supplies of Rand concentrates" because it had "instead obtained the supplies [it] required from France, without safeguard inspection requirements."[24]

In tackling the problem of how to sell more uranium abroad, South Africans at home found themselves arguing about the relationship between uranium's status as a nuclear thing, its status as a commodity, and the effect of apartheid on its marketability. These debates played out in institutional tensions between the Atomic Energy Board and the mining industry (represented by the Chamber of Mines).

Until 1961, South African uranium sales had remained under the sole control of the AEB, the state's guardian of things nuclear. But the AEB didn't produce uranium oxide directly—that was done by the mining industry, which treated uranium as a by-product of gold. So long as the American and British bomb programs were the only consumers of South African uranium, this arrangement had worked well enough. But with the end of those contracts, the mining industry wanted to take charge of marketing. In 1961 the Chamber of Mines appointed Hugh Husted as its own "uranium adviser," and the Chamber and the AEB began discussing how to erect a uranium sales organization.[25]

Husted's very first report challenged the AEB's long-standing policy of keeping nuclear things secret. He suggested publishing more detailed information about South African uranium production for the benefit of potential buyers.[26] South Africa's commercial consul in New York agreed, noting that statistics "reflecting imports of Rand Concentrate" had already been published in the US and arguing that

> it may also be opportune to review the arrangements in regard to the treatment of uranium oxide as an export commodity. At present information relating to uranium is treated as secret and is handled by the Department of Foreign Affairs, but there appears to be no reason why it should not in future be treated *as a normal export commodity*.[27]

The mining industry and the Department of Commerce and Industry saw uranium as a commodity in search of customers. One might debate about whether uranium belonged in mineral markets or in fuel markets. Either way, however, publishing statistics could demonstrate South Africa's strength and reliability as a supplier to potential customers.[28] Effective capitalism, after all, relied on persuasion: that's what marketing was.

But the AEB and the Department of Foreign Affairs objected forcefully to publishing production statistics for South African uranium. The secrecy of all matters nuclear—which in South Africa still meant all matters uranium—could not be violated. Anxious to retain the final word on such matters, the AEB insisted that

> uranium concentrates not be treated as a normal export commodity, because this material continues to have immense strategic significance and its transfer from one country to another is consequently fraught with political implications, also because countries like the United Kingdom and other possible buyers of South African uranium concentrates might not wish to have statistics published regarding sales of this material to them.[29]

By suggesting that the preservation of secrecy might become a marketing tool, the AEB sought to preserve uranium's exceptionalism—and thereby its nuclearity, and the AEB's ultimate power to pronounce its fate.

AEB experts thus framed the importance of secrecy in nuclear terms. For some "possible buyers," however, discretion was motivated at least as much by the provenance of the uranium. Consider France.

In 1963, Bertrand Goldschmidt, director of external relations for France's Commissariat à l'Energie Atomique (CEA), traveled to South Africa to discuss possible uranium purchases and scientific exchanges. He wooed South Africans by conveying personal sympathy with their "race problem":

> M. Goldschmidt was most favourably impressed by South Africa and said that its economic strength and degree of development far surpassed his expectations. A few years previously he had spent some months in Kenya where his sister and her husband were now obliged to contemplate the abrupt loss of a large estate which they had

built up over the past 20 years; he was therefore well able to comprehend the point of view of the Europeans in Africa and the complex nature of our multi-racial problem.[30]

Times were changing, however, and such matters now required tact.[31] France needed to preserve a privileged relationship with its former colonies. For this it would gladly pay a premium. The CEA's contracts with Gabon, for example, "could not easily be reduced as they constituted an essential form of economic subsidy with political implications." None of France's uranium sources could match South African prices, "but the invisible cost to France of cutting down her own sources of production would have to be very carefully weighed in the balance."[32] Precisely because uranium contracts always had implications beyond those of ordinary market transactions, Goldschmidt suggested, the CEA's willingness to purchase yellowcake depended on just how low the South African bid could go.

In the meantime, the CEA offered to help South African experts experiment with processes to convert their ore into uranium tetrafluoride. This offer of collaboration emphasized the benefits of thinking beyond mere market considerations. Goldschmidt's expressions of colonial solidarity, meanwhile, seduced the South Africans and gave them confidence that France, at least, would not let racial politics trump nuclearity. Hugh Husted argued that "the value of the deal in terms of money, while important, lies more in the long-term implications."[33] The French approach persuaded the South Africans to offer the CEA rock-bottom prices, and a substantial uranium contract resulted.[34]

Before concluding the deal, however, Goldschmidt reiterated that France could in no way jeopardize its relationship with other African nations:

[T]here were international political considerations, notably the open objections of the Afro-Asian bloc to countries trading with South Africa. . . . [B]ecause of this latter implication, Mr Goldschmidt made it quite clear that any deal that might be concluded should be regarded as confidential.[35]

Could anyone imagine a better vindication of the AEB's refusal to publish uranium production statistics? To be sure, the fact that France's motivation for secrecy stemmed from apartheid's threat to its postcolonial relations—not nuclearity—was a glitch. But this could be made to vanish easily enough. The AEB warned that the contract had to remain "strictly confidential, since the French authorities regarded all government contracts for nuclear materials as secret."[36] In a typical apartheid syllogism, secrecy was redistributed onto the nuclearity of uranium, the pesky matter of racism swallowed whole by Cold War reasoning.

Notwithstanding the AEB's rhetorical sleight of hand, the mining industry continued to fight for greater independence in uranium production. It achieved some measure of this in 1967 with the creation of the Nuclear Fuel Corporation (NUFCOR), structured as a consortium of uranium-producing gold mines. NUFCOR coordinated the uranium output of the mines, operated the plant that processed this output into uranium oxide, and marketed the oxide overseas. The arrangement gave the mining industry greater commercial autonomy while (in principle) preserving close technical collaboration with the AEB.

At first, relations between NUFCOR and the AEB went smoothly enough, thanks in no small part to the French connection. When the CEA's test conversion of South African ore to uranium tetrafluoride (UF_4) proved successful, NUFCOR placed an order for a French UF_4 plant. This pleased the AEB, which promptly suggested that NUFCOR now investigate the next stage of the industrial process: converting UF_4 into uranium hexafluoride (UF_6), the feed material for enrichment plants.[37]

Some South African contracts specified uranium oxide, in which case the customer arranged conversion of the South African material separately, but some were for UF_6. To satisfy these, NUFCOR had arranged for toll conversion at a British facility. But what if the new Labor government in Britain decided to block South African material? France had proved friendly to South Africa, and the French conversion process carried considerable commercial promise. Perhaps NUFCOR should switch to the CEA once its contract with Britain expired? Or even build a hexafluoride conversion plant in South Africa?

AEB metallurgists seemed utterly persuaded that selling a more highly processed product would inevitably yield greater profits. Raw materials, they argued, were what colonized, backward nations produced. A highly developed metallurgical industry placed South Africa ahead of the rest of the continent, assimilated it into the industrialized West. The nation practically had a duty to produce the most highly processed form of uranium. Besides, NUFCOR's profit margin would surely increase if it could sell UF_6 directly.

NUFCOR obviously found the prospect of a larger profit margin tempting. But it also knew that profits did not flow inevitably from processing. Conversion plants entailed high capital costs. Profits depended on economies of scale, trends in the price of uranium, finding enough willing customers, ore grade, efficient removal of impurities, the particular conversion process, and much more besides. So before committing to a UF_6 plant, NUFCOR commissioned a series of feasibility studies and market forecasts.[38]

Meanwhile AEB officials squirmed impatiently. They had invoked the language of profit, but grander things were at stake. Sometime in the 1960s, the AEB had secretly begun research into a full-blown enrichment plant (hence its desire for conversion plants to produce the feed material).[39] Initial results seemed promising. This time, peeling off the first layer of secrecy governing nuclear activities might bolster South Africa's weakening international status—especially if framed in terms of the ideological edifice of nationalism, capitalism, and developmentalism with which the "Western powers" justified their nuclear hegemony. Prime Minister B. J. Vorster decided to go public with the fact (though not the details) of the enrichment project. In 1970 he revealed to Parliament the existence of a pilot plant, calling it an "obvious" step in the (white, industrial) history of the nation.[40] Enrichment would allow South Africa to market uranium more profitably, and eventually to supply its own nuclear power program. Vorster lauded South African scientists for bolstering "the prestige of their country. In the past they have made lasting contributions to science, but perhaps the achievement that I am announcing today is unequalled in the history of our country." He went on:

> The South African process, which is unique in its concept, is presently developed to the stage where it is estimated that under South African conditions, a large scale plant can be competitive with existing plants in the West. . . . South Africa does not intend to withhold the considerable advantages inherent in this development from the world community. We are therefore prepared to collaborate in the exploitation of this process with any non-communist country(ies) desiring to do so, but subject to the conclusion of an agreement safeguarding our interests. However, I must emphasize that our sole objective in the further development and application of the process would be to promote the peaceful application of nuclear energy—only then can it be to our benefit and that of mankind.[41]

Vorster's insistence on the "unique" character of the South African enrichment process would quickly pervade official discourse on this topic, serving as an affirmation of national(ist) technological prowess. So would the more veiled remark that a large-scale plant would prove competitive "under South African conditions." That phrase encoded two things. First, the South African process was not really unique. It closely resembled the German "jet-nozzle" enrichment process. (Indeed, the ANC would later argue that South African experts had copied that process directly.[42]) But high energy costs made the jet-nozzle process uneconomical in Europe. Hence the second bit of code: "South African conditions" referred to cheap energy, via cheap black labor. Official descriptions of the enrichment program elided these "conditions," remaking them into an apolitical, technical trait of industrial development.

The prospect of commercial enrichment in South Africa also derived legitimacy from "existing plants in the West." In 1970 the United States had a monopoly on the actual provision enrichment services, but pilot plants in France and the Netherlands heralded the arrival of a commercial enrichment market. South African efforts seemed in line with these, especially as apartheid leaders asserted that they would sign the NPT as soon as a few thorny matters of sovereignty were resolved.

In the event, South Africa didn't sign on to the NPT until 1991. We know today that Vorster didn't merely elide apartheid repression in this speech. Like most political leaders of nuclear weapons states, he also lied about the intent of the enrichment program. The lie was deeply entwined with an two-pronged assertion about South Africa's strong "non-communist" affiliation: South Africa would act like a responsible Western nation in matters both nuclear (by not proliferating) and capitalist (by sharing its technology under proper commercial conditions). A few days later, the Minister of Mines developed this theme by congratulating parliamentarians on "the insight shown . . . in not asking unnecessary questions"[43] about nuclear development over the years. This helped ensure the secrecy and security of atomic matters, which in turn bolstered South Africa's credibility as a responsible nuclear state. Maintaining secrecy not only proved the nation's nuclear modernity, it also helped South Africa weather geopolitical storms. In a typically oblique reference to international opposition to apartheid, one member asked:

> What could it mean if the knowledge underlying this discovery [of a new enrichment process] became of more general knowledge throughout other countries of the world who are to-day looking with very suspicious eyes, shall I say, on South Africa? The security attached to this particular discovery to my mind transcends the need for security in any other matter at our disposal here in South Africa.[44]

Secrecy thus legitimated South African nuclearity both inside and outside national borders.

For NUFCOR, the announcement of an enrichment program cast the AEB's push for a UF_6 plant in a new light. Clearly, the AEB needed a conversion plant to feed its enrichment plant. Feasibility studies showed that a hexafluoride plant would not be commercially viable below a minimum annual capacity of 4,000 tons of uranium.[45] Market forecasts suggested that demand would not support this capacity anytime soon. Yet there were other considerations: "As a financial proposition [the production of UF_6] might not be attractive to Industry but [it] might be considered as being in the national interest."[46] So surely the state should subsidize the plant?

Whatever the case, NUFCOR refused to build a commercial-scale conversion plant on its own. The AEB and the mining industry compromised on a pilot plant, funded primarily by the state with a little help from NUFCOR's research budget.

As plans for South African enrichment capacity progressed, relations between the AEB and the mining industry deteriorated. The AEB's veil of secrecy thickened. NUFCOR received no information on the potential feed requirements for an enrichment plant. It began withholding information from the AEB, citing commercial confidentiality. AEB president A. J. A. Roux grew livid, noting acidly in mid 1974 that "in terms of the Atomic Energy Act the Board was entitled to ask that all information on uranium research should be disclosed to the Board." He hoped the situation wouldn't come to that.[47]

The central objects of tension were the hexafluoride plant, the uranium market, and the relationship between them. Roux desperately wanted the hexafluoride plant so that South Africa could feed its own enrichment plant. It wasn't merely that the enrichment plant would in turn would fuel South Africa's atomic bombs (none of Roux's correspondence with the mining industry even hinted at this, and if NUFCOR directors suspected anything at this stage, they kept it out of their written records). Roux believed that the ability to produce enriched uranium—a product legitimated by Cold War capitalism, whose manufacture and flow powered neocolonial circuits—would protect South Africa's international standing:

The production of uranium in South Africa is a matter of great importance to the State, quite apart from the significant economic benefits that arise from the export of uranium concentrates. South Africa's position in international affairs and its prominent status as a foundation member of the International Atomic Energy Agency are very largely due to the fact that this country is one of the top three uranium producers in the world.[48]

This position absolutely had to be preserved because the IAEA "was perhaps the last international body where South Africa was permitted to make a contribution and where South Africa's viewpoint was given respectful attention."[49] The solution to South Africa's political isolation lay in the uranium market. And there, Roux believed, prospects looked good: the spot price had begun to climb after an all-time low, and orders for reactors were up. "I am sure you will agree," Roux wrote to NUFCOR chairman A. W. S. Schumann, "that the present marketing situation indicates that a far greater quantity of uranium can be sold in future years than is currently being produced by the mining industry."[50]

Schumann wasn't so sure. The spot price bore little relationship to real prices fixed in long-term contracts. Roux accused the mining industry of holding back crucial market forecasts. Schumann replied that "the whole subject of uranium marketing was very much more tentative than the Board might think and it was difficult to know how to give more information than was already supplied which could be useful." AEB director of extractive metallurgy R. E. Robinson chimed in: he needed NUFCOR's projections because he wanted to establish an econometric model to predict the uranium market. Surely, Schumann replied testily, this was beyond the scope of a metallurgical laboratory. On the contrary, Robinson affirmed, such knowledge was "essential . . . in formulating [a] research programme."[51]

As for Roux, he thought that "a projection of future uranium prices and their possible effect on the Industry's uranium production capability" was crucial to the decision about whether to build a commercial-scale enrichment plant. As documents, market forecasts could help to create the markets themselves. Could industry ramp up to this level of production? If so, would it? Schumann replied that "since the Industry had no official information on the prospects of for uranium enrichment in South Africa its interest in possible UF_6 production had declined."[52] Without feed contracts—or, at the very least, more information about enrichment—industry had no incentive to increase production.

In the end, South Africa's conversion and enrichment plants did not help South Africa maintain its foothold in the IAEA. The apartheid state was voted out of the agency's Board of Governors in 1977. But its uranium flowed north long after the establishment of sanctions, and well after the emergence of concrete evidence of a bomb program. The apartheid state never did build a commercial-scale enrichment plant. But the facility that had begun its life as a pilot plant went on to enrich enough fuel for at least six atomic bombs. It was decommissioned, along with the bombs, during the death throes of apartheid—which were also the dying days of the Cold War.

Conclusion

Since the advent of democracy in 1994, South Africa has become the poster child for non-proliferation. Proliferation experts invoke the dismantling of South African bombs as evidence that atomic weapons development can be rolled back. The eradication of archival records by state officials is even, sometimes, presented as a virtuous act: destroying thousands of papers documenting apartheid's arsenal prevented atomic secrets from falling into "unreliable" hands. Not, many would hasten to add, that the African Nation Congress itself posed a threat. But it had, after all, received support from countries like Libya during the liberation struggle. Who knew how former freedom fighters might repay such debts?

Such rhetoric renders nuclear renunciation as redemption, the eradication of history as an act of global citizenship. The moral of the story? If South Africa could give up the bomb, so can others.

But if we look beyond "the bomb," if we push past the despair that scholars feel when facing large-scale archival annihilation, we can see that the history of South African uranium production carries deeper and more complex implications for the nuclear world we inherited from the global Cold War.

South African uranium networks involved wide-ranging accommodations between the Cold War, colonial power, and capital flows. American and British atomic arsenals in the early Cold War gave birth to large-scale uranium production in South Africa. The global commercialization of nuclear power enabled and structured by the IAEA, Euratom, the NPT, and other supranational institutions sustained such production. The South African uranium industry operated in a space delineated by entanglements between the politics of market capitalism and those of global Cold War. Representatives of the apartheid state helped to shape that space through their interventions in uranium's nuclearity. France's uranium purchase, for example, was driven by its own atomic arsenal, structured by Euratom's safeguards, inflected by its ties to postcolonial Africa, and subsidized by its search for conversion business.

Cold War nostalgia trades on the notion that even though nuclear relations "back then" posed the prospect of planetary annihilation, bipolarity made their control relatively straightforward; the proof is that we're all still around to talk about it. But the messiness of nuclear diplomacy today flows directly from Cold War structures and practices. Safeguards regimes and the uranium market emerged from the same historical processes. Uranium became a legitimate commercial product thanks to safeguards; the definitions and practices that constituted safeguards were circumscribed by the push to make uranium into a commodity. Thus South Africa could credibly claim to pursue commercial enrichment because the NPT and the IAEA defined a series of frameworks under which enriched uranium was a legitimate commercial commodity. The more other nations built on these frameworks, the more enrichment—and earlier stages of uranium production—became market activities. And the more the United States tried to dominate those markets, the larger and more unwieldy they became. Over the course of the 1970s and the 1980s, the US implemented a series of measures designed to protect its domestic uranium production from foreign competition. In the long run, these only generated more competition, as producers elsewhere found niches for non-US uranium. From the perspective of "pure" market capitalism, the geographic spread of uranium production became more and more justified.

Safeguards measures did seek to keep pace with this expansion. In 1971, shortly after the NPT came into force, an international committee was formed to devise a "trigger list"—a list of things nuclear enough to trigger safeguards.[53] At that time, safeguards were understood as export controls: states in a position to export these things should somehow ensure that purchasers not divert them for military purposes. The first trigger list, published by the IAEA in 1974, was both brief and general. Even still, not all "nuclear exporters" agreed with its specifications. Competing lists developed, grew longer, became more detailed.[54] By the late 1990s, the list of enrichment plant components supposed to trigger safeguards exceeded 20 pages and specified the tolerances and diameters of shut-off valves and rotary shaft seals "especially designed or prepared" to handle uranium hexafluoride gas.[55] Yet not even these minute details resolved everything. Did uranium ore count as a source material? It depended which IAEA document you read. Did yellowcake count as "natural uranium" for export purposes? Also unclear. Until the passage of yet another protocol in 1997, uranium mines and mills were not subject to the ritual practices that certified the separation between civilian and military domains and established the relative nuclearity of things. (Even then, adherence to the protocol was voluntary.)

The end of the Cold War did not interrupt these dynamics between safeguards and uranium markets. If anything, it intensified them. Most immediately, the collapse of the Soviet Union enabled uranium producers in the former republics to sell their product to Western utilities at rock-bottom prices. Western uranium producers cried foul and lobbied for protectionist measures. Uranium producers in southern Africa (including Namibia) were especially distressed: they had just broken loose from the anti-apartheid sanctions that had hampered (though by no means stopped) their commercial activities in the mid 1980s, and had been looking forward to flooding the market.

The legitimacy of uranium as a commercial commodity has always had a geography. The end of the Cold War just made its contours more visible, and more extensive. It's not that the end of the Cold War changed nothing, but rather that the change has not been one from simplicity to complexity. Nor has it enacted a rupture. Current tensions between the spread and containment of nuclear things have deep roots in the Cold War. They emerge directly from institutions, practices, and meanings generated during that time. At the dawn of the twenty-first century, Iranian leaders cite the "inalienable right" clause of the NPT to legitimate their nuclear efforts, arguing that any attempt to stop them would constitute imperialism. They use the discovery of large uranium deposits in Iran to argue for enrichment as the natural next step in their nation's development. In its nationalism and its technopolitical logic, President Mahmoud Ahmadinejad's announcement of Iran's ability to enrich uranium eerily echoed Vorster's declarations concerning South African enrichment.

In the 1970s and the 1980s, the prospect of an apartheid bomb was at least as frightening to postcolonial Africa as the prospect of an Iranian bomb is to twenty-first-century North America and Europe. The leaders of the African National Congress were not fooled by Vorster's attempt to legitimate uranium enrichment in commercial terms. South African enrichment efforts were but the tip of the iceberg, they argued; the apartheid state was building a bomb. At the 1979 launch of the World Campaign against Military and Nuclear Collaboration with South Africa, director Abdul Minty led off UN hearings with a phrase that would resonate for years to come: South Africa was the "nuclear Frankenstein" of the West, and only through sanctions could the West redeem itself and hope to control its monster.[56] Sanctions were eventually imposed, but only after enormous resistance from the West, and not before the apartheid state built its bombs.

Some things are different. It matters that apartheid South Africa was deeply embedded in American and European nuclear networks, while post-revolution Iran is embedded in Russian networks and especially in black markets seeking to subvert dominant nuclear and uranium markets (we need to note that Iran under the Shah got its nuclear start from the United States and its uranium from southern Africa). In the case of Iran, it's the West pushing for sanctions, the non-West resisting them. (Funny how these categories of Cold War geography continue to hold sway, even as their contents change.) Ahmadinejad has dismissed the power of sanctions with a wave of his hand, proclaiming them a thing of the past. (Debates still rage over whether economic sanctions against South Africa did anything to hasten the end of apartheid.)

Nevertheless, the pressures exerted by commodification remain strong. Russia resisted sanctions against Iran: it didn't want to lose its multi-million dollar investment in the Bushehr reactor site, nor did it want to lose face by yielding to US pressure. The IAEA too has sought to protect its investments: it took Iran to task for not permitting inspections, but kept 55 projects of "technical assistance" to Iran active until March 2007, when it suspended 22 of them (leaving the remaining 33 on the grounds that they had medical, agricultural, or humanitarian purposes).[57]

As for the dream of postcolonial nuclearity, it's more powerful than ever. After the end of apartheid, South Africa regained its seat on the IAEA's Board of Governors. At a meeting in 2004, its representative to that august body vigorously defended Iran's right to enrich uranium, declaring that South Africa associated itself with the Non-Aligned Movement and citing a statement by President Thabo Mbeki:

> The imposition of additional restrictive measures on some NPT States, while allowing others to have access to these capabilities, only serves to exacerbate existing inequalities . . . already inherent in the NPT and undermines one of the central bargains . . . contained in the Treaty.[58]

The representative in question? None other than Abdul Minty, the very man who in earlier decades had led the charge against South African enrichment. Was it only because of concern about the intersections between nuclearity and imperial power that Minty defended Iran? Or do we need to bring the market back in for a fuller explanation? Because the ironies are endless. In March 2007, the ANC government declared that the time had come for South Africa to master the nuclear fuel cycle. Feasibility studies for commercial conversion and enrichment of uranium in South Africa are now underway. Again.

Acknowledgments

Portions of this text have been extracted and revised from Hecht 2006a, Hecht 2006b, and Hecht 2007. My thanks to the publishers for permission to re-use this material, and to John Krige, Fred Cooper, Hans Weinberger, and participants in both workshops on "Bodies, Networks, Geographies" for their comments on previous drafts.

Notes

1. Quoted on p. 41 of Cawte 1992 and p. 62 of Hecht 1998/2009.

2. Krige 2006.

3. Emphasis added. In 1956, members of the first category were the US, the USSR, the UK, France, and Canada; members of the second were South Africa, Brazil, Japan, India, and Australia. See Fischer 1997.

4. South Africa's uranium was located in the same mines that produced its gold. In the decade after World War II, supplying uranium to the US and Britain saved many of these mines from economic collapse and served as a conduit for massive foreign investment in South Africa's industrial infrastructure. See Borstelmann 1993 and Helmreich 1986.

5. "International Atomic Energy Agency," annex to South Africa minutes no. 79/2, 28/7/56, pp 10–11, National Archives of South Africa (hereafter NASA), BLO 349 ref. PS 17/109/3, volume 2. The position of South Africa relative to the IAEA is thoroughly documented in the BLO 349, BVV84, and BPA 25 series of these archives.

6. Ibid.

7. Forland (1997) sketches out how commercial considerations shaped the ways various nations approached safeguards. She also discusses the differences between bilateral and international safeguards, and conflicts within the US Atomic Energy Commission over what safeguards should consist of and how they should be implemented.

8. For more details, see Hecht 2006a.

9. Forland 1997; Krige 2008.

10. Scheinman 1987; Abraham 1998, 2006, 2009; Perkovich 1999; Forland 1997.

11. Article IV of Treaty on the Non-Proliferation of Nuclear Weapons (signed at Washington, London, and Moscow July 1, 1968) (emphasis added). For the full text of the treaty, and the US State Department's triumphalist version of its history, see http://www.state.gov/t/np/trty/16281.htm.

12. Between 1958 and 1993, the IAEA gave out $617.5 million in "technical assistance." The top ten recipients were Egypt, Brazil, Thailand, Indonesia, Peru, Pakistan, the Philippines, Bangladesh, South Korea, and Yugoslavia (Office of Technology Assessment 2005).

13. IAEA, INFCIRC/66/Rev.2, September 16, 1968.

14. IAEA, INFCIRC/153 (corrected), June 1972.

15. "Joint declaration by a group of Members in Africa and Asia regarding South Africa," general debate and report of the Board of Governors for 1962–63, agenda item 10, October 1, 1963, NASA, BVV84 13/1, volume 7 and annex.

16. "Statement by South Africa," general debate and report of the Board of Governors for 1962–63, agenda item 10, October 1, 1963, NASA, BVV84 13/1, volume 7 and annex.

17. "Report of the South African Delegation to the Fifth General Conference of the IAEA," October 12, 1961, p. 32, NASA, BVV84 13/1, volume 7 and annex.

18. Donald Sole, "Uranium Sales Survey: Interim Report on Continental Western Europe," June 8, 1959, NASA, HEN 2756 ref. 477/1/17.

19. AEB Sales Committee, minutes of 5th meeting, February 24, 1961, NASA, HEN 2756 ref. 477/1/17.

20. See H. McL. Husted, "Uranium Sales," report no. 7, November 8, 1963, NASA: HEN 2757 ref. 477/1/17/2; A. J. Oxley to Acting Secretary for Foreign Affairs, "UK/Canadian Agreement for the Supply of 12,000 tons of Uranium," August 1, 1962, NASA, BLO 40 ref. 64/237.

21. See Forland 1997 for a detailed discussion of Japan's push to dispense with bilateral safeguards agreements in favor of IAEA oversight.

22. And, as a reminder, consisted merely of pledges: there was never any question of verification by inspections.

23. Forland 1997, p. 11.

24. A. J. Brink to H. R. P. A. Kotzenberg, "Sale of Uranium by France," March 14, 1962, NASA, HEN 2756 ref. 477/1/17.

25. AEB Sales Committee, minutes of 5th meeting, February 24, 1961, NASA: HEN 2756 ref. 477/1/17; AEB Marketing Advisory Committee, minutes of 1st meeting, February 2, 1962, NASA: HEN 2756 ref. 477/1/17; H. McL. Husted, "Uranium Sales Organization, Report No. 1: 1962, Mission to Europe 15th October to 30th November, 1961," NASA: HEN 2757 ref. 477/1/17/2.

26. Husted, "Uranium Sales Organization."

27. Consul (Commercial) New York to Secretary for Commerce and Industry in Pretoria, November 30, 1962, included in AEB Marketing Advisory Committee, Supplementary Agenda for the 1st meeting for February 2, 1962, NASA: HEN 2756 ref. 477/1/17 (emphasis added).

28. Ambassador (Brussels) to Secretary of Foreign Affairs, "Visit to Brussels of Mr. Husted," September 28, 1962, NASA: BLO ref. 64/237.

29. AEB Marketing Advisory Committee, minutes of 1st meeting, February 2, 1962, NASA: HEN 2756 ref. 477/1/17 (originally all in capitals).

30. J. R. Jordaan (Ambassador to Paris) to Secretary for Foreign Affairs (Pretoria), "Uranium Sales to France," May 9, 1963, NASA: BPA, volume 8, ref. 18/25 part 1.

31. Moments like this—when Europeans would find ways to express support for South African racial policies in order to continue doing business—abounded. Another example: In 1964 Euratom's Belgian director-general, F. Spaak, traveled to South Africa in order to assess the sophistication of its industrial economy and discuss a possible uranium supply chain. Spaak's report did admit that black mine workers were paid one-fiftieth as much as whites, but his explanation mimicked that

of the Transvaal Chamber of Mines: "Wage levels have been and still are strongly attractive to Africans when the normal conditions of existence of the Africans in Central and South Africa are considered." (Was Spaak thinking of Belgian colonial mining companies as he wrote?) The sheer magnitude of the mining industry impressed him with its stability, power, and ability to shield South African uranium from market vagaries. Spaak concluded that South Africa had "considerable elasticity in uranium production based on the peculiar mining conditions applicable to the country." In the long term, he expected that South Africa would remain "an important supplier of limited quantities of uranium until the end of the century and perhaps until deep into the next century," and that Euratom should therefore maintain active contracts with South Africa. F. Spaak and J. Brinck, Report on South Africa and Its Uranium Supplies (A Translation from a German Version of a Report Originally Written in French), July 17, 1964, p. 4, 24, 30, NASA, MMY65 M3/7.

32. Ibid.; J. R. Jordaan (Ambassador to Paris) to Secretary for Foreign Affairs (Pretoria), "Uranium Sales to France," May 9, 1963, NASA: BPA, volume 8, ref. 18/25 part 1.

33. Husted, "Uranium Sales," p. 10.

34. The contract specified a firm delivery of 1,300 tons of uranium oxide between 1964 and 1968, an option to double the quantity during that period, and an option on another 1,000 tons between 1969 and 1973. AEB Marketing Advisory Committee, minutes of 6th meeting, March 23, 1964, p. 5, NASA: HEN 2757 ref. 477/1/17/2.

35. Ibid, p. 5.; Husted, "Uranium Sales," p. 10.

36. AEB Marketing Advisory Committee, minutes of 6th meeting, March 23, 1964, p. 5, NASA: HEN 2757 ref. 477/1/17/2.

37. NUFCOR, Record of discussions held in Johannesburg among representatives of SUCP, CEA, NIM, and NUFCOR on November 7, 8, and 12, 1968, Goldfields Archives (Rhodes University, Grahamstown, South Africa): records of the Nuclear Fuel Corporation (NUFCOR) Uranium Technical Advisory Committee (UTAC).

38. The material on these debates is too voluminous to cite individually. Records of meetings, economic and technical feasibility studies, and other correspondence relating to a potential UF_6 plant for NUFCOR can be found in the UTAC papers from 1968 to 1973 (Goldfields archives).

39. Some archival documentation concerning the enrichment project survives in the National Archives of South Africa; see, for example, papers in EAE 143 ref EA 2/2/13, volume 1 and MEM 1/590, ref. 121/2.

40. Newby-Fraser 1979, p. 91. This was the official history of the AEB, written by its public relations director; it's full of nationalist assertions of South African technological prowess.

41. "Verklang deur Sy Edele die Eerste Minister," 20 Julie 1970 (n.a.), NASA, MEM 1/590, ref. 121/2.

42. Cervenka and Rogers 1978; African National Congress 1975; Dan Smith, "South Africa's Nuclear Capability" (World Campaign against Military and Nuclear Collaboration with South Africa; UN Centre Against Apartheid, February 1980).

43. Republic of South Africa, Assembly Debates, July 27, 1970, p. 476.

44. Republic of South Africa, Assembly Debates, 27th July 1970, p. 475.

45. See, for example, J. C. Paynter and H. E. James, "Feasibility Study of the Commercial Production of Uranium Hexafluoride in South Africa," National Institute for Metallurgy Research report no. 924, March 16, 1970, Goldfields archives, UTAC papers.

46. NUFCOR, report of 18th meeting of UTAC, May 14, 1970, p. 4, Goldfields archives, UTAC papers.

47. R. E. Worroll, report on Interview with Dr. A. J. A. Roux in Pelindaba, June 17, 1974, part of NUFCOR/UTAC Circular no. 26/74, July 5, 1974, Goldfields archives, UTAC papers.

48. A. J. A. Roux to A. W. S. Schumann, liaison between AEB and NUFCOR, 3.6.74, part of NUFCOR/UTAC Circular no. 26/74, July 5, 1974, Goldfields archives, UTAC papers.

49. R. E. Worroll, report on Interview with Dr. A. J. A. Roux in Pelindaba, June 17, 1974, part of NUFCOR/UTAC Circular no. 26/74, July 5, 1974, Goldfields archives, UTAC papers.

50. A. J. A. Roux to A. W. S. Schumann, Liaison between AEB and NUFCOR, 3.6.74, part of NUFCOR/UTAC Circular no. 26/74, July 5, 1974, Goldfields archives, UTAC papers.

51. R. E. Worroll, report on Interview with Dr. A. J. A. Roux in Pelindaba, June 17, 1974, part of NUFCOR/UTAC Circular no. 26/74, July 5, 1974, Goldfields archives, UTAC papers.

52. R. E. Worroll, report on interview with Dr. A. J. A. Roux in Pelindaba, June 17, 1974, part of NUFCOR/UTAC Circular no. 26/74, July 5, 1974, Goldfields archives, UTAC papers.

53. This was called the Zangger committee, after its chairman, Claude Zangger. It was initially composed of 15 states that were "suppliers or potential suppliers of nuclear material and equipment." IAEA, INFCIRC/209/Rev. 1, Annex. See Hecht 2007 for more details.

54. Notably, two trigger lists developed in parallel: one under the rubric of INFCIRC/209, and another under the rubric of INFCIRC/254. Different nations

adhered to different lists; the two streams were brought into synch in 1977, but continue to develop separately.

55. IAEA, INFCIRC/209/Rev. 1/Mod. 4, April 26, 1999.

56. *Nuclear Collaboration with South Africa: Report of United Nations Seminar*, London, February 24–25, 1979. See also Smith 1980.

57. Molly Moore, "UN Nuclear Agency Curtails Technical Assistance to Iran," *Washington Post*, March 9, 2007.

58. South Africa Statement by Abdul Samad Minty, IAEA Board of Governors, Vienna, November 29, 2004, re: Item 4 (d): Implementation of the NPT Safeguards Agreement in the Islamic Republic of Iran: Report by the Director General (GOV/2004/83), available at www.dfa.gov.za.

5 Rare Earths: The Cold War in the Annals of Travancore

Itty Abraham

The Political Economy of Rare Earths

In a way it begins with semantic confusion. So-called rare earths were "rare" because it was assumed that these naturally forming mineral-laden compounds were scarce and hard to find.[1] It didn't mean they were valuable—though economists are quick to assume the identity of scarcity and value—at least not until two independent transformations of rare earths took place, each giving new meanings and value to particular rare earths. But before we get there, it also turns out that rare earths aren't really that scarce after all. This nomenclature is a symptom of the geo-historical origins of modern chemistry. In Europe in the nineteenth century, when rare earths were named, these compounds were found in only a few places, notably Sweden, hence they were assumed to be rare. Later on, their rarity could be associated with their place in the periodic table, since most rare earths are in the lanthanide group, which were among the last elements to be identified. Semantic confusion also doesn't imply unimportance. Once named rare it was assumed that they were, in terms of value; perhaps counter-intuitively, this feeling was most pronounced in places where such minerals are relatively abundant.

The southwest coast of India is one such place. Between the coastal towns of Kollam (formerly Quilon) and Nagercoil in the former princely state of Travancore (today's Kerala) lies a huge concentration of monazite beach sands. Monazite is a phosphate compound composed of a number of rare earth elements. Among the elements found in monazite is thorium, which is radioactive.[2] The official history of rare earths in Kerala begins innocently in 1908–09 when a German chemist, C. W. Shomberg, "chanced upon a few yellowish-green particles in some material from South India."[3] The first mining lease for monazite was granted to the London Cosmopolitan Mining Company, which, notwithstanding its name, was controlled by

German interests. By 1911, the bulk of extraction activities were carried out by two British companies: the Travancore Minerals Company and Hopkins and Williams Limited.[4]

Travancore's rare earths first became valuable when they moved from scientific curiosity to industrial use. Monazite was used in the manufacture of gas mantles, a form of incandescent lighting based on chemical rather than electrical power. The mantle is composed of thorium nitrate and other salts: when heated by a flame, these salts oxidize and give off an intense white light. The gas mantle lamp was invented by the Austro-Hungarian chemist Carl Auer von Welsbach, (1858–1929), a student of Robert Bunsen, after his chemical studies of the properties of rare earths in the late nineteenth century.[5] The gas-mantle lamp, which was in competition with early forms of electrical lighting, received a patent at the time when municipalities in Europe were beginning to invest in better street lighting. This ensured the commercial success of this product and set off a global scramble for rare earths that would eventually lead to the realization that these minerals were not so rare as once thought.

The demand for monazite rose steadily through the 1910s. Thanks to their political control of India, with its huge reserves of monazite, British firms had a lock on the supply of raw monazite to lighting companies all over Western Europe and North America. German firms, on the other hand, controlled the basic patents to manufacture thorium nitrate and "limited [the industry] to a very narrow circle in order to prevent competition."[6] The production of monazite rose steadily until the end of World War I, when the market crashed. The reason is explained by a technical advisor to the British Ministry of Supply: "It is doubtful that there can be any great expansion in the output of monazite until some [other] outlet is found. . . . The chief cause of the slump may be put down to the general decline in the use of gas as an illuminating agent."[7] Electricity and electric light had taken the place of gas and incandescent lamps far quicker than had been anticipated. The demand for monazite as an industrial input would be marked by acute fluctuations for the rest of the century, the first indication of its unreliability as a material ally.

The second transformation of rare earths came a half-century later when it became known that thorium was a potential nuclear fuel. Thorium's radioactive properties had been known since 1898, owing to the independent researches of Marie Curie and Gerhard Carl Schmidt. The importance of thorium as a potential nuclear fuel, however, only became widely realized after the dawn of the nuclear age in 1945. The isotope thorium-232, when bombarded by slow-moving neutrons, decays into uranium-233, which is fissionable and can be used as a fuel in a nuclear "breeder" reactor. Breeder reactors, so called because they are supposed to breed more fuel in the course of burning it up, represent one possible future of the nuclear power industry. Far from yet being commercially viable, breeder reactors are the subject of much ongoing research in India.

Although rare earths are not rare in the sense of quantity or scarcity, the name is still consequential. Focusing on the shifting value and meaning of rare earths in different contexts and time periods allows me to map a shifting terrain of political power in late colonial and early post-colonial India. Giving voice to rare earths enables me to discuss multiple histories that are usually situated in isolation from one another and certainly outside the ambit of Cold War studies. Rare earths, I find, are a fickle ally. From originally being a means to a grand political end, i.e., independence for the princely state of Travancore (a course of action that depended on the insatiable global demand for atomic fuels by the United States), thorium becomes an instrument by which Indian territorial sovereignty was reinforced. In the remainder of this essay, I explore three geopolitical scales across which the technopolitics of rare earths in Travancore become visible. Rare earths, like all material agents, are polyvalent: new meanings are produced through transformations that come from technological change but which become significant only when understood as political relations. Their embeddedness across multiple scales of political relations enables rare earths to become an instrument in more than one political struggle. To be successful as a political instrument, however, requires establishing a dominant meaning: in effect, stripping rare earths of polyvalence. This reduction of rare earths to a single dominant meaning would not be possible without the intervention of the Cold War scale.

The Cold War in this story is thus both a backdrop and a globally powerful index of meaning. In relation to struggles taking place at local and regional scales, it appears to have far less agency than ostensibly mute rare earths; this repositioning is in itself a useful way of claiming the relative autonomy of the region and its histories from the tyranny of the global. The Cold War scale, however, is critical to this story: by establishing a dominant meaning to hitherto polyvalent rare earths, it conditioned Travancore's future irrevocably.

Travancore and India

Travancore was one of the more progressive Indian princely states.[8] It was located on the southern Malabar coast, with Mysore to the east and the

Arabian Sea to the west. Along with its neighbors to the north, the smaller kingdom of Cochin and the Malabar region of British India, it would become the Indian state of Kerala in 1956. In 1937, a well-known lawyer and experienced administrator, Sir C. P. Ramaswamy Aiyar, former legal advisor to the royal family, was appointed Dewan or Prime Minister of Travancore. Sir C. P., as he was usually known, would rule Travancore for the next decade. He resigned from this position when the Maharaja of Travancore agreed to merge his state with India unconditionally.

Starting from 1942, in the midst of the huge "Quit India!" political uprising led by the Congress to demand freedom from colonial rule, a series of plans were advanced to discuss the eventual form of a politically independent India. Two basic problems were addressed in these schemes. The first considered the future of the princely states and their relationship with the political entities that would emerge after colonial rule ended; the second concerned the possibility that British India would be divided so as to establish a homeland for the subcontinent's Muslim population: Pakistan.

Although it was legally possible that the princely states could seek formal independence (or, as they preferred to put it, restore or resume independence), such a possibility was eliminated as a practical outcome, in private and in public, both by the Congress leadership and the last British Viceroy. In effect, the princely states had to choose whether to belong to India or to Pakistan. The two largest princely states found this choice unbearable. The Nizam of Hyderabad tried to get the United Nations to accept his plea for independence, to little avail.[9] In September 1948, the Indian military would intervene in what was called a "police action," leading to the end of Hyderabad's nominal independence. The Maharaja of Kashmir dithered until Pakistani irregular forces and Pathan tribesmen invaded the state and military conflict broke out between India and Pakistan, creating a territorial dispute that has yet to be resolved. The Muslim ruler of tiny Junagadh, which today lies in the Indian state of Gujarat, also chose to join Pakistan. This decision did not last long. In all three cases, the new Indian government sent troops in to take over, resolving by force what was a contested and legally murky transition of political power and legal authority.

In the traditional recounting of the events of this period, the story ends here. By and large, nationalist historiography sees the problem of accession as a political issue that has been long resolved. The credit for this relatively peaceful transfer of power is typically attributed to Deputy Prime Minister Sardar Vallabhai Patel and his chief assistant, the civil servant V. P. Menon. Menon's memoir of that period and his role in it sets the template for the consolidation of this conventional wisdom.[10] His story effaces how contentious and complicated the politics of this period were, and how many states and communities were willing to consider independent political existence as a practical possibility.

In recognition of their military and political weakness, the princely states especially focused on the legal case for their independence. The minutes of a meeting between the British Viceroy, Mountbatten, and the (Princely) States Negotiating Committee in June 1947 show how keen the princes were to get Mountbatten to agree that, if the relation of paramountcy between the British Crown and the Native States lapsed at the moment of the creation of India and Pakistan, it would mean that the princely states were in effect independent political entities.[11] Mountbatten did agree that this was might be so, but went on to make clear that, regardless of possible legal arguments for independence, "the first step" was for each princely state to enter into "practical negotiations" with the successor states.[12]

There were few officials in the princely states as vocal about independence as Travancore's Dewan, Sir C. P. Ramaswamy Aiyar.[13] An avowed conservative, an avid supporter of the monarchy,[14] and deeply distrustful and dismissive of the Congress' leading lights, including Gandhi, Nehru, and Patel,[15] C. P. Ramaswamy Aiyar spent the period from 1942 to 1947 engaged in making as strong a case as he could for Travancore's right to regain its independence after the British left. His argument was variously historical, legal, and geo-economic, and used all manner of forums, from closed meetings and public speeches to letters and press conferences, to press the case for Travancore's independence.

In his view, Travancore was quite unlike most other states.[16] Apart from the state's unique history, its location and unfettered access to the sea,[17] and the legal case (whose validity, if not relevance, even the British accepted), Sir C. P. would return constantly to the economic conditions allowing for a viable independent existence. Travancore's economic resources, especially its valuable export commodities (including tea, rubber, and coffee), allowed a favorable comparison with other, similar countries. Belgium and Thailand, both small but independent constitutional monarchies, were his favorite examples. Then why not Travancore?

It remains a matter of historical dispute whether C. P. Ramaswamy Aiyar intended to force Travancore's outright independence, as is strongly indicated by a literal reading of much of his writing and correspondence,[18] or whether his efforts were merely a carefully calibrated ploy to ensure protection for Travancore's special interests within an independent India, including de facto autonomy and continuation of the monarchy.[19] The answer also depends on the historical moment about which the question is being asked: the closer to political independence, the more intractable the Dewan seemed.[20]

If calling for Travancore's independence was only a bargaining ploy, C. P. Ramaswamy Aiyar was certainly willing to go quite far to make the bluff look credible.[21] In June 1947, the government of Travancore announced the appointment of Khan Bahadur Abdul Karim Sahib, former Inspector General of Police, as the representative of Travancore to the Dominion of Pakistan "to take charge of duties when Paramountcy lapsed." The Congress reacted harshly to this step. Nehru would write a furious letter to the Viceroy.[22] C. P. Ramaswamy Aiyar was "warned that Travancore would be starved out," and that there would be economic reprisals that would lead to the "elimination of independence in three months." In response to this "pistol at my head," C. P. Ramaswamy Aiyar raised the stakes still further: he began negotiations for the purchase of rice from the Pakistani province of Sindh to break a possible embargo on food grains and planned to do the same with Burma and Siam. With the textile mills of Ahmedabad and Bombay refusing to make sales to Travancore at the behest of the Congress, C. P. Ramaswamy Aiyar proposed to approach American and British suppliers, "until the time we establish our own textile mills." With the expected closing of the Indian market to Travancore's exports, other arrangements were being made with Pakistan and Australia for the sale of copra and coconut oil products.[23]

Clearly, all Travancore's resources, real and imagined, were being mobilized to push the case for independence. Under these circumstances, given how high the stakes were and facing the loss of much more than just his credibility were he to fail in this gamble, what is most surprising is the Dewan's seeming reluctance—or inability—to use a powerful weapon in his state's arsenal: thorium. At just this moment, the United States was scouring the world looking to purchase or control all sources of atomic energy; this was no secret to Travancore's rulers.[24] In turn, the United States was equally well aware of Travancore's strategic value. After all, more than two-thirds of the US's supply of rare earths came from this region.

The Global Atomic Monopoly

Once the United States dropped two nuclear explosives on Japan in August 1945, one war officially ended and another became public knowledge. The so-called Cold War between the US and the USSR had many manifestations. One of the most important early struggles was the US's effort to ensure its military dominance over all comers through a continued monopoly over all atomic expertise and materials.[25] This desire was expressed in a variety of forums, including introducing resolutions at the United Nations seeking to establish a global regime controlling all aspects of atomic research and production (the Baruch Plan).[26] The US was even willing to distance itself from its wartime allies Canada and Britain, both of whom played important roles in the making of nuclear weapons.[27] At the same time, the US was engaged in a furious effort to control all sources of nuclear fuel worldwide, either by outright purchase or by seeking a veto over sales to third parties. Initially uranium was the main object, but when, in 1944, it was realized that thorium too could be used to fuel nuclear reactions, it was added to the list of strategic commodities the US sought to control. Along with these nuclear fuels, mineral ores that had military applications, including beryl and cerium, became strategic objects to seek to control. Travancore's monazite sands, along with thorium in Brazil and South Africa, uranium in the Congo, and smaller quantities of strategic minerals in western Europe, became the focus of a global effort to ensure that these materials did not end up in the hands of the enemy and that they would be made available to the United States as needed.[28] The window of opportunity was small, and the US did its best to capitalize on it. At times because of ignorance on the part of sellers, at others because of its willingness to offer large sums of ready cash, the US was quite successful in getting control over much of the world's known supplies of uranium. However, what was out there proved to be quite underestimated. Rare earths turned out not to be so rare after all, especially after a massive search for them began all over the world. Even for the United States, with its deep pockets and unquestioned military might, a global monopoly would soon prove to be impossible. Jonathan Helmreich treats Eisenhower's Atoms for Peace initiative in 1954, which completely reversed the policy of denial, as the symbolic end of the American effort to obtain this monopoly.[29]

The original search-and-control effort operated through the Combined Development Trust (CDT), a joint US-UK endeavor. By now, the US and the UK's had effectively divided up the "free" world into independent spheres of influence.[30] The UK was "responsible" for its African colonies and much of Asia up to Hong Kong, the US for everything else. This complicated matters with respect to Travancore, which was ostensibly in the British zone. Since it was widely known to have the largest and richest holdings of monazite in the world, the stakes were too high to leave Travancore to its impoverished and weakened ally. Were it a matter involving only the US Department of State and the UK Foreign Office, US interest in Travancore's mineral resources might have been held in check. But atomic diplomacy, as played by General Leslie Groves and the CDT, often led to unilateral action against British advice and bypassed as well the Department of State, leading to inter-agency fury when the extent of these efforts was finally revealed.[31] Foreign partners would find that the outcomes of dealing with the US depended on the agency involved. In general, the US Department of Defense had the deepest pockets and was most susceptible to being played off against other potential international competitors, especially if they were alleged to be "Reds." The Department of State was in the middle. The Atomic Energy Commission had the least available funds, knew the most about the relative value of various atomic materials, and was the least concerned about other countries getting their hands on them.

Travancore's Rare Earth Reserves

Travancore's rare earth reserves were in the form of surface deposits, and, as a result, could be extracted relatively easily. Further, the presence and value of the monazite sands had been known since the beginning of the century. Four private companies, Indian and European, had been awarded exclusive contracts to mine and export monazite, mostly to Germany, France, the United States, and England. As it had done during World War I, the government of India sought to control foreign access to monazite, seeking particularly to ensure that it did not fall into the hands of German industry during World War II. Travancore obliged by placing an embargo on the sale of all rare earths, creating a considerable stockpile at home and shortages around the world, especially in the United States.[32] In June 1945, with the war winding down, Sir C. P. announced that he intended to embargo all "exports of ilmenite and monazite except those essential for war needs."[33] For a while, the situation was so dire that the Illinois-based Lindsay Lighting and Chemical Company, seeking to restore access to the 3,000 tons of monazite that it used to import annually from Travancore, threatened "to turn off the lights" by blocking sales of all thorium-related products to India.[34]

What Travancore wanted was an end to the sale of raw sands and access to the capital and expertise necessary to set up domestic processing plants. Sir C. P. had long wanted to strengthen the economic base of his state through industrialization, and saw the rare earths industry as an obvious place to start.[35] Among the most vexing issues in Travancore's relationship with the four mining companies was the companies' reluctance to build a processing plant locally. Hopkins and Williams and Thorium Limited, both British firms, much preferred the cheaper and more profitable option of shipping unrefined sands to be processed overseas. C. P. Ramaswamy Aiyar made no bones about his desire to find mining companies that would be more amenable to his designs, based on his confidence that the acknowledged value of monazite would attract other, better offers. In particular, he hoped to attract Lindsay Lighting, whose owner had powerful connections in Washington and who was already a regular and substantial purchaser of Travancore's monazite. This hope was not far-fetched. Between 1937 and 1945, 73 percent of US monazite came from Travancore; in the opposite direction, one-third of Travancore's monazite exports went to the US.[36]

Not surprisingly, Andrew Corry, an extremely able geologist attached to the Foreign Economic Administration in New Delhi (and later to become minerals attaché), kept a close eye on minerals-related developments around the country. Corry conducted, at the Travancore government's request, a survey of the monazite sands, and kept his superiors well informed of changes in India's export policies.[37] As long as the US's closest ally formally controlled India, there was an in-built limitation on the freedom of maneuver for US diplomats and expression of its interests. Although, strictly speaking, the US government could not make any official approach to the princely states without first getting the permission of the government in New Delhi, US diplomats soon found covert ways to begin direct talks with the Dewan of Travancore.[38]

From as early as 1944, the United States had been aware that Travancore harbored visions of independence. "As far as it is possible," one official noted, "[Dewan] intends to run Travancore (which has about the area and population of Belgium) as an independent nation analogous to those making up the British Commonwealth of Nations."[39] As late as a month before India would formally become independent of Britain, the official US policy on the future of the princely states was still wait-and-see. Though in principle US policy favored a united India "as far as possible," such an outcome was not written in stone. In correspondence with Ray Bower, the US consul in Madras, in mid July 1947, the Department of State advised: "We must be careful to avoid giving the Dewan any opportunity to claim a special relationship with us, *unless it should actually be our intention to establish one.* At this particular time, I should imagine he would be particularly eager to see in the Consulate's dealings with him some explicit or tacit approval of his idea of Travancore independence, in advance of the time when we can take any decision on that question."[40] William Pawley, head of the Inter-Continent Corporation, who had spent much time in India during and after World War II, would argue that if monazite was important enough, the US should support Travancore's independence, as it was a viable state "politically and economically."[41]

In sum, Travancore was in direct contact with the US, and the US was aware of Travancore's desire for independence. As a matter of policy, the US was trying to control all sources of atomic fuel worldwide, and Travancore's reserves of thorium were the largest in the world. All that remained was making the link between thorium and independence.

Rare Earths and Atomic Energy in India

After Hiroshima, when the strategic value of thorium came to light, Travancore's monazite sands became valuable in an entirely different way.[42] This realization had the potential of transforming Travancore's relations with the Indian government as well as the rest of the world. C. P. Ramaswamy Aiyar was not uninformed about this transformation of rare earths from commercial mineral to strategic resource. Only days after Nagasaki, he wrote to the Maharaja: "If thorium can be utilized for the manufacture of atomic bombs (there is no reason why it should not be), Travancore will enjoy a position very high in the world."[43] His official biographer reports: "C. P., who did a lot of soul searching about [the atomic revolution] wrote to the Maharaja that the atom bomb would revolutionize all industry and power production. If the cost of breaking up the atom could be reduced by further research, all steam engines and power projects would become unnecessary and the launching of new power projects would stop across the world. . . . That the manufacture of the atomic bomb was based on the disintegration of uranium awakened him to the fact that Travancore with its uranium and thorium enjoyed a very special position in this field."[44] Even if he did not fully understand the new science of the atom, Sir C. P. was aware of the international interest in Travancore's thorium—"which is used for atomic energy"—as early as March 1946.[45]

In April 1946, the newly formed Board for Atomic Energy Research of the Council of Scientific and Industrial Research (CSIR) announced they would soon begin "intensive geological and physico-chemical surveys" of the Travancore thorium deposits.[46] This drew a spirited response from C. P. Ramaswamy Aiyar. He made it clear that Travancore was the sole owner of the mineral sands and was not willing to "surrender control of thorium deposits to any outside agency—the British Government included."[47] In September 1946, Travancore concluded negotiations with the monazite extracting companies, re-establishing the state's sole ownership of the mineral sands and making the companies mere "agents and contractors" of the state of Travancore.[48] In response to questions in the Legislative Assembly about the Travancore government's policy on thorium, C. P. Ramaswamy

Aiyar falsely stated that no foreign companies had contracts to extract the sands, that there was a ban on exports of monazite, and that all decisions would be made in close consultation with the government of India.[49] In fact, Travancore was already engaged in secret negotiations with the British government for exporting monazite to the UK. The negotiations were concluded in early 1947 with an agreement to supply 9,000 tons of monazite to Britain over the next three years.[50] In return, the British government promised to "contribute their good offices" to encourage Thorium Limited to construct a processing plant in Travancore.[51] Although this arrangement was publicly announced only in May 1947,[52] the British informed the interim government of India as soon as the agreement had been signed. The information was shared with the prominent physicist Homi J. Bhabha, chairman of the CSIR's atomic energy board. British records of this conversation show he is reported to have "expressed satisfaction at having been informed in advance." Most important, "[Bhabha] did not consider that the action of Travancore conflicted . . . with the interests of India."[53]

Rumors that some arrangement had been made with the British began to circulate in early 1947. According to Jawaharlal Nehru, there was "consternation" among scientists at the annual meeting of the Indian Science Congress in January 1947 about possible exports of monazite. This led to the passage of a special resolution that "the State should own and control all these minerals. . . . [It] referred to all minerals, and more especially and specifically to those minerals which are necessary for the production of atomic energy."[54] Speaking to the Indian cabinet in April 1947, Nehru is reported to have said that he "would approve the 'use of airpower against Travancore, if necessary, to bring them to heel.'"[55] Although Nehru was prime minister of the interim government, he was not aware of the full extent of the agreement between Travancore and the British. Gravely concerned about the precedent and its implications, Nehru was in no doubt that the government of India should be involved in any dealings with foreign parties regarding the sale of India's mineral resources.[56] He then asked the head of CSIR, Sir Shanti Swarup Bhatnagar, to travel to Travancore to obtain more information about the nature of their arrangement with the British.

Bhatnagar was accompanied by Dr. Bhabha on his visit. Their trip seems to have borne fruit rather quickly. In early June 1947, the creation of a [Travancore-India] Joint Committee on Atomic Energy was announced. Of the nine members of the committee, six would be appointed by the CSIR and three by Travancore. If the balance of power expressed through ratios on the joint committee was not statement enough about who would control India's thorium, the press release issued following this meeting makes it clear that widely expressed fears of the loss of rare earths could be put to rest. As C. Rajagopalachari, Minister of Industries and Supplies, put it: "The public may rest assured that the atomic energy resources of India will not be frittered away or go to waste."[57] He would add: "I am grateful to Sir C. P. Ramaswamy Aiyar, Dewan of Travancore, for the cooperation he has extended in this matter. We ha[d] deputed Sir S.S. Bhatnagar and Professor Bhabha to go to Travancore to discuss matters with him, and the present arrangement is the result of these negotiations."[58]

Bhabha had been thinking of the potential of atomic energy since 1944.[59] His message to the Indian political leadership was simple: atomic power could solve the problem of lack of electric energy and would do so cheaply and by exploiting local resources. For politicians looking for distinct and genuinely new ways of thinking, as befitted a post-colonial state, a vision that combined the latest of modern technologies and a boundless source of electrical energy was too good to be true.

The furor over Travancore's effort to export rare earths had provided Bhabha with a unique opportunity, which he began slowly to realize. After his visit to Travancore and the creation of the Joint Committee, he approached Prime Minister Nehru and explained to him the importance of rare earths as a state resource. Bhabha no longer thought it was perfectly fine to allow Travancore to export monazite. He now presented thorium as a national asset that would allow India to obtain scarce foreign technologies and expert assistance in return. Given its potential, thorium was too valuable to be controlled by any agency other than the national state.

Bhabha took this opportunity to ensure that he would have complete operational control of the country's atomic energy program. In a note written to Nehru soon after independence, he wrote: "The development of atomic energy should be entrusted to a very small and high-powered body composed of say three people with executive power, and answerable directly to the Prime Minister without any intervening link. . . . The present Board of Research on Atomic Energy cannot be entrusted with this work since it is an advisory body which reports to the governing body of the Council of Scientific and Industrial Research, composed of 28 members including officials, scientists and industrialists. Secret matters cannot be dealt with under this organization."[60] In this note Bhabha strikes themes that would characterize the growth of atomic energy programs worldwide: secrecy, centralization, and unaccountable executive power.

Nehru bought into this vision entirely and gave Bhabha all he wanted. The first step was the dissolution of the Atomic Energy Board, with its wider representation and access to diverse political views, and its replacement with a three-man Atomic Energy Committee. The new committee had Bhabha in the chair and two of his closest scientific allies (Bhatnagar, head of the CSIR, and another eminent physicist, K. S. Krishnan) as members. Notably absent was any explicit political representation or anyone representing the ministry of finance, a startling exception from established government rules and procedures. The next step was the legal creation of an atomic energy establishment insulated from public scrutiny and accountability "for reasons of secrecy" and reporting directly to the prime minister. Direct access to the highest political power in the country also opened the door to state resources on a grand scale. The urgency of the matter required immediate action. Without even waiting for a properly elected parliament to come into being, the Indian Constituent Assembly passed legislation to create the Indian Atomic Energy Commission in 1948. The first chairman of the Commission was Homi Bhabha, the first members Bhatnagar and Krishnan. Thorium and all atomic materials were now legally under the control of the new state, and Bhabha was in charge.

The Last Days of Travancore

On June 3, 1947, it was announced that India would be divided into two states, India and Pakistan; the date of the transfer of power and territorial partition was set for August 14 (Pakistan) and 15 (India). Two days later, Hyderabad announced its intention to remain independent of India. On June 11, 1947, C. P. Ramaswamy Aiyar announced that Travancore would do the same and refused to enter into any interim arrangement with the government of India that did not recognize Travancore as an equal political entity. He dissolved the legislature (the Sri Mulam Assembly) and banned the political party known as the Travancore State Congress. A week later, the Maharaja of Travancore publicly confirmed his support for this decision. On hearing the news, Gandhi is said to have responded that this was "tantamount to a declaration of war against the free millions of India."[61] A series of meetings between the Dewan, the Viceroy, and V. P. Menon of the Ministry for States followed in Delhi. The Dewan remained firm in his decision, though he agreed to carry back to Travancore a copy of the Instrument of Accession and a letter to the Maharaja from the Viceroy. Travancore appeared firmly on the path to political independence.

The Maharaja seemed to be quite unaware of popular sentiments in his state. Though there still may have been respect for the monarchy, the local population hated the autocratic and bigoted Dewan with increasing ferocity. Leading the charge were the Travancore State Congress and the

Communist Party, both of whom had faced state repression for years. As independence grew closer, the political parties in opposition were joined by a variety of communal, service, and professional groups, including former military personnel, the Bar Association, and students. The Dewan responded by sending groups of armed thugs around the state to disrupt political meetings and attack prominent politicians.[62] Faced with this overt repression, the public mood in Travancore swung in favor of union with India and against independence. Eventually even the royalist Nair Service Society would join the opposition. Travancore was in political ferment. A variety of plans for overthrowing the government and acceding to India were forged by a variety of actors: some were violent, others not; some involved the support of the political forces from outside the state, others were quite local.[63]

On July 25, 1947, the Viceroy spoke to the princes at a meeting in New Delhi, enjoining them to accede to the successor states at the end of British rule and assuring them of protection of many of their rights. As one commentator put it, the meeting "was a personal triumph for [Mountbatten]. He could charm the birds off the trees, and the Princes into their constitutional cages—cages that later were gilded with substantial pensions."[64] That evening, C. P. Ramaswamy Aiyar, who had chosen not to attend the Viceroy's meeting as he had nothing to discuss, attended a concert at the Music Academy in Trivandrum.

In a letter to the Maharaja of Bhopal, Sir C. P. describes what happened at the end of the concert:

Your Highness, it is needless to recount the dramatic and carefully calculated plot that led . . . to an attack on my life. The lights were put out just before I was getting into the car (with the evident connivance of some of the employees of the Electricity Department), another person intercepted the Inspector General on the pretext of talking to him urgently, the Police were either taken by surprise or tacitly permitted a crowd of about 200 to rush the gates sometimes before my departure and I was attacked by a butcher's knife and got four gashes; one hit entered my skull and one stroke gashed my cheek and but for my having warded off the attack with my left hand my throat would have been cut. . . . All things considered, [the maharaja of Travancore] has, after much hesitation, decided to accede subject to the conditions discussed, and has sent a letter a copy of which I enclose. If I thought that I could, in the critical period between now and the end of August, fly from place to place conducting propaganda for strengthening the loyal and putting down the disloyal elements, the case might have been different. But I am crocked up and disabled and for a fortnight (perhaps more) shall not be able to move around freely. Now that I am writing candidly (propped up on cushions to give your Highness, as your loyal friend, a picture of the situation), let me ask your Highness to ponder over the situation in your own state. My advice given in Bhopal to your Highness to [accede to India] straight away be carefully considered. With Kindest Regards and deep regret at the turn of events.[65]

The violent attack on C. P. Ramaswamy Aiyar seems to have finally demonstrated to the Maharaja where the passions of the people lay. In something of an anti-climax, the very next day he informed the government of his decision to accede to the Indian Union. On July 30, 1947, the palace announced that Travancore would accede unconditionally to India. The battle for Travancore's independence was over.

Sir C. P. Ramaswamy Aiyar appears to have given up control over Travancore and thorium rather easily. In a few short months he had gone from insisting on complete ownership and control of rare earths to "cooperation" with the government of India. He meekly allowed the creation of a joint committee that was to be "the [sole] authoritative advisory board" to both governments and on which his own representatives would always be in the minority.[66] Whether Ramaswamy Aiyar was serious about taking Travancore in the direction of Kashmir and Hyderabad and demanding outright independence, or even if he was settling for the lesser objective of delaying accession in order to negotiate from strength with the Congress, it is difficult to understand why he would not have acted to maintain control over the most strategic and globally significant resource in his possession. From the evidence we have, we know that Ramaswamy Aiyar understood, however incorrectly, the transformation of Travancore's commercial rare earths into an entirely new strategic material. Then why did he not act on that knowledge? At the same time as he was backing away from deploying thorium as a strategic resource, he showed no reluctance in confronting the Congress in other ways, including arranging to send an ambassador to Pakistan and negotiating for imports of food grains to break a possible embargo.

The most obvious interpretation that seems possible under these circumstances is that Travancore's rare earths were simply not endowed with the significance that we now assign to them. Rare earths were appreciated in terms of their exchange value, but only as a commercial product not fundamentally different from tea, coffee, rubber, and the other high-value agricultural products the state traded in. (And, it is important to remember, Professor Bhabha thought no differently about thorium's value at this time). Perhaps Sir C. P. could envisage keeping more of the value-added that would come from increased local processing, as indicated by his efforts to get extracting companies to build monazite processing plants in Travancore, but could not see beyond that. Had Sir C. P. failed to realize the newly added value of this resource? This seems unlikely. Given the range of evidence that identified the monazite sands as "associated with atomic energy" and his own correspondence with the Maharaja, Sir C. P. clearly knew the potential significance of rare earths beyond mere commercial value.

Conclusion

Travancore's defiance of New Delhi, via its threat to become independent, indicated serious trouble for India on many fronts. In the first instance, this was a Hindu-majority state, which, if allowed to become independent, would provide additional proof of the "two-nation theory" that justified the creation of Pakistan. Hence Nehru's threat even to bomb Travancore if it did not "come to heel." But further, the terms of accession, which constituted the basis of state formation, needed to be absolutely clear about where sovereignty lay. New Delhi was willing to allow the princes a certain amount of superficial autonomy: continued use of princely titles and entitlements, sweetened by a generous privy purse. It stopped well short of allowing a constituent part of the country to sell off its strategic resources. Asserting New Delhi's veto over Travancore's desire to sell its mineral resources was a decisive step in the larger political battle of establishing territorial sovereignty over the constituent parts of India. The struggle over Travancore was an object lesson for all princely states with visions of independence to note. Rare earths had become a means of sovereign control.

Professor Bhabha too saw rare earths in terms of political independence. Though at first Bhabha understood thorium only in terms of its exchange value, he soon realized that it also gave him a means to get rid of the unwieldy official structure that monitored India's interest in atomic energy and thereby regulated his own activities. If thorium's strategic value was to be realized, he argued privately to the political leadership, it required the centralization of decision making, the removal of public accountability, and the privileging of specialized, expert knowledge. This was the minimum, Bhabha argued, that foreign states would demand. Above all, it would have to be done without answering to a committee composed of "officials, scientists, and industrialists," who would neither understand what was at stake nor be able to keep these matters appropriately sub rosa. To realize the value of rare earths called for a new kind of state organization—one constituted in the recognition that in this new frame of meaning no effort could be spared to control this most rare and powerful of new substances. Atomic energy required no less.

Sir C. P.'s problem was that he could not find a means to deploy the "surplus value" of rare earths. This was not for lack of ability. Rather, he calculated that, precisely because of their new value, denying Travancore's thorium to India would be going too far. Denial of a strategic resource would open the door to a military response against which he could not defend, and which he was desperately trying to avoid. A military intervention would ensure the annexation of Travancore by India, sans the guarantees that Sir C. P. had spelled out in his negotiations with the Viceroy and the States Ministry. Without access to the bigger guns of US Department of Defense (the only agency that conceivably might have been willing to fight this battle for him), Sir C. P.'s hands were tied. It appears that the transformation of rare earths into atomic energy had made them altogether too rare for Travancore to control and use. Almost overnight, the variety of meanings that had accrued to rare earths had reduced to a single one: rare earths now meant atomic energy. This transformation of meaning came about only because of the Cold War and its "flashy flagship" (as the introduction to the present volume puts it): the over-determined condition of nuclearity.[67]

The mutation of this commodity, rare earths, from beach sand to commercial raw material to strategic resource describes the life course of a number of similar minerals inserted into the circuits of modern technological capitalism. What makes rare earths different are the associations of value and dominant meaning that bring rare earths into direct correspondence with state imperatives. The identity between rare earths and atomic energy had to be established before it could be understood that Travancore's sale of monazite was not in the interest of the new state. This particular mutation, in other words, was only possible as a result of the interpellation of a particular frame of meaning that belonged, uniquely, to the Cold War. Centralization, secrecy, and lack of popular accountability are institutional products of this meaning. The new meaning of rare earths is reinforced by removing it from public purview, by limiting access to it in all ways, and by asserting the commonsense of rare earths as a national strategic resource. Once the new value of rare earths was known, the state's response was determined. Once rare earths became atomic energy, Travancore had to submit to India. Rare earths may have belonged to Travancore, but after the onset of the Cold War atomic energy belonged to the state.

By the early 1950s, India was busily selling its great strategic resource, thorium, to the United States. But so were Brazil and South Africa, where significant quantities of rare earths were now being mined. Thorium was not as rare, and hence valuable, as had been hoped, but in India that mattered little: rare earths were now, thanks to the Cold War, atomic energy, a state resource.

Notes

1. For a useful introduction to rare earths, see http://www.du.edu/~jcalvert/phys/rare.htm.

2. The black sands also contain quantities of ilemenite (iron-titanium oxide); titanium oxide is of considerable value in the paint and pigments industry.

3. "Souvenir of the opening of the factory of Indian Rare Earths Limited at Alwaye by Pandit Jawaharlal Nehru on Dec. 24, 1952" (Bombay: Tata Press, n.d.[1952]). The casual tone adopted by the souvenir describing how the sands were found to be valuable stands in stark contrast to contemporary efforts being made in Europe to control the lucrative mantle industry and its various material inputs. Schomberg's firm was scouring the globe looking for these materials.

4. Government of India, Foreign and Political Department. Secret (Internal B). Proceedings,1926. File 713-I/26. "Reconstitution of Travancore Minerals Company."

5. For more on Welsbach, see http://www.althofen.at/AvW-Museum/Englisch/menue_e.htm.

6. Government of India, Foreign and Political Department, Secret Proceedings,1916, no. 1–42, "Proposed grant by the Travancore Durbar of a monazite sand concession in the state to Mssrs. Hopkins and Williams (Travancore) Ltd of London."

7. Government of India, Foreign and Political Department. Secret (Internal B) Proceedings,1926, File 713-I/26, "Reconstitution of Travancore Minerals Company." Memo by J. Coggin-Brown, 8-8-1927.

8. Professor A. Sreedhara Menon's voluminous writings, going back many decades, offer the most comprehensive political history of Travancore. See, in particular, Menon 2001.

9. For a contemporary account of these well-known events, see Talbot 1949: 321–332.

10. Menon 1956.

11. For a critique of the legal case, see Noorani 2003.

12. Minutes of meeting of Mountbatten with States Negotiating Committee, 3 June 1947, Folder 63, microfilm roll no. 1405, C. P. Ramaswamy Aiyar Papers, National Archives of India.

13. For a very sympathetic biography, see Sundararajan 2003.

14. Among many examples in his private papers, see C. P. Ramaswamy Aiyar to the ruler of Gwalior State: "I am fundamentally and intensely interested in the preservation of the best features of monarchy that I feel it incumbent on me to tender these

observations and suggestions. I trust I will not be misunderstood; even if I am, I feel I shall have done my duty to Indian monarchy by writing this letter to one like your Highness." December 11, 1946, microfilm roll 1391, C. P. Ramaswamy Aiyar Papers, National Archives of India.

15. "For an Indian, who after the Quit India and the Objective Resolutions, to have voted for a non-Indian as Governor-General is to furnish proof of the lowest possible moral cowardice and degradation and to manifest a lack of self-respect, which disentitles men like Gandhi, Patel and Nehru from any real part in Indian affairs." C. P. Ramaswamy Aiyar to Nagaraj, July 11, 1947, roll 1397, C. P. Ramaswamy Aiyar Papers, National Archives of India.

16. "The special history and background of Travancore have alone inducted the Government to keep out of a Union in which her special interests may not have full scope (e.g., the decisions of the Union Powers Committee and even the Fundamental Rights Committee.)." C. P. Ramaswamy Aiyar to Jawaharlal Nehru, June 2, 1947, microfilm roll 1397, C. P. Ramaswamy Aiyar Papers, National Archives of India.

17. "In the nature of things, and quite apart from the historical past, the position of maritime States with plenty of resources, agricultural and industrial, is different from [landlocked] States like Hyderabad and Mysore, however wealthy and powerful these may be." C. P. Ramaswamy Aiyar to T. R. V. Shastri, April 8, 1947, microfilm roll 1383, C. P. Ramaswamy Aiyar Papers, National Archives of India.

18. In calling for outright independence, Sir C. P. would have had, at a minimum, the support of the Hindu extremist right-wing group Rashtriya Sevak Sangh (RSS). Veer Sarvarkar, founder of the RSS, wrote to Sir C. P. in June 1947 supporting "the Maharaja and the far sighted and courageous determination to declare the independence of our Hindu state of Travancore." Sarvarkar's reasons for supporting Travancore was easily explained by his Manichean view of the world. "The Nizam (Muslim ruler of Hyderabad) has already proclaimed his independence and other Muslim states are likely to do so. Hindu states bold enough to assert it have the same rights." From C. P. Ramaswamy Aiyar to the newspaper *Hindu*, quoting a telegram sent to him by Sarvarkar, June 19, 1947, Folder 63, roll 1405, C. P. Ramaswamy Aiyar Papers, National Archives of India.

19. "Sir C. P. thought that if they could have a measure of independence coupled with close contacts with India, that would be best for Travancore. What is more the Maharaja made it clear that he entirely supported Sir C. P.'s policy in this matter and that they had changed their minds only when it became clear that the Congress Party leaders intended to give the State and Princes a chance, in July." Note of an interview of HE the Governor General with HH the Maharajah of Travancore on the future of Travancore, March 22, 1948, Ministry of States P[olitical] Branch, No. 546-P/48 (1948), National Archives of India.

20. Nehru saw Sir C. P.'s behavior as a manifestation of "vanity [gone] paranoiac": "Even if he had been treated curtly by any of us, there are some things which are not done by any decent and patriotic individual." Letter to G. P. Hutheesing, July 14, 1947, in *Selected Works of Jawaharlal Nehru*, Second Series, volume 3 (New Delhi: Jawaharlal Nehru Fund).

21. The similarity with M. A. Jinnah's approach to the Congress is too obvious not to mention. Ayesha Jalal (1985) persuasively argues that Jinnah's call for Pakistan began as a tactic to force the Congress to consider a confederated, weakly centralized India, but was blocked by Congress' intransigence and hatred of him.

22. Letter to Mountbatten, June 22, 1947, in *Selected Works of Jawaharlal Nehru*, Second Series, volume 3 (New Delhi: Jawaharlal Nehru Fund).

23. All quotes from C. P. Ramaswamy Aiyar to Chinni, July 12, 1947, microfilm roll 1397, C. P. Ramaswamy Aiyar Papers, National Archives of India.

24. Travancore's rulers must have also been aware that the United States had used its military power to create the country of Panama in order to serve US strategic interests.

25. "Two days after the Quebec Agreement was signed [August 19, 1943], the Military Policy Committee approved [Gen. Leslie] Groves' recommendation that the United States allow nothing to stand in the way of achieving as complete control as possible of world uranium supplies." Helmreich 1986: 10.

26. Broscious 1999.

27. Gowing 1974.

28. Helmreich 1986.

29. Ibid.: 226–231.

30. Fraser 1994.

31. Helmreich 1986: 97–113.

32. Memorandum, Wendel to Marks, "Travancore Monazite Exports," November 14, 1946, National Archives and Records Administration, State Dept., Atomic Energy Files, RG 59: 1.

33. Ibid.

34. See Helmreich 1986 and correspondence in National Archives and Records Administration, State Dept. Files, RG 59, Decimal Files 1945–49, Box 6104.

35. Sundararajan 2003: 409–444.

36. Wendel to Marks, "Travancore Monazite Exports," p. 2; Briefing memorandum for Ambassador Loy Henderson, Sept. 17, 1948, FRUS, 1948, volume 1: 758–759.

37. File L/P&S/13/1286, 17 (AB), Public Records Office, Kew, UK. Report on Mineral Sands Industry of Travancore, South India, by Andrew Corry, Sept. 8, 1945. National Archives and Records Administration, State Dept. Files, RG 59, Decimal Files 1945–49, Box 6103.

38. "Now it so happens that the Dewan of Travancore is impatient with these restrictions and to a considerable degree is ready and willing to establish direct contact with the consulate. The consul knows the Dewan personally. It would be indiscreet to give details, but the consul has means of exchanging messages verbally." Roy Bower, American consul in Madras, to Special Adviser, Supplies and Resources, Dept. of State, September 8, 1944, National Archives and Records Administration, State Dept. Files, RG 59, Decimal Files 1940–44, Box 5090.

39. "The present Dewan, Sir C. P. Ramaswamy Aiyar, is one of the most able administrators in the country. He speaks of Travancore as a 'country,' and all other parts of the world including parts of India as 'foreign.' Despite the fact that he owes his tenure of office to the British Government, he is extremely assertive of State's rights, and frequently risks (or seems to risk) the displeasure of the Government of India in so doing. He has declared that Travancore will resist any form of Pakisthan, Hindusthan, Dravidisthan [sic], etc. As far as it is possible, he intends to run Travancore (which has about the area and population of Belgium) as an independent nation analogous to those making up the British Commonwealth of Nations." Roy Bower, American consul in Madras, to Special Adviser, Supplies and Resources, Dept. of State, September 8, 1944, National Archives and Records Administration, State Dept. Files, RG 59, Decimal Files 1940–44, Box 5090.

40. Emphasis added. E. A. Guillon, Memorandum (Secret), July 17, 1947, National Archives and Records Administration, State Dept., Atomic Energy Files, RG 59.

41. E. A. Guillon, Memorandum (Confidential), July 28, 1947, National Archives and Records Administration, State Dept., Atomic Energy Files, RG 59.

42. Sundararajan states that uranium was discovered in Kerala by "Masillamani, a Geologist" in June 1939. Since in her text the value of uranium was imputed by its association with radium (which makes little sense from a scientific point of view, but certainly radium would have been, in 1939, the material most easily associated with radioactivity thanks to Madam Curie's fame), this claim seems dubious. More important than the degree of veracity, however, Sir C. P. was informed about the discovery and he reported the same to the Maharaja. See Sundararajan 2003: 427.

43. Sundararajan 2003: 429 and footnote 64 on 744.

44. Sundararajan 2003: 429.

45. Letter from Kanji Dwarkadas to C. P. Ramaswamy Aiyar, marginal note: "you must have seen questions in the House of Commons about thorium in Travancore—thorium which is used for atomic energy," February 2, 1946, microfilm roll 1402, C. P. Ramaswamy Aiyar Papers, National Archives of India. In May 1946, the Viceroy

Lord Wavell also communicated with him about Travancore's thorium deposits (Noorani 2003: 7).

46. "Government Must Encourage Thorium Research," *Morning Standard* (Bombay), May 21, 1946.

47. Wendel to Marks, "Travancore Monazite Exports," November 14, 1946, Attachment 2: 2.

48. Quote from C. P. Ramaswamy Aiyar remarks at press conference, June 11, 1947, Folder 63, microfilm roll 1405, C. P. Ramaswamy Aiyar Papers, National Archives of India.

49. Wendel to Marks, "Travancore Monazite Exports," November 14, 1946. National Archives and Records Administration, State Dept., Atomic Energy Files, RG 59, Attachment 2: 2.

50. In April 1947, 700 tons of monazite was shipped to the UK. The remaining amount was never delivered. "Travancore Monazite," *Foreign Relations of the US*, 1 (1948): 777.

51. "Outline of Indian atomic energy situation," (briefing of Ambassador Henderson, Sept. 17, 1948). *Foreign Relations of the US* 1 (1948): 702.

52. "The Dewan disclosed that Trivandrum had entered into an arrangement with the Government of India for the purpose of conjoint research on mineral sands and atomic research. We have also entered into certain arrangements with the British Government for joint research and exploitation of the possibilities of the mineral sands of Travancore, atomic fission and the production of atomic energy." API report of a press conference held in Trivandrum, May 17, 1947. Folder 63, roll no. 1405, C. P. Ramaswamy Aiyar Papers, National Archives of India.

53. Sundararajan 2003: 430.

54. "State Control of Monazite and Thorium Nitrate," February 27, 1947. File no. 17 (4) 47-PMS. From *Selected Works of Jawaharlal Nehru* (New Delhi: Jawaharlal Nehru Memorial Fund), volume 2, pp. 604–607.

55. Sundararajan 2003: 431, drawing on the Viceroy's (Mountbatten) notes.

56. "State Control of Monazite and Thorium Nitrate," p. 607.

57. "Atomic Research Board Set Up," *Hindustan Times*, June 6, 1947. Folder 63, microfilm roll 1405, C. P. Ramaswamy Aiyar Papers, National Archives of India.

58. "Atomic Research Board Set Up," *Hindustan Times*, June 6, 1947. Folder 63, microfilm roll 1405, C. P. Ramaswamy Aiyar Papers, National Archives of India.

59. This section draws on the discussion in chapter 1 of my book *The Making of the Indian Atomic Bomb* (Abraham 1998).

60. "Note on the Organisation of Atomic Research in India," dated April 26, 1948, republished in "The Architects of Nuclear India," *Nuclear India* 26, no. 10 (1989): 3. Only much later would thorium would be defined as the critical substance in a long term three-stage plan leading to breeder reactors, which would obviate any fears of Indian dependence on the outside world for energy resources.

61. Quoted in Rangaswami 1981: 218.

62. Telegram from Pattom Thanu Pillai to Vallabhbhai Patel, July 10, 1947, *Sardar Patel's Correspondence, 1945–50*, ed. D. Das (Ahmedabad: Navjivan), volume 5: 446.

63. Ouwerkerk 1994: 260–277.

64. Ouwerkerk 1994: 251.

65. July 30, 1947, microfilm roll 1391, C. P. Ramaswamy Aiyar Papers, National Archives of India. Another version of the same event is offered by Ouwerkerk, based on interviews with the conspirators. "They found a young man to do the deed. . . . Suffice it to say that he was a Brahman and he had a sword-stick. . . . That evening the young man dressed himself in black and blacked his face and hands, and set off on his bicycle, prepared to die in the attempt. He leaned his bike against the high wall that surrounds the Music Academy, and climbed over. After the concert, as Sir C. P. came down between the two rows of chairs to get into his waiting car, he leapt forward and attacked him. The brave little man defended himself and his devoted secretary interposed himself between the Dewan and his assailant. Just then Trivandrum was plunged into darkness as all the lights went out (not part of the plot but just an ordinary failure not unfamiliar to Trivandrum residents) and the young man escaped, unaccountably leaving behind a pair of khaki shorts as well as his knife. He scaled the wall again, got on his bicycle and pedaled off to a junction where a lorry was waiting for him. He and his bicycle were hauled on board the lorry and they roared off through the night to Madurai. There he took the train to Bombay and turned up at Communist Party headquarters, where he had an old classmate who he was sure could arrange for him to go underground." (Ouwerkerk 1994: 271–272).

66. "Atomic Research Board Set Up," *Hindustan Times*, June 6, 1947. Folder 63, microfilm roll 1405, C. P. Ramaswamy Aiyar Papers, National Archives of India.

67. Also see Hecht 2006a: 25–48.

6 Nuclear Colonization?: Soviet Technopolitics in the Second World

Sonja D. Schmid

In 1990, the government of reunited Germany had every Soviet-designed nuclear power reactor in the former German Democratic Republic shut down, ostensibly for safety reasons. Shortly thereafter, Czechoslovakia split peacefully into two republics. A decade later, the Czech Republic and Slovakia each launched two Soviet-designed reactors that had been completed with substantial contributions from Western firms. These reactors brought the nuclear share of their electricity generating capacity to 31 percent and 55 percent, respectively.[1] How can we explain these remarkably different developments among post-communist states? The former East Germany and the former Czechoslovakia had both relied on Soviet assistance to create their domestic nuclear industries; in fact, they had been the first East European states to import Soviet nuclear technology. Both had been members of the Warsaw Pact, and of the Council for Mutual Economic Assistance (CMEA), the common market of the socialist world. And yet, the fate of their respective nuclear industries after the collapse of the Soviet bloc indicates that East European nuclear industries were anything but monolithic. Whereas the East German nuclear program struggled almost from the outset, and eventually stalled completely in the early 1980s, Czechoslovakia not only managed to manufacture its own nuclear reactors but became the monopolist producer of Soviet-designed pressurized water reactors for the entire Soviet bloc.

This essay focuses on the Soviet Union's transfer of nuclear technology to Eastern Europe during the Cold War.[2] Many aspects of Soviet rule in Eastern Europe suggest a colonial relationship, as scholarship on Soviet imperialism within the USSR and in the Eastern bloc countries has shown.[3] Rather than analyzing traditional forms of exerting political power such as military force or economic pressure, however, I focus here on Soviet technopolitics. Science and technology have played fundamental roles in many, if not all, colonial settings. The concept of technopolitics allows us to examine how the practice of designing and implementing technologies is used to enact political goals.[4] In particular, the institutional and cultural framework in which scientific knowledge and technological artifacts are embedded have often determined the success or failure of a given political agenda. Soviet nuclear technology transfer to Eastern Europe is a case in point. In addition to the mining of East European uranium for the Soviet nuclear weapons program, technical cooperation in the nuclear energy sector began in the mid 1950s. This cooperation started as bilateral technical assistance, but over the course of the Cold War it developed into complex multilateral collaboration. Here, I reconstruct Soviet attempts to establish the rules of the game through technical designs and management structures, and show how these attempts ultimately failed. Early technical and organizational choices in each state, combined with the changing character of cooperation, produced unexpected and sometimes contradictory outcomes, as individual East European states successfully enacted goals quite different from those of the initial Soviet plans. The Soviet modernizing mission relied on the idea of science and technology as the motor of progress. But the Soviet commitment to progress was more than a powerful rhetorical resource. Soviet technological assistance materialized in the form of nuclear reactors, and, even more importantly, in indigenous nuclear expertise.[5] The stability of this technopolitical regime depended on the degree and kind of cooperation between the Soviet Union and individual states, and ultimately affected the success of post-Cold War East European nuclear industries.[6] The claim that science (and to a lesser degree technology) was politically neutral allowed for more than one interpretation of what exactly constituted desirable social transformations. The post-1989 disintegration of the Soviet bloc has shown that even a tightly knit technopolitical arrangement allowed alternative narratives to emerge—for example, entitlement in the face of hegemony, or national achievement.

In the following, I first lay out the changing characteristics of the colonial relationship between the Soviet Union and Eastern Europe, then identify the specifics of inner-bloc trade and their effects on Soviet control. I address the special role of science and technology in the communist context before turning to the analysis of nuclear cooperation in the Soviet bloc. Was Soviet technology transfer to Eastern Europe a colonial practice? If so, was the *nuclear* cooperation any different? Using the examples of the German Democratic Republic and Czechoslovakia, I illustrate the variation in Eastern Europe's colonial experience in the area of nuclear energy. Ultimately, my argument qualifies the idea that Soviet imperialism in Eastern Europe was homogeneous, and emphasizes the interpretative flexibility of technopolitical regimes.[7]

Soviet Imperialism in Eastern Europe

A growing body of literature addresses the Soviet Union as an empire, focusing on its territorial expansion through invasion or subversion; its strong, central administration controlling the governments of all subsidiary and satellite territories; and its interference (including the use of military force) in the internal politics of its allies.[8] In particular, scholars have begun to analyze former Soviet republics as postcolonial spaces.[9] Their attention has focused on how the Soviet Union subjected, developed, and modernized these republics, and on how the introduction of a uniform system of education, infrastructure, and industry resulted in the peculiar mix of promoting indigenous, "national" identities and an envisioned universal Soviet (or, even broader, Communist) identity.[10]

However, the Soviet Union not only succeeded the Russian Empire, it also de facto controlled the territories occupied by the Red Army during World War II.[11] Soviet rule in Eastern Europe bore a number of features characteristic of a colonial relationship, such as "lack of sovereign power, restrictions on travel, military occupation, lack of convertible specie, a domestic economy ruled by the dominating state, and forced education in the colonizer's tongue."[12] The changing face of Soviet imperialism in Eastern Europe has been the subject of a broad debate at the boundary of political science, history, and area studies. In her 1985 essay "The Empire Strikes Back," for example, Valerie Bunce analyzed the evolution of the Eastern bloc from its Stalinist origins to the rise of Mikhail Gorbachev.[13] Bunce argued that Eastern Europe, created as a political entity in only five years (from 1944 to 1949), initially fulfilled the classic functions of an empire: it increased Soviet national security, yielded economic benefits, and enhanced domestic stability.[14] After a period of ruthless exploitation and relocation of assets from the occupied territories to the homeland, the Soviets gradually switched their policy to one of support for individual East European economies.[15] Stalin ultimately based his decision to keep control over Eastern Europe on the idea of a defensive shield.[16] In 1949, in a typical coupling of political interests and economic policies, he set up the CMEA and installed Moscow-trained Communists as rulers of the East European People's Republics.[17]

As Palladino and Worboys argue, imperialism comes in many different forms, "sometimes cultural, sometimes economic, or political, or social, or scientific . . . [or] . . . in changing combinations of any of these."[18] East European states differed from the non-Russian territories within the Soviet Union. In contrast to these provinces that the Soviets considered "backward" and in need of "civilizing," the Soviet satellites in Eastern Europe were in many respects ahead of their colonizer.[19] The Soviets "avoided governance over internally weak colonial states," one of the classic problem in other empires, and instead sought to enhance regime stability by supporting economic prosperity in the colonies.[20] Inner-bloc relations, according to the Soviets, involved the idea of "socialist internationalism," a form of government in the spirit of "socialist harmony" that did not impair the sovereignty of individual states, while critics countered that East European policy making was in fact Soviet domestic imperial politics.[21] Like other empires, the Soviet regime in Eastern Europe went through significant changes. These changes were effected not only by shifts in Soviet leadership, but also by geopolitical concerns and by revolts in East European states.[22] Eastern Europe was the "rare example of an ideal colony" under Stalin (1945–1953), significantly deteriorated in colonial usefulness during Khrushchev's regime (1953–1964), and had become a serious burden on the Soviet Union by the end of the Brezhnev era (early 1980s).[23]

Throughout this history, the Soviet Union weighed national security interests and concerns about the proliferation of nuclear weapons against the need to keep its East European allies economically prosperous and thus politically stable. These goals were to be accomplished first and foremost through modernization, which in practice meant Soviet technological assistance. The nature and extent of Soviet assistance, however, often remained ambiguous, which caused both unrealistic expectations, as well as concerns about the degree to which this aid was tied to political machinations. East European states bargained over the terms of economic and political dependence with increasing proficiency, alternately invoking Soviet "generosity" and deploring Soviet "exploitation."

Trade within the Soviet Bloc

The Council for Mutual Economic Assistance (CMEA) was established on January 25, 1949. Its founding members were Bulgaria, Romania, Czechoslovakia, Hungary, Poland, and the Soviet Union; East Germany joined in 1950.[24] Starting in 1953, CMEA members set up an institutional framework with an executive committee, a secretariat, and various specialized commissions. Regular conferences brought together party and government representatives of member states. In 1955, after the first treaties on nuclear cooperation had been signed, the Warsaw Pact military alliance was founded.[25] Together, these two organizations completed Soviet–East European political and economic integration.

Trade within CMEA was divided into three broad categories: energy and raw materials, machinery and equipment, and consumer goods.[26] CMEA goods were priced according to a formula that was supposed to eliminate the unpredictable fluctuations of the world market.[27] Most importantly, energy and raw materials tended to be traded below world market prices, while machinery and equipment, although usually of inferior quality, were traded above world market prices. The Soviet Union mostly exported raw materials and energy and imported machinery and equipment, which amounted to a net subsidy for Eastern Europe.[28] For example, after a crash program in oil field development in the late 1950s, the Soviet Union was able and willing to deliver increasing amounts of raw materials to Eastern Europe at prices below those of non-CMEA suppliers.[29] At the time, this seemed advantageous to both sides. East Europeans received cheap energy, and the Soviets got a buffer zone between the capitalist West and the homeland. Over time, however, the costs of such "empire maintenance" became a substantial burden on the Soviet economy. Soviet leadership repeatedly attempted to reduce this liability, especially since East European trade played only a minor role in the Soviet economy. By contrast, East Europeans relied almost exclusively on cheap Soviet oil, and on subsidized sales of equipment and machinery.[30]

In 1959, Khrushchev called for increased economic integration within CMEA, following the example of the European Economic Community (which had been established in 1957). But CMEA member countries rejected his plans.[31] Their economies benefited from the existing arrangement to such an extent that, instead of trying to reduce their trade dependence on the Soviet Union, they sought to increase it.[32]

Why did the Soviets continue their trade subsidy instead of integrating the East European economies into a larger system? Was Khrushchev distracted by the Sino-Soviet split, or was he reluctant to risk the bloc's political stability for economic benefits?[33] Especially after the cataclysmic events of 1956, the Soviets were certainly aware that their trade subsidies to Eastern Europe were a powerful "institutional mechanism for linking trade and foreign policy variables."[34] But several scholars have found that, despite the Soviet Union's obvious leverage, CMEA members behaved surprisingly autonomously in their economic strategies.[35] No doubt East Europeans were getting better at taking advantage of poorly motivated Soviet negotiators, loopholes in legal documents, and the inefficient monitoring of how agreements were actually implemented.[36] But at the same time, East European leaders learned to use their own weakness to enhance their bargaining position. Only economic success (that is, tangible improvements in living standards) would ensure the public's compliance, whereas an economic downturn—the certain result of reduced Soviet subsidies—would threaten the authority of loyal East European elites.[37] By emphasizing the relevance of strong economies for political stability in the bloc, they successfully transformed the question over assistance to Eastern Europe into a matter of Soviet security.[38] This strategy also allowed East Europeans to put pressure on the Soviet Union in the nuclear sector.

The Role of Science and Technology

In addition to the fusion of economics and politics, scientific and technological cooperation was a critical feature of the interaction between the Soviets and their East European allies.[39] Marxism, and accordingly Soviet doctrine, saw science and technology as the driving force of a historical process that would eventually culminate in communism. This legitimated the emphasis that Soviet leaders placed on science and technology as the vehicle for the communist civilizing mission. The economic and military ties among socialist states, the Soviet ban on participation in American-sponsored post-World War II reconstruction programs, and the North American Treaty Organization's 1949 embargo on sensitive technologies, limited the available options for Eastern Europe to advance science and technology.[40] The transfer of any kind of technology, let alone nuclear technology, was closely watched by intelligence agencies, and was increasingly regulated by export controls. These rules and regulations created and reinforced specific economic and technopolitical Cold War geographies as the superpowers carved out their spheres of influence.

During the Stalinist period, rapid socioeconomic transformation opened up new opportunities for East European citizens. The Soviet emphasis on education, modernization, and progress was compatible with hopes and needs of many East Europeans elites, who, as Bunce put it, gained "the role they had always sought—teleological power brokers—and the system they had always wanted—rational redistribution."[41] The deep-seated belief in the modernizing power of science and technology was widespread beyond the socialist camp, however. Nuclear energy in particular emerged as the symbol of progress and modernity in the 1950s, and capitalist and communist leaders alike pursued nuclear energy even before the idea of an independent energy supply became a dominant concern.[42]

East European leaders embraced nuclear energy as an innovative technology that promised accelerated economic growth and social development; strategic considerations regarding their countries' energy policies followed suit. The structure of CMEA trade also shaped the cooperation in the nuclear energy sector, in the sense that the Soviets controlled materials, prices, and barter agreements. But the underlying belief in technological progress made this cooperation special: nuclear cooperation was perceived as part of a shared modernization effort. East European elites did not perceive modernization as an inherently political concept. Whether or not this idea of technological progress as an avenue relatively free of ideological baggage obscured the potential implications of technology transfer, it facilitated parallel narratives of dependency and independence.[43] It allowed East Europeans to criticize the practice of Soviet nuclear assistance, *and* to uphold an ideology-free ideal of science and technology as the foundation for progress and modernization.[44]

President Eisenhower's Atoms for Peace proposal invigorated nuclear enthusiasm around the world. Announced at the United Nations General Assembly in Geneva only nine months after Stalin's death, the program resulted in the First UN Conference on the Peaceful Uses of Atomic Energy, held in August 1955 in Geneva.[45] Specifically for this conference, the United States and the Soviet Union amended laws protecting state secrets to allow the declassification of documents on nuclear energy.[46] The Geneva conference allowed for an unprecedented level of international exchange of information, and the Soviets impressed the rest of the world. Not only did they announce the launch of their first nuclear power plant; it also became clear that the level of Soviet nuclear expertise was very high, and that the Soviets' uranium enrichment capacity matched that of the United States and Britain.[47]

The Soviets were deeply ambivalent about the Atoms for Peace proposal, since they recognized civilian nuclear reactors as a potential proliferation risk.[48] They expressed concern that even the "peaceful" generation of electricity in nuclear reactors would inevitably create fissionable material, and thus add to the stockpiles of weapons-grade materials.[49] They initially made their participation contingent upon a complete ban on nuclear weapons. Only when it became clear that the United States would go ahead with the program regardless of whether the Soviets joined did they revise their position.[50] They were unwilling to leave the initiative entirely to the West, and they feared that Western nuclear assistance to Eastern Europe would compromise Soviet economic control, which in turn could become a threat to political unity within the bloc. Ironically, under these circumstances the Soviets' best option was to intensify cooperation with Eastern Europe in the nuclear energy sector, despite their concerns about proliferation and despite the fact that they would almost certainly not profit from this cooperation.

Technology Transfer—A Colonial Practice?

As a colonial practice, technology transfer has mostly been studied for "Third World" contexts.[51] Postcolonial theorists as well as scholars in Science and Technology Studies have emphasized that the introduction of Western science and technology transformed colonized societies more profoundly "than the political systems and social struggles that had all but monopolized the attention of historians of European imperialism until the 1980s."[52] While technologies were often used to discipline the colonized, they also sometimes served, or at least corresponded to, the interests of the colonized population.[53] The transfer of nuclear technologies from the Soviet Union to Eastern Europe was no exception.

The Soviets agreed to supply East European states with nuclear reactors at least in part to keep control over materials and designs, but they also fulfilled East European aspirations to train scientists and engineers in this new area. Especially Germany and Czechoslovakia, whose industries had been among the most advanced in Europe before the war, and which had accomplished scientific communities, pushed for Soviet scientific and technical assistance. The fact that science and technology were considered politically neutral facilitated the success of the Soviet technopolitical regime in this early period, when the exact nature of Soviet rule was still uncertain.

One aspect of Soviet–East European nuclear cooperation that indicates rather unambiguously a colonial relationship was uranium mining. The Soviets started mining East European uranium in the immediate postwar period, and uranium imports from East Germany, Czechoslovakia, and Hungary contributed significantly to the Soviet nuclear weapons program.[54] In 1945, General Leslie Groves had turned over to the Soviet occupation authorities the German Erzgebirge region, close to the border with Czechoslovakia, whose uranium reserves he considered insignificant.[55] The Soviets quickly proceeded to exploit the mines. Deliveries from East Germany to the Soviet Union were treated as part of reparations payments, and by the early 1950s the "State corporation of non-ferrous metals" (code name Wismut) employed over 100,000 people.[56] Probably in reaction to workers' revolts in June 1953, East Germany became a legal co-owner of its mines in 1954.[57] Wismut remained under Soviet management, and the entire German uranium production was still delivered to the Soviet Union, but in exchange the Soviets agreed to fulfill East Germany's uranium needs.[58]

Some scholars have argued that the Soviet Union dominated nuclear cooperation in every detail, from selecting the sites for nuclear power plants to choosing the type of reactor that would be installed. Others have

Nuclear Colonization? 133

emphasized the relative autonomy granted to CMEA members through the "interested country" principle. The Soviets did in fact limit their exports to Eastern Europe to one type of reactor, a pressurized light water reactor quite similar to the preferred design for US exports. However, the fact that Czechoslovakia and Romania were able to choose the type of reactor for their nuclear power stations over Soviet objections challenges the argument that East Europeans were just Soviet puppets.[59] East Europeans hoped that nuclear power would decrease their dependence on Soviet oil and gas, while the Soviets saw nuclear technology as a means to strengthen inner-bloc ties and to further demarcate their sphere of influence. These competing goals joined in the conviction that science and technology would assure progress and modernization. Belief in the universality of science and in the direct link between technological advances and economic prosperity was central to Soviet doctrine, but was also pivotal in any East European resistance ideology.[60] It could justify both unprofitable exports and increased economic and political dependence on the Soviet hegemon. Thus, the Soviet Union's technological cooperation with Eastern Europe hardly fits a straightforward colonial model. East Europeans not only initiated and kept this cooperation running, they also benefited from Soviet expertise while pursuing rather idiosyncratic design preferences.

Peaceful Nuclear Energy in the Early Cold War

The conditions of the emerging Cold War determined which technologies were available to Eastern bloc countries. These available options, in turn, confirmed the political order that materialized in postwar Europe. East European nuclear industries originated in the 1950s and the 1960s, when enthusiasm for nuclear energy prospered throughout the industrialized world. The American offer for nuclear assistance under the Atoms for Peace program was not extended to East European countries, though. In 1950, in addition to the 1949 NATO embargo on sensitive technologies, the United States and its European allies set up the Coordinating Committee (CoCom) to control and constrain technology exports to the Eastern bloc.[61] Though "meant to strike at the power of the USSR," these measures "strengthened the Soviet nuclear stranglehold over the East European countries."[62] Eastern Europe's only remaining nuclear option was Soviet assistance, but the Soviets were dragging their feet.[63] Soviet leaders feared that exporting civilian nuclear technology to Eastern Europe would entail the spread of sensitive, dual-use materials and know-how. Only when East European leaders began to demand Soviet nuclear assistance in exchange for many years of uranium deliveries did the Soviets hesitantly agree to provide nuclear aid to "friendly" countries.[64] Over the years, this assistance would range from the supply of research reactors and the training of nuclear specialists to the delivery of turn-key power plants. In an attempt to secure all potentially dual-use materials and technologies, many agreements came with sophisticated guarantees of fuel supply and arrangements for the return of spent fuel.[65]

Hailed as one of the most successful initiatives in CMEA history, nuclear cooperation in the Soviet Bloc remained bilateral for many years. In 1957, West European countries formed Euratom, an organization intended to strengthen multilateral cooperation in the civilian nuclear sector. East European states, however, continued to rely on exclusive agreements with the Soviet Union.[66] As we have seen above, this type of trade agreement was profitable for East European states, but there were lingering concerns about dependence on Soviet materials and expertise. This explains why most East European states readily agreed to create their own national nuclear industries. In a conscious effort to increase their technical, economic, and political autonomy, some states, like Czechoslovakia, chose reactor designs that would not require continued Soviet enrichment services. By the time these initial plans gave way to a more sober assessment of the magnitude of the investments required for the development of a domestic nuclear industry, East European were up to their ears in debt, which ruled out alternatives to Soviet designs.[67] Therefore, even after the NATO embargo was gradually alleviated in the late 1960s and the early 1970s, the Soviet Union remained the dominant source of nuclear technology within CMEA.[68]

In the following section, I consider more systematically the main phases of Soviet–East European nuclear energy cooperation. Restrictions on cooperation in the nuclear energy sector delineated clear technopolitical boundaries: while Western assistance remained beyond reach for East European states, Soviet aid became available and was often welcomed. Using East Germany and Czechoslovakia as examples, I trace important stages of nuclear cooperation and illustrate the variation in Eastern Europe's colonial experience in the area of nuclear energy. Despite the fact that the Soviet technopolitical project fused nuclear reactor designs and economic mechanisms in an effort to coordinate East European societies, I show that the Soviet satellites had room to maneuver, and that they used this leeway ever more skillfully. I conclude that nuclear technology transfer—a potential instrument of technopolitical power—developed in remarkably diverse ways across Eastern Europe. The broader context of the Cold War continued to affect the bargaining leverage of East European governments and resulted in quite uneven results of scientific and technological cooperation.

Initial Soviet Commitments: The 1950s

In the spring of 1955, the Soviet Union signed a series of brief bilateral treaties with its East European allies about nuclear assistance.[69] Although the details of these arrangements were to be determined in further bilateral negotiations, the Soviets committed to supplying experimental research reactors, particle accelerators, isotopes, technical documents, and relevant expertise, as well as to training local specialists at Soviet universities and nuclear research centers.[70] The recipient states, in turn, pledged that no technical information, equipment, or fissile material would be transmitted to a third party.[71] Recipients of Soviet nuclear technology had to bear the full cost of the reactors; payments were to be made in cash, commodities, or short-term loans. Technical documents and relevant expertise, by contrast, were provided free of charge.[72] Similar to the United States, the Soviet Union insisted that spent fuel from its research reactors abroad be returned.[73] One immediate result of these agreements was the establishment of the Joint Institute on Nuclear Research in Dubna on March 26, 1956 by CMEA members and China.

In the wake of the Atoms for Peace proposal, many states felt entitled to nuclear expertise, but most underestimated the new webs of political and economic dependencies such technical cooperation would entail. East European states were eager to participate in nuclear research, and eventually to build up their own nuclear industries, while the Soviets worried about the spread of dual-use technologies. Their concerns conflicted with the modernization imperative central to communist doctrine, a tension East European leaders quickly learned to exploit. They argued that nuclear technologies were particularly suitable to ensure progress in their respective states, and underscored the link between technological progress, economic prosperity, and political stability to appeal directly to Soviet security interests.

In the following years, the Soviet Union signed bilateral agreements with East European states about the export of industrial-scale nuclear reactors. The Soviets continued to train scientists and engineers from East European countries at Soviet institutions and facilities, and in addition sent teams of Soviet experts abroad to supervise construction and to help with the reactor start-up. But the actual administration of the nuclear assistance program proved to be challenging: the civilian nuclear industry within the Soviet Union was still in its infancy, and an effective organizational framework was not yet in place. In the spring of 1956, the Soviets created the Chief Administration for the Use of Atomic Energy to handle all activities related to the peaceful use of nuclear energy, including international cooperation.[74]

At this early stage, many questions about the imminent nuclear cooperation remained unanswered. Would the Soviet Union deliver turn-key nuclear power plants, or would East European countries develop their own nuclear industries? Who would train the nuclear workforce needed? Would nuclear power become economically competitive with conventional power sources? Would Eastern Europe form an association similar to Euratom? Who would be able to manufacture nuclear fuel and extract plutonium from spent fuel? Would East European states produce their own nuclear weapons?[75] Over time, expectations and preferences of individual states changed as they acquired experience with their first nuclear reactors and as they developed their own nuclear industries. East Germany and Czechoslovakia were the first East European states to request Soviet assistance with the development of industrial-scale reactors.[76] The terms of these agreements limited Soviet assistance to technical know-how and training. After an initial period of extensive assistance, East Germany and Czechoslovakia would develop and rely on their own domestic nuclear industries. The Soviets implemented a comprehensive system for supplying fresh fuel and returning spent fuel, which testifies to their commitment to nuclear nonproliferation (and to a closed nuclear fuel cycle). This system, however, showed the underlying problems of exporting dual-use technologies.[77]

A Model Ally: Czechoslovakia

Czechoslovakia's substantial uranium deliveries to the Soviet Union and its non-enemy status during World War II had singled it out as an ally from the outset.[78] In addition, Czechoslovakia had a significant tradition of nuclear research. In 1955, Czechoslovak engineers concluded that the country's natural resources were insufficient to satisfy the country's growth plans, and determined that the unaided construction of a research reactor would take too long.[79] That same year, they decided to purchase a Soviet-designed reactor, and opened a technical college for nuclear technicians.[80] A nuclear physics faculty for undergraduate studies was established at the Charles University in Prague around the same time, and graduate training in nuclear physics and engineering began at the University of Brno and at the Plzen Engineering Institute. In addition, Czechoslovak specialists benefited from CMEA scholarships for training at Soviet technical universities and nuclear research facilities.[81]

Having been a "model satellite" for more than seven years, Czechoslovakia was the first country to conclude an agreement with the Soviet Union about a full-scale power reactor.[82] The Czechoslovak nuclear program was ambitious: by 1970, nuclear energy was supposed to provide one-third of the country's electricity.[83] Czechoslovakia intended to rely on Soviet assistance for speeding up the time-consuming development process, but from the outset it sought to build its nuclear industry with substantial contributions from Czechoslovak engineering.

In 1956, Czechoslovakia signed an intergovernmental agreement with the USSR on the construction of a nuclear power plant at Bohunice, in western Slovakia; construction started two years later.[84] Czechoslovak engineers chose a heavy water design that operated on natural uranium, a design Czechoslovak scientists had studied for many years and one that would render them independent of costly uranium enrichment services.[85] This reactor, built as a joint CMEA project by Soviet and Czechoslovak engineers,[86] is a remarkable technological artifact because it was the first and only time the Soviets agreed to build an industrial-scale reactor type outside the Soviet Union before it had been tested domestically.[87]

Construction of the Bohunice reactor proceeded slowly, and the delays eventually prompted Czechoslovak scientists to accuse the Soviet Union of deliberately hampering nuclear development in Czechoslovakia.[88] The Soviet invasion in 1968, moreover, provoked fears that the Soviets would abandon the reactor altogether.[89] But the long delays are better explained by the difficulties the civilian nuclear program was undergoing within the Soviet Union and by the lack of experience Soviet engineers had with this type of reactor. Ultimately, the Bohunice reactor was completed in 1972, fourteen years after construction had begun. Only four years later, it suffered its first serious incident, and on February 22, 1977, after a major accident occurred during refueling, the Czechoslovak government decided to decommission the reactor and abandon the heavy water design.[90]

Soviet Super Domino: East Germany

The German Democratic Republic (GDR) was officially established in early October 1949, soon after the founding of the Federal Republic of Germany and just over a month after the Soviet Union detonated its first atomic bomb.[91] When Stalin died, in 1953, the future of the East German state was still uncertain. Its economy was in shambles because of early Soviet spoliation, but also because of the ongoing loss of skilled workers through the open border in Berlin. The GDR was quickly becoming an expensive liability for the Soviets. The idea of a united Germany remained on the table. Georgii Malenkov, one potential successor to Stalin, publicly declared the GDR superfluous to the Soviet Union. Nikita Khrushchev, by contrast, considered the GDR of utmost importance, a "Soviet super domino" in the Cold War. He intended to turn this outpost of socialism into a "showcase of the moral, political and material achievements of socialism."[92] It was only in 1957 that the East German regime was finally confirmed, and German reunification definitively ruled out.[93]

The GDR, like Czechoslovakia, had a long scientific tradition, and German industry excelled in the area of precision instruments. Some German scientists had been taken to the Soviet Union and had worked in Soviet research establishments during the war. Upon returning to Berlin, they became leading figures in the GDR's nascent nuclear industry, even before their state's political fate was settled.[94] Once they had received a Soviet research reactor, the Germans started creating their own domestic nuclear industry, with the goal of a uniquely German reactor. It soon became clear, however, that this was far beyond the country's scientific and economic capacities. Just as in Czechoslovakia, the leadership consequently opted for the purchase of a Soviet-designed power reactor.

According to a contract signed on July 17, 1956, the Soviets agreed to deliver a small pressurized water reactor operating on enriched uranium for a nuclear power plant near Rheinsberg, some sixty miles north of Berlin. German industry would assume responsibility (including financial) for equipment and construction work, while the Soviets would deliver the nuclear fuel for free and would take back the spent nuclear fuel.[95] But construction did not proceed as planned. In October 1957, the Soviet minister of energy and electrification, Konstantin Lavrenenko, informed the Germans that Soviet industry would not deliver any equipment for the nuclear plant before 1960, owing to massive backlogs at home. Meanwhile, the costs of the project exploded, and skepticism about the economic viability of nuclear power grew among German scientists and engineers. This in turn prompted Soviet concerns about the country's political loyalty. The Rheinsberg plant had been marketed as a showcase for the entire socialist camp. Questioning it cast doubt not just on nuclear energy, but on the political system.[96] Realistically, the Germans had little choice but to hold out until the Soviets finally delivered. The Rheinsberg reactor, the first Soviet-designed power reactor outside of the USSR, would not be started up until May 1966, ten years after the initial agreement had been signed.[97]

Bottlenecks, Delays, and Emerging Structures of International Regulation
The significant cost overruns and delays in completing both the East German and the Czechoslovak plants were not only frustrating, they also exacerbated serious economic stagnation.[98] Disappointed East Europeans tended to attribute the problems to a lack of Soviet commitment, and to see them as indicative of Soviet imperialism.[99] The development of the nuclear

industry within the USSR, however, suggests that the Soviets faced similar delays and bottlenecks at home. Furthermore, despite claims to the contrary, Soviet scientists and economists were still hotly debating the viability of nuclear power plants.[100] In 1959, Khrushchev prematurely abandoned the sixth Five Year Plan as too ambitious—a measure without precedent in Soviet history. The revised Plan drastically reduced funds for the domestic civilian nuclear program, which would make a second start only in 1962.[101] The immature state of the Soviet program explains why the Soviets had trouble providing as much support as their allies had initially hoped for.[102]

Throughout the 1960s, concerns about the proliferation of nuclear weapons intensified as structures of international nuclear governance were slowly taking shape. The Soviet Union's nuclear policy was less than clear at that point. The Soviets' recent experience with China had made them hesitant to provide nuclear technologies to their East European allies.[103] The International Atomic Energy Agency, established in 1957, administered an inspection system intended to ensure that nuclear reactors were used only for civilian purposes. But countries with developing nuclear industries felt that these inspections did not "adequately protect commercial interest in civilian nuclear power."[104] Acutely aware of the political relevance of controlling enrichment technologies, these countries criticized the nuclear weapons states' monopoly over them.[105]

East Europeans had hoped that signing the Treaty on the Non-Proliferation of Nuclear Weapons (NPT) would alleviate Soviet concerns, but to no avail.[106] This led some observers (both in the West and in Eastern Europe) to accuse the Soviet Union of using its nuclear expertise as a bargaining chip to "extract political concessions from recalcitrant states."[107] But by the late 1960s, Soviet policy toward the capitalist world had changed, and now explicitly encouraged East European states to trade with the West.[108] On the other hand, the Soviets knew that East European states had little spending power and would need their approval for hard-currency loans.[109] Ultimately, the Soviets established extensive controls to safeguard nuclear materials and insisted on not transferring enrichment or reprocessing technologies even to their East European allies.

Changes in Soviet Technopolitics: From Artifacts to Practices

After Khrushchev's ouster, Soviet–East European agreements from the mid 1950s were renewed. Beginning in the late 1960s, the Soviet Union provided technical documentation, equipment, and fuel for nuclear reactors, and also specialists who helped set up nuclear power stations. By the mid

1970s, 1,000 Soviet specialists had been sent to work in Eastern Europe, and over 3,000 East European specialists had been trained in the Soviet Union.[110]

In 1965, the East German government decided to abandon domestic nuclear R&D, and to import "turn-key" nuclear power plants from the Soviet Union instead. The Soviets had promised to supply nuclear fuel for the first and second East German nuclear power plants at no charge, and to take the spent nuclear fuel back. Still, abandoning the development of a domestic nuclear industry was a momentous decision, because it led to total reliance on Soviet production, expertise, and safety concepts.[111] Although the goal of importing turn-key plants was never accomplished in full, the GDR's contribution to its nuclear industry was comparatively small. Czechoslovakia took the opposite route. Czechoslovak engineers drew heavily on Soviet assistance with their first reactor, but they also continued developing their domestic nuclear engineering capabilities.

An increasing trend that started in the late 1960 was multilateral financing of CMEA energy projects. For example, East European states funded the construction of two new nuclear power plants in Ukraine, along with power lines to transmit electricity from those plants to the East European grid.[112] This seemed to be a rational division of labor, especially since creating a nuclear industry from scratch was beyond the capabilities of these comparatively small states. From the perspective of CMEA members, participation in Soviet projects ensured that no individual state would gain inner-bloc leverage through developing a monopoly over crucial commodities.[113] But shifting the emphasis from raw material imports and technology transfer to joint financing of Soviet resource development was also part of a Soviet strategy to revise the terms of trade with Eastern Europe. The Soviets knew that East European economies relied on growing amounts of energy imports, but the trade subsidies to Eastern Europe were more and more difficult to justify domestically.[114] This development marks a remarkable shift in the Soviet technopolitics vis-à-vis its bloc allies: rather than ensuring technopolitical dominance through the export and operation of material artifacts (reactors), the Soviets increasingly relied on administrative, economic, and organizational technologies (structures of trade interaction, scientific and technological cooperation). Technopolitics went from providing artifacts to controlling practices—that is, ways of doing things.

Expansion and Control: The Early 1970s
As the subsidized inner-bloc oil trade deteriorated, 1970 and 1971 brought significant changes in nuclear cooperation within CMEA. Tempted by

hard-currency sales, and with energy demands rising within the Soviet Union, Moscow began to curtail oil exports to Eastern Europe and to replace them with non-subsidized gas deliveries. The imperative to diversify their energy mix thus took on a new urgency for East European states, even before the oil crisis.[115] Developing nuclear energy in CMEA also promised to solve problems with regional energy supplies, and cost projections for nuclear electricity looked favorable.[116] In Czechoslovakia, for example, the cost prognosis for nuclear energy was encouraging even when compared to that for domestic coal or that for imported Soviet oil and gas.[117] Planned increases in the percentage of electricity generated in East European nuclear plants by 2000 ranged from 50 percent to 72 percent.[118] East European nuclear expansion was to be based on Soviet-designed pressurized water reactors.[119] Following an agreement with the Soviet Union in 1970, Czechoslovak factories started producing all but the primary equipment for these reactors.[120] At that time, the reactor vessel, the steam generators, the circulation pumps, and the measurement and safety systems were still manufactured in the Soviet Union; only in subsequent years was the Soviet-Czechoslovak cooperation agreement extended.

Still, nuclear energy cooperation in CMEA in many ways resembled the oil regime: East European states depended on the Soviet Union not only for technical expertise and specialized equipment, but also for enrichment and fuel take-back services. This made them highly vulnerable to unilateral Soviet decisions. In 1970 the Soviets sold large amounts of heavy water to Canada, and in 1971 they began offering commercial uranium enrichment services to capitalist countries, breaking the US monopoly.[121] This policy change affected East European states severely: once the Soviets could get hard currency for these services, not overpriced and low quality goods, they had more leverage to streamline multilateral cooperation within the bloc. In contrast to Khrushchev's unsuccessful earlier proposal, this time the Soviet leaders could even dictate the terms.[122] And so, in 1971, CMEA members signed a "Comprehensive Program for the further intensification and improvement of cooperation and the development of socialist economic integration."[123] In the nuclear sector, this led to the creation of two specialized organizations to ensure a reliable infrastructure for the nuclear industry. Interatominstrument, founded in 1972, coordinated the manufacturing of high-tech equipment for nuclear power plants. Interatomenergo, founded in 1973, managed all other tasks involved in the transfer of nuclear power technology.[124]

When the 1973 oil crisis hit, East European leaders became aware that without curtailing their plans for growth significantly they could afford neither to pay world market prices for oil nor to purchase more un-subsidized Soviet oil. In the interest of domestic stability, however, curtailing their plans for growth was not an option.[125] Apart from contracting Western loans on a massive scale, the only bargaining chip left for East European leaders was to use their countries' economic weakness to press for Soviet aid in the interest of political stability.[126] Likewise, the Soviets continued to export enriched uranium to Eastern Europe because of the military-strategic significance of bloc cohesion.

Gaining Momentum: The Late 1970s and the 1980s

By 1974, Czechoslovakia had mastered significant challenges in reactor engineering, East Germany and Poland had specialized in precision instruments, and Hungary and Bulgaria had concentrated on other nuclear equipment.[127] However, it was not until 1979 that an agreement was signed on "the joint production and mutual deliveries of nuclear power plant equipment for the period 1981–90."[128] This agreement was the biggest ever signed in any field of engineering in CMEA.[129] But problems with innerbloc cooperation persisted. Earlier fears about monopoly formation proved well founded: the division of labor had the effect that certain nuclear equipment was produced in only one country, sometimes by a single enterprise, for the entire bloc. The absence of competition led to poor quality, which in turn raised concerns about nuclear safety. Also, suppliers of indispensable components, most prominently Czechoslovakia, used their inner-bloc monopoly position to raise prices to world-market levels, forcing CMEA client states to scale down their nuclear programs.[130] Integration also enforced dependency, since the Soviets kept control over the bloc's specialization program. They assigned projects to individual states, set prices for manufactured goods, and were able to modify the terms of a contract at any time.[131]

In 1982, 1984, and 1987, the first jointly financed nuclear reactors started operation in Ukraine. Half of the electricity produced by these reactors was earmarked for the Mir system, a new set of high-voltage transmission lines.[132] As a consequence, the Soviet Union's exports of electricity to Eastern Europe nearly doubled in the period 1980–1988.[133] By 1986, CMEA's nuclear program had become so successful that, despite the Chernobyl accident, the member states agreed to intensify cooperation in the period 1991–2000.[134] Czechoslovakia's nuclear engineering sector, in particular, was prospering. In the six years from 1979 to 1985 alone, Czechoslovak engineers launched four pressurized water reactors at the Bohunice site (after decommissioning the original heavy water reactor), two of them manufactured domestically. In the period 1985–1987, they started up

Nuclear Colonization? 143

another four domestically engineered reactors at Dukovaný (near Brno), and built, with Soviet support, eight more reactors (four at Mochovce, in the Slovak part of the country, and four in Hungary). Thus, while the GDR and other CMEA countries were struggling to gain a foothold in the Soviet bloc's nuclear market, Czechoslovakia effectively dominated it.[135]

Conclusions: From Soviet Security to European Safety

The announcement of the Marshall Plan, the escalating events in Hungary and later Czechoslovakia, and the Sino-Soviet split were among the factors that shaped the Soviet Union's technopolitics vis-à-vis its East European satellites. Stalin's death marked the end of the first, "imperialist" phase of Soviet rule in Eastern Europe, and the Soviet empire degenerated further during Khrushchev's tenure.[136] Trade with the Soviet Union was economically beneficial for Eastern Europe, and by and large these considerations outweighed concerns about Soviet military and political domination.[137] Only a stable economy guaranteed the public's compliance, which in turn ensured the regime's survival. Czechoslovakia and the GDR competed for Soviet assistance, rather than avoiding it. It is hard to believe, however, that successful East European negotiating can be attributed to their de facto bargaining power and to consistently mismanaged Soviet representation in CMEA, as Randall Stone has argued. I find it more convincing that East European states strategically used the threat of social unrest as a result of economic crisis to garner Soviet assistance. Above all, appeals to the universal nature of science, and the commitment to progress that was so pivotal for Communist doctrine, swayed these negotiations. The satellites' function as a "security belt" for Soviet national security was guaranteed only as long as their economies were stable. And only scientific modernization and technological progress could provide the basis for economic prosperity, and thus Soviet national security. Over time, the satellites learned to use the language of "brotherly cooperation" to their advantage.

The Soviet technopolitics of nuclear energy were designed to ensure modernization, progress, and development. But they proved unexpectedly flexible on ideological grounds. East Europeans repeatedly "pushed the envelope" politically and economically, but also technologically (for example, by requesting specific reactor designs). Sometimes, most prominently in Hungary and Czechoslovakia, the Soviet Union reacted with military invasion. But at other times, the Soviets granted unconventional requests, or at least did not interfere with individual states' initiatives.[138] The Soviets did not block the decision by Czechoslovakia to build a heavy water reactor, nor did they stop Romania's negotiations with the West. They did, however, remain extremely concerned about nuclear weapons proliferation. This explains their sophisticated system of nuclear fuel supply and take-back, and their choice for export to other countries of the pressurized water reactor design—a design that makes it more difficult to extract weapons-grade plutonium than it is with other designs. Perhaps this cautious management of dual-use technologies explains, at least in part, why Soviet technopolitics ultimately failed on the technical level. Turn-key plants in Eastern Europe awaited Soviet deliveries for years, and the system of spent fuel returns suffered serious backlogs by the 1970s.

How could the nuclear power cooperation within CMEA become one of the most successful initiatives in its history, despite these manifest complications? I suggest that the emphasis on modernization through scientific and technological progress was the one part of the Soviet civilizing mission that was compatible with other ideologies. Technical elites in Eastern Europe benefited tremendously from scientific and technological cooperation with the Soviet Union, as it earned them international prestige and recognition.[139] And while the general resentment of Soviet political domination grew, especially among intellectuals, Soviet technical assistance eventually came to support post-Cold War national interests. When the Soviet bloc collapsed in 1989–1991, what proved most valuable for the former bloc countries was the technical training East European specialists had received in the Soviet Union, and the scientific and technical expertise Soviet specialists had shared on site with their East European colleagues. Soviet technopolitics thus failed on the political level too, as even successful Soviet nuclear assistance fell short of ensuring the satellites' political allegiance.[140]

In spite of these failures, Soviet nuclear technopolitics has left a lasting legacy for post-Cold War Europe, however contradictory. Nuclear reactors were created with one set of norms and design decisions, even as they got disentangled from their Soviet origins. By the late 1980s, many reactors and most of the equipment for East European nuclear plants were manufactured outside the Soviet Union, even though they were still based on Soviet technological standards.[141] These Soviet-style artifacts became bargaining chips for newly independent East European states in the confusing post-1989 political landscape. For these states, accession to the European Union threatened to come at the price of shutting down their nuclear stations, which had become primary energy sources. To appease the European Union, East European reactors now had to comply with yet another technopolitical regime: that of Western safety standards.[142] Western analysts who studied Soviet-designed nuclear reactors in post-communist East European countries were no doubt aware of the complex processes that had shaped the East European nuclear reactors. They noted that in some respects the Soviet designs were superior to comparable Western ones, but they also discovered that there were significant differences among individual East European countries with regard to nuclear safety.[143] Where national industries had participated actively in the development of nuclear energy, as Czechoslovakia had, nuclear safety was generally higher than in countries with more limited domestic nuclear expertise, such as the GDR.[144]

This difference of involvement in CMEA's nuclear cooperation thus affected the post-Soviet fate of individual East European industries. After the peaceful breakup of Czechoslovakia, the governments of the Czech and Slovak republics continued to pursue their nuclear energy program. Although both republics have increasingly turned to Western suppliers for equipment and apparatus, nuclear power has become a defining element of a revitalized national identity in each of them.[145] The leaders of the Czech Republic, for example, responded to the European Union's concerns about the safety of these plants by invoking a successful and uninterrupted tradition of Czech engineering, by emphasizing their economy's need for a reliable supply of electricity, and by defending their country's authority to choose a suitable energy policy. By contrast, in the former GDR all Soviet-designed nuclear reactors were shut down, even the small research reactor at Rossendorf. After reunification, the German government also halted the construction of, and eventually demolished, all new reactors in the former GDR.[146]

The degree of engagement in CMEA's nuclear cooperation was not the only factor that shaped the post-Cold War industries of Eastern Europe, but it is one piece of the puzzle. East European states were distinct entities before the Soviet empire absorbed them. Remarkably, this variation persisted, to varying degrees, throughout Soviet rule. The Soviets' emphasis on modernization through science and technology was compatible with post-World War II European aspirations to resurrect the war-torn economies. But the start of the Cold War prevented Eastern Europe from participating in the Atoms for Peace program, and the Soviets were reluctant to risk the proliferation of nuclear weapons material and know-how. So East European leaders learned to appeal to the belief in scientific and technological progress that they shared, and to invoke Soviet security, as they initiated bilateral cooperation agreements with the USSR. The Soviets' desire for bloc stability prompted their engagement in ever more costly cooperation with their satellites, despite increasing criticism from economists at home. Soviet technopolitics in the nuclear sector were characterized by creative ideas for securing a complex, dual-use technology, and by a lack of industrial, economic, and political stamina to make these ideas work. As a result, East European leaders were able to make quite autonomous decisions with regard to their nuclear research and development. Many of these early decisions, however, carried high stakes, as became clear even before the fall of the Berlin Wall. Combined with the specifics of CMEA trade, they generated unexpected asymmetries in the distribution of nuclear expertise. Furthermore, emerging Cold War geographies gave East European states the leverage to negotiate highly idiosyncratic deals with the Soviet Union. An unfaltering belief in the universality of science and technology allowed flexible interpretations of Soviet technopolitics. Hence, nuclear technopolitics could have been a powerful instrument of Soviet power. But they ultimately failed on both the technical and the political level, leaving unprecedented challenges for post-Soviet Europe.

Acknowledgments

I thank Gabrielle Hecht, Vincent Lagendijk, Martha Lampland, William Gray, Bill Leslie, and Matthew Wisnioski for helpful comments on earlier versions of this essay.

Notes

1. World Nuclear Association, country briefings on Czech Republic and Slovakia, available at www.world-nuclear.org.

2. Throughout this essay, "Eastern Europe" refers to the former Soviet satellites. In contrast to today's preferred label, "Central and Eastern Europe," "Eastern" in this essay should be read as a political rather than a geographical category.

3. On the Soviet empire in Eastern Europe, see Seton-Watson 1962; Bunce 1985; Rowen and Wolf 1987; Reisinger 1992; Rubinstein 1992; Brown 1966; Brzezinski 1967; Harrison 2003; Narkiewicz 1986; Polach 1968; Roskin 2002; Simons 1993; Young 1996. The term "imperialist" in Communist doctrine was reserved exclusively for the United States and its allies (Seton-Watson 1962: 121; Oldenziel, this volume). Significantly less post-colonial analysis has been conducted on Eastern Europe; an exception is Moore 2001.

4. Hecht and Edwards 2008. See also the introduction to this volume.

5. This is contrary to Yuri Slezkine's argument (2000: 228) that the Soviet Union had nothing but its commitment to progress and modernity as its claim to legitimacy. See also Slezkine 1994: 421.

6. Westad 2005; Seton-Watson 1962; Lampland, this volume.

7. Cooper 2004.

8. Moore 2001: 120; Pearson 2002: 48; Lieven 2000.

9. E.g. Slezkine 1994, 2000; Baberowski 2003; Moore 2001; Northrop 2004; Hirsch 2000, 2005; Lapidus, Zaslavsky, and Goldman 1992.

10. E.g. Slezkine 1994.

11. Seton-Watson 1962: 87.

12. Moore 2001: 121.

13. Bunce 1985, 1999.

14. On the rushed construction of the Soviet empire, see Pearson 2002: 46–47.

15. Brown 1966: 34–35; Pearson 2002.

16. Roskin 2002: 56. Roskin (78) claims that Stalin was determined to stay in Eastern Europe since the Yalta conference in 1945. By contrast, Pearson (2002: 27–30, 46–47) argues that there was no pre-designed master plan for Eastern Europe.

17. Pearson 2002: 50–51.

18. Palladino and Worboys 1993: 96. See also Seton-Watson 1962; Cooper 2005.

19. Lieven 2000; Cooper 2004.

20. Bunce 1985: 5.

21. Reisinger 1992: ix–x; Fleron, Hoffmann, and Laird 1991.

22. As Frederick Cooper summarizes, "the long-term survival of empires depended on their rulers limiting their transformative ambitions even as they extended their power" (2004: 247).

23. Bunce 1985; Seton-Watson 1962: 95–96.

24. Other countries joined, or attended sessions; China was involved in CMEA trade from 1950 until 1961.

25. The "Warsaw Treaty on Friendship, Cooperation, and Mutual Assistance" was signed in May 1955. That same year, Moscow committed to global decolonization at the Bandung conference, and agreed to an allied peace treaty with Austria. See Schiavone 1981: 20; Pearson 2002: 56–57; Holloway and Sharp 1984; Mastny and Byrne 2005; Selvage 2001.

26. Stone 1996: 50.

27. Levcik and Skolka 1984: 23. According to the "Bucharest Formula," CMEA prices were based on average world market prices over a five-year period. Initially, changes

could only be made every five years, but this policy was modified in the 1970s to account for extreme oil price fluctuations (Simons 1993: 162–163; Trend 1976).

28. Scholars have debated who benefited more from this arrangement, the Soviet Union or Eastern Europe. See Levcik and Skolka 1984; Wilczynski 1974: 74–75; Stone 1996: 5–6.

29. Reisinger 1992: 20; Simons 1993: 110. With the exception of Poland and Romania, no East European country had significant energy resources.

30. Harrison 2003: 221; Reisinger 1992: 17; Bunce 1985: 12, 21; Brown 1966: 85–123.

31. Most importantly, the "interested country" principle allowed CMEA members to abstain from participation in any given project. This principle remained in force even after the adoption of the "Comprehensive Program for Socialist Economic Integration" in 1971.

32. Schiavone 1981: 27, 40; Stone 1996: 9, 70, 240; Brown 1966: 169–170.

33. Brown 1966: 127–129, 169. That still would not explain why significant reforms of inner-bloc cooperation were postponed until 1971, and even then remained half-hearted.

34. Reisinger 1992: 84; Modelski 1959: 180; Pearson 2002: 67–68; Gittings 1964; see Stone 1996 for a contrary interpretation. After the 1956 events, the Soviets tried to consolidate their position in the socialist world, most notably by making concessions to China. Lewis and Xue 1988: 61–62. See also Kramer 2003, 2004, 2005.

35. Stone 1996: 88; Wallace and Clarke 1986: 74–75.

36. Stone 1996.

37. Bunce 1985: 30, 41–45; 1999; Reisinger 1992: 21.

38. As Hope Harrison put it (2003: 139, 94), "The tail boldly wagged the dog."

39. The link between economic progress and communist victory through science and technology goes back to Lenin, who also emphasized per-capita electricity output as a measure of development. See Polach, 1968: 841–842; Koriakin 2002; Kramish 1963; Trotsky 1960.

40. Young 1996: 16–17. Czechoslovakia in particular was interested in Marshall Plan aid, but the Soviets claimed that Czechoslovakia's participation would violate the 1943 Treaty of Friendship (Wallace 1976: 233–234, 262–263; Crampton 1997: 236–237). Czechoslovakia withdrew. Czechoslovakian and Polish industry suffered most from the technology embargo. See Narkiewicz 1986: 153; Wilczynski 1974: 164, 331.

41. Bunce 1985: 8. East Europeans resented the assumption that they needed "civilizing," as they often felt culturally superior to the Soviet Union. See also Palladino and Worboys 1993: 99–101.

42. Josephson 1996. At the Communist Party Congress in 1956, Soviet Premier Nikolai Bulganin linked nuclear energy directly with the construction of Communism and thus gave the program priority. He declared that communists "must fully harness atomic energy—that tremendous discovery of the twentieth century—in the service of the cause . . . of building communism. . . . Our country leads other countries in the peaceful use of atomic energy. We must maintain this lead in the future." (Modelski 1959: 114–115).

43. Schmid 2005: 39–40; Adas 2006; Bunce 1999.

44. Levkic and Skolka 1984: 29.

45. Krige 2006b; Wilczynski 1974: 585.

46. Weiss 2003.

47. Modelski 1959: 111–117; Fermi 1957.

48. Bunn 1992.

49. Modelski 1959: 126; Polach 1968: 851; Weiss 2003: 40.

50. Fermi 1957: 16; S. M. 1955.

51. Headrick 1988.

52. Adas 1997: 478–484.

53. Palladino and Worboys 1993: 97.

54. Müller 2001: 221.

55. Müller 2001; Stokes 2003.

56. Müller 2001; Stokes 2003; Newman 1978: 38; Engeln 2001.

57. Harrison 2003: 94–95; Crampton 1997: 279–280. The charter of the 1956 Hungarian revolution explicitly demanded access to Soviet records concerning Hungarian uranium (Modelski 1959: 138), and the Czechoslovak public learned for the first time in 1968 that the Soviets had exclusive rights to the country's sizeable uranium deposits (Polach 1968: 848). Both countries remained important producers of uranium throughout the mid 1980s. Today the mines are exhausted and remediation work is in progress.

58. Although Wismut partly relied on prison labor in the early years, it was no East German version of the GULAG system. Workers were recruited through incentives (e.g., exceptional health care services), and by 1951 essentially the entire workforce

was voluntary. See also Stokes 2003; Newman 1978: 38; Mueller, 2001: 220–226; Modelski 1959: 127; Engeln 2001; Karlsch 2007.

59. Wilczynski 1974: 67, 71.

60. Often the best way to modernization and technological progress was thought to lead through scientific planning—another ideologically promiscuous set of ideas (see Lampland, this volume).

61. Young 1996: 17; Narkiewicz 1986: 153; Wilczynski 1974: 164, 331.

62. Wilczynski 1974: 588.

63. Weiss 2003; Bunn 1992; Hsieh 1964.

64. Fermi 1957; Modelski 1959: 127; Newman 1978: 38–40.

65. Headrick (1988) distinguished "geographic relocation" and "cultural diffusion," labels that could not hold up in practice but which no doubt served as powerful rhetorical resources for both Soviets and East Europeans to justify, or constrain, nuclear technology transfer in the context of the emerging Cold War.

66. Modelski 1959: 143; Scheinman 1987; Newman 1978.

67. Stone 2008.

68. Wilczynski 1974: 588; Newman 1978: 58.

69. In 1955, agreements were signed with Romania on April 22, with Czechoslovakia and Poland on April 23, with East Germany on April 28, and with Hungary on June 13; in 1956, Yugoslavia followed (Newman 1978: 38; Ginsburgs 1960; Modelski 1959: 127). China signed the first of several agreements with Moscow in 1955; that agreement involved uranium deliveries in exchange for a research reactor (Lewis and Xue 1988: 41).

70. Wilczynski 1974: 585; Newman 1978: 38.

71. Newman 1978: 39–40.

72. By contrast, in US bilateral agreements half the cost of the research reactors was financed through grants (Newman 1978: 40; Modelski 1959: 128). Some Western observers condemned this arrangement as lacking generosity, and as an attempt to make a profit—a concept apparently deemed incompatible with socialism. Conversely, some deplored the lack of Soviet business acumen regarding the creation of large future markets—clearly the explicit goal of the US initiative (Modelski 1959: 129–130; Ginsburgs 1960; Lewis and Xue 1988: 61–62).

73. Like the US, the USSR offered financial incentives to CMEA countries to return spent fuel (Müller 2001: 251).

74. Sidorenko 2001: 217–18; Emelyanov 1958; Kramish 1959: 178; Modelski 1959: 11–14.

75. In 1959, this was apparently still considered a legitimate pursuit by some (Modelski 1959: 144).

76. Polach 1968: 834; Tatarnikov 2002: 348.

77. This fuel supply and take-back system had an advantage over Western bids at the time: it effectively "closed" the nuclear fuel cycle for their customers and relieved them of "the plutonium problem." Although this feature was generally advertised as economic convenience, the probable underlying reasons were concerns about nuclear proliferation and industrial espionage.

78. Hungary and Poland became suspect allies in 1956, making Czechoslovakia an even more important partner (Modelski 1959: 154). Because Czechoslovakia had been a Nazi protectorate, its potent industrial base had not been destroyed (Wallace 1976: 223; Brown 1966).

79. Polach 1968: 835, 840.

80. Modelski 1959: 152–157, 171.

81. Emelyanov 1958: 360; Polach 1968: 833.

82. Brown 1966: 23.

83. Polach 1968: 842; Modelski 1959: 172–173.

84. Polach 1968: 842–843.

85. Wilczynski 1974: 67.

86. The Institute of Theoretical and Experimental Physics (ITEF) in Moscow had operated a heavy water research reactor since 1949. The first Bohunice reactor, referred to as KS-150, had a power output of 150 MW and posed a series of challenges for its designers (Goncharov 2001: 29).

87. Modelski 1959: 113; Goncharov 2001: 33–34, 58; Schmid 2005: 83.

88. Polach 1968: 848.

89. In Hungary, the delivery of the promised research reactor was canceled after 1956 (Ginsburgs 1960).

90. Asmolov et al. 2004: 66; Petros'iants 1993. This accident was rated 4 on the International Nuclear Event Scale (INES-4, on a scale of 7, labeled "Accident without off-site risk"). The Soviets subsequently abandoned this design, either because of the failed Czechoslovak "test case" or for economic reasons (Polach 1968: 834–851; Modelski 1959: 218).

91. Roskin 2002: 71; Brown 1966: 33.

92. Harrison 2003: 139–142, 178, 227.

93. Brown 1966: 34.

94. Modelski 1959: 157–161; Albrecht, Heinemann-Grueder, and Wellmann 1992.

95. Müller 2001: 127, 157.

96. Modelski 1959: 217; Müller 2001: 174.

97. Müller 2001: 167–171; Reisinger 1992: 43.

98. Brown 1966: 85–123.

99. Polach 1968: 843–845.

100. Goncharenko 1998: 148.

101. Goncharov 2001: 49; Kramish 1959: 145.

102. The difficulties the Soviets encountered with expanding their own nuclear industry also help understand why they advocated a more rational division of labor ("integration") within the bloc. See Modelski 1959: 130–131; Reisinger 1992: 43; Schiavone 1981: 27; Brown 1966; Stone 1996: 157.

103. Westad 1998; Lewis and Xue 1988.

104. Gilinsky and Smith 1968: 817.

105. "For the distant future, control of plutonium implies control of electrical power, which implies control of industry." (Gilinsky and Smith 1968: 820).

106. Czechoslovakia, East Germany, Hungary, Poland, and Bulgaria ratified the treaty in 1969, Romania in 1970 (Goldschmitt 1980: 75).

107. Newman 1978: 57; Polach 1968: 850–851.

108. This policy change explains in part Soviet non-interference when Romania negotiated with France, with Great Britain, and later with Canada, about a reactor running on natural uranium (Finch 1986).

109. Polach 1968: 849; Müller 2001: 75; Stone 2008.

110. Reisinger 1992: 51; Wilczynski 1974: 585.

111. Müller 2001: 9, 81, 135, 187.

112. Reisinger 1992: 58.

113. Reisinger 1992: 50; Bunce 1985: 15.

114. Stone 1996: 36–38; Bunce 1985: 13; Brown 1966: 86, 123.

115. Pearson 2002: 91.

116. Modelski 1959: 169.

117. Wilczinski 1974: 571.

118. At the time, the United States and Britain were expecting 70 percent of their electricity to be nuclear by 2000 (Wilczynski 1974: 577).

119. The VVER-440 became the standard Soviet export reactor.

120. Sobell 1984: 151; Hewett et al. 1986.

121. Wilczynski 1974: 590.

122. Stone 1996: 156; Müller 2001: 86.

123. Schiavone 1981: 32.

124. Wilczynski 1974: 71; Office of Technology Assessment 1981: 295; Sobell 1984: 155.

125. Newman 1978.

126. Pearson 2002: 91–92. Newman 1978; Bunce 1985.

127. Wilczynski 1974: 581–582.

128. Schiavone 1981: 39.

129. Sobell 1984: 153. On Czechoslovakia, see Levcik and Skolka 1984: 21.

130. Stone 1996: 167; Wallace and Clarke 1986: 82; Stone 2008.

131. Müller 2001: 87.

132. Mir connected the Khmel'nitskaia station with the Polish city of Rzeszów, and from there with Czechoslovakia, and the Southern Ukraine station near Odessa with Isaccea in Romania and Dobrudzha in Bulgaria (Reisinger 1992: 55–59; Stone 1996: 153).

133. Stone 1996: 165–166.

134. Reisinger 1992: 66.

135. On the GDR's trouble fulfilling its obligations, see Müller 2001: 237–243.

136. Pearson 2002: 50–51, 88–91; Brown 1966: 230.

137. This statement obviously does not apply to 1968 Czechoslovakia or to 1956 Hungary; it also ignores less visible resistance movements. The point here is that these were notable exceptions to a generally conformist (because Moscow-installed) East European leadership elite.

138. Examples include the Berlin Wall (which was built at East Germany's request), the tolerance of Romania's eccentric and provocative political course, and the settlement of Poland's reform wishes in 1956 through negotiations (Harrison 2003: 184–187. See also Brown 1966: 36; Pearson 2002: 64).

139. Pfaffenberger 1993: 344; Brown 1966: 39; Bunce 1985: 8.

140. The fuel supply and take-back system deteriorated early on (Müller 2001: 253, 236).

141. Levcik and Skolka 1984: 64.

142. The idea of so-called safety upgrades for Soviet-designed reactors rested on two contradictory assumptions. On the one hand, it implied that these reactors were somehow imbued with Communist ideology, therefore inherently inferior, and in dire need of a Western overhaul. On the other, it suggested that these reactors were just like any other technological artifact, and as such almost certainly could be improved.

143. Sidorenko 1997.

144. Müller 2001: 135–136.

145. Hecht 1998; Dawson 1996; Hezoucký 2000.

146. Haenel 1998. These actions no doubt were motivated by economic considerations, but there were also concerns about West Germany's particularly vocal and influential anti-nuclear movement.

7 The Technopolitical Lineage of State Planning in Hungary, 1930–1956

Martha Lampland

The battle to gain allies and control territory in the Cold War was orchestrated in large part by experts in economic development promising technological innovation to usher in a modern future. The epochal break symbolized by the Iron Curtain was rendered visible in geographic polarities of East and West, and ever-shifting lines on the outdated maps of formerly colonial hinterlands. The strategies and techniques experts deployed to promote economic development, however, did not follow the same geographical contours, with similarities crisscrossing political divides. Nor were the policies adopted novel approaches to economic progress, as the techniques and practices deployed were already well in use before the Cold War had begun. Unfortunately, studies of the transition to socialism in the early years of the Cold War, most notably in Eastern Europe, have been colored by the scholarship chronicling the rapid escalation of hostilities, overshadowing the continuities in expertise and statecraft in the region. This neglect has thwarted our ability to appreciate the specific dynamics of regime change in the late 1940s and the early 1950s, and led us to overlook valuable comparisons with developing economies in other parts of the world. In short, both the temporal and the geographical boundaries characterizing scholarship on the Eastern European socialist transition have been distorted, a problem I wish to remedy in small measure here.[1]

The transition to Stalinism in Hungary in the late 1940s has long been portrayed as a swift and radical economic about-face, ripping Central Europe from its capitalist moorings and consigning the region to the dark ideological irrationalities of the East. Well-schooled, tempered experts were sacrificed to political expediency; alien practices were foisted on a conquered people by imperialist forces. This depiction is inaccurate, fostered as much by the impossibility of primary research as by Cold War ideological struggles.[2] These constraints no longer hold. My approach, based on research in archives and interviews, broadens our perspective, situating the early Cold War era of Eastern European economies in relation to the global history of developmental economics in the mid twentieth century.[3]

Studies of the socialist transition have not paid sufficient attention to the similarities between European states and states in other regions, most notably East Asia, during the 1930s and the 1940s. As Johnson has argued,[4] recent histories of market-driven developmental states in East Asia have ignored the early gestation of these states in state-directed state-managed growth, leading to a misunderstanding of the conditions under which thriving economies were initially created. Notable among these innovations was an economy of experts, a wide-reaching network of social scientists committed to a grand restructuring of states according to the principles of scientific management and business administration.[5] The promise of technological solutions to the intractable social problems of the 1920s and the 1930s appealed to a wide range of actors—from businessmen to politicians to bureaucrats—even though their specific explanations for social ills varied enormously. This explains how regimes with widely differing political aims could embrace similar state policies. Indeed, during this period planned economies were found in capitalist and socialist societies, in fascist and liberal regimes, and in colonial states as well as in sovereign states. Further investigation is required to discern how these developmental states differed—most notably, in this case, how a planned capitalist economy in Hungary was transformed into a socialist economy.[6]

My approach to the transition to socialism in Hungary is informed by Hecht's notion of "technopolitical rupture-talk," i.e., "the rhetorical invocation of technological inventions to declare the arrival of a new era or a new division in the world."[7] Hecht's analysis focuses on how the technology, the infrastructure, and the materiality of nuclear weapons were deployed to inaugurate a new geopolitical order that relied extensively on perpetuating colonial relations and sustaining pre-existing political networks. The intimate relationship between colonial resources and the technopolitical development necessary for joining the "nuclear club" had to be denied, since the ability to engineer nuclear weapons constituted the crucial political distinction between the developed and the not-developed countries. I would like to extend the notion of "technopolitical rupture-talk" to an analysis of the transition to socialism and of the early Cold War. The "originary" technology in this case is the socialist state, portrayed as a novel configuration in the postwar political economy of Eastern Europe. Missing from this history are the substantial constraints that social science theorizing and political practice of preceding decades exercised on the new regime, i.e., the technopolitical lineage of state planning. Ignoring the importance of epistemology and disciplinary formations to economic policies and state formation in the twentieth century has had unfortunate consequences. Blinded by the technological rupture-talk of ideologues on both sides of the Cold War (and of their compatriots in the academy), we have overlooked two simple questions: How and to what degree was the postwar Hungarian economy altered in the transition to socialism? Who in fact fashioned policies in the early years of the socialist state, and why?

Deprovincializing Planning

Strong state involvement in the economy was widespread between the two world wars. Sweden, Nazi Germany, and the United States shared many features of a state-led economy in the 1930s, having chosen a similar configuration of policy options to cope with economic distress.[8] Japan, Korea, and Taiwan also adopted strong interventionist policies during the 1930s, joining the ranks of countries struggling to prevent economic collapse.[9] The search for pragmatic solutions to the economic crisis was worldwide; economists and policy makers studied the different strategies governments attempted, rejected, and advocated.[10] Japanese rationalization demonstrated a particularly lively mix of influences.

Japan's specific conception of [industrial rationalization] originated as a poorly digested amalgam of then current American enthusiasm ("efficiency experts" and "time-and-motion studies"), concrete Japanese problems (particularly the fierce competition that existed among the large number of native firms and the consequent dumping of their products), and the influence of Soviet precedents such as the First Five Year Plan (1928-33) and the writings of the Hungarian economist and Soviet advisor Eugene Varga.[11]

Economic policy making and theoretical elaboration entailed more than simply reading up on government programs or debating the merits of new ideas, such as the fascist model or the pioneering work of Keynes. The pursuit of viable institutions led officials to study political programs in action. Japanese and Soviet officials traveled to Germany seeking advice and policy templates in the late 1920s, and Germans traveled to Detroit for inspiration.[12]

This focus on policy as problem-solving, however, tends to underestimate the degree to which theories of economic development in this era were seen as *scientific instruments*. This is significant, as it directly influences the specific character of rationality envisioned, i.e., relying on experts in economics and management sciences to modernize bureaucratic practices.

Advocates of the new field of administrative science drew inspiration from economic policies adopted during World War I, and from the rationalization movement within industry. Zoltán Magyary, a staunch advocate for administrative modernization in Hungary, praised Herbert Hoover's 1921 report "Waste in Industry" as the manifesto of the rationalization movement.[13] Such efforts were especially salient in Central Europe, where the death of imperial regimes and the birth of new states required the building of new administrative structures. Taking a page from Hoover's report, Magyary declared: "Our [state] administration may . . . be considered a large factory, the management of which bare 'empiria' or dilettantism is no longer sufficient."[14] The era of random and impulsive statecraft was over.

Despite the widespread interest in planning, it is important to underscore the variety of means by which states pursued a more ordered and vibrant economy. There was no clear consensus among its proponents about what rationalization actually entailed.[15] "Neither rationalization nor efficiency were clear and concise concepts. Indeed a large part of their popularity lay in their elasticity, in their ability to encompass so many phenomena achieved or desired."[16] In light of the diversity of application, Rabinbach goes so far as to say that productivism, "the common coin of European industrial management and of the pro-Taylorist technocratic movements across the European landscape between the wars," was "politically promiscuous."[17] Even though the left and the right espoused different values in scientific management (e.g., social harmony vs. factory autocracy[18]), both camps assumed that decisions were to be made by trained specialists and experts—technocrats and bureaucrats. Scientific remediation of social dislocation and of economic hardship rested upon basic principles of exclusive knowledge and omnipotent vision. The apparent neutrality of technique appeared to sever means from ends, obscuring the relations of authority and dominance achieved with these new policy tools. This politics, often termed managerialism, did not solicit public opinion or submit policies to processes of democratic review. In some versions of this approach, of course, parliamentary procedures or other interest-based means of adjudicating policy decisions were eliminated entirely as superfluous or intrusive.

The Technopolitical Lineage of State Planning in Hungary, 1920–1947

During the 1920s and the 1930s, many Hungarian economists and policy makers promoted planning, discussing the variety of ways in which planning would enhance productivity and efficiency. They often pointed to policies they admired, most notably policies developed in Germany and

Italy.[19] Prominent social commentators—from both the right and the left—kept their eyes on events developing in Soviet Russia, expecting to learn valuable lessons from the grand experiment under way.[20] The enthusiasm for innovation in some quarters was met with intense skepticism by others, who questioned the motives of planning advocates. Example: "Today across the world large enterprises, large industrialists and wealthy landowners are enthusiastic for Soviet plans. . . . As long as they could earn [money], free competition was good; now that one must pay, then the public should pay for it. Planned economy is 'the socialization of the losers.'"[21] Yet whatever position one took on the balance between markets and planning, there was strong agreement that experts in the new fields of economics and business sciences would have to be involved in decision making. This required a united front of economic specialists to break the legal profession's monopoly on positions in government administration.

The Hungarian government's policies toward the economy in the interwar period entailed extensive state participation, ranging from legal measures and regulations to direct control of nationally owned concerns. Intervention preceded the Depression, but the role of the state in protecting Hungarian businesses and the health of the economy overall increased between 1931 and 1938.[22] Enterprises were regularly bailed out by the state, aided by provisions in the bankruptcy law (1924. évi IV. t.c.), although the law had been intended to facilitate the demise of unprofitable enterprises.[23] Monopolies were given special consideration, becoming the focus of state assistance as of 1931.[24] The state set specific taxation policies, or engaged in the setting of prices (often with the collusion of particular cartels) to dampen competition. The state also invested heavily in a number of private concerns, although the ties were often difficult to discern from official statistics:

> Firms established in Hungary before the First World War resisted liquidation, existing firms expanded, indeed, entirely new kinds of corporations started to appear, which appeared to be private businesses (e.g. joint stock company, cooperative) but which in fact represented a unique combination of state capital and state intervention. Therefore the development of hidden state capitalism intensified, with the state acquiring stakes in private enterprises, either in the form of the majority of shares, or actually the entire enterprise.[25]

As was true in many nations at the time, monetary policy stood at the center of state policy, specifically foreign exchange restrictions introduced at the beginning of the currency crisis in 1931, but maintained as "valuable tools for extricating themselves out of the crisis."[26] Protectionist tariffs were levied to protect industry.[27] Another crucial measure, directly tied to foreign exchange restrictions, was the introduction of premiums on exchange.[28] This placed the control of currency exchange directly in the hands of the National Bank.[29] In Hungary, long-brewing animosities between agricultural interests and financial and industrial interests played themselves out in struggles over the particular schedule of premiums, which had consequences for import and export policies.[30]

The role of monopolies in the economy grew in the 1930s. Approximately 60 percent of agricultural exports were handled by monopolies.[31] A number of crops were regulated by price as well as by trade. Ihrig provides a list of the crops under state regulation in 1935: "wheat, sugar beets, tobacco, linseed and flax, hemp, milk, firewood, wool, paprika, alcohol and potatoes. One must also consider that the price trends of wheat more or less influence the other grains, indeed seed for fodder as well. Thus in the area of crop production the zone which is managed is larger than that of production which is not."[32] Laws preventing the establishment of new firms eliminated competition, strengthening already existing monopolies.[33] The argument the Minister of Industry in 1935 made against new firms was simply "the danger of spreading limited capital too thin."[34] The law eventually passed initiated a quota (or *numerus clausus*), forcing cartelization within industry. This was considered necessary in the increasingly state-managed economy, as a 1935 article in the newspaper *Pesti Tőzsde* (*Stock Market of Budapest*) made all too clear: "There is no doubt that . . . promulgating the industrial *numerus clausus* represents a newer and forceful step toward a managed economy. . . . It is obvious that this condition is not a temporary world phenomenon, but an overture toward a newer chapter of the economy."[35] In the banking sphere, increased concentration was evident. By 1938, 72 percent of all capital stock was in the hands of eight of the largest banks.[36] All these measures supported the interests of finance capital, industry, and (to a lesser degree) large-scale agriculture. As such, they demonstrated the power and efficacy of state intervention in the economy, constituting positive evidence for those favoring greater state management.

The state also was actively engaged in labor policy, passing regulations on the length of the work day, setting minimum wages, and, by the war years, setting maximum wage levels. Throughout the interwar period, and during World War II, the state consistently acted to moderate extremes in the price of agricultural labor, either by reigning in workers' demands in times of labor scarcity by setting an upper limit or by forcing employers to pay a base minimum when the fortunes of workers had drastically worsened. The state also initiated public works projects in rural communities facing substantial shortages in winter supplies.[37] Regulations issued by the Ministry of Agriculture in 1926, in 1928, and in 1930 stipulated more carefully the exact role of national and county authorities in supervising the movement and the contractual obligations of labor. Advocates of a better-managed economy lamented the shortcomings of these agencies, most notably their inadequate methods of gathering information. During the war, Béla Reitzer complained that the consistent absence of proper numerical data impeded "the continuation of planned labor market policy."[38]

State intervention increased substantially with the introduction of what came to be known as the Győr Program of 1938. Following Germany's lead in ignoring the ban on militarization set by the Treaty of Versailles, revisionists keen to enhance Hungary's power and reputation in the region began to call for serious plans to prepare for war in 1937. Elements of the military began to lobby the government to invest in the armed forces. Designed to assist rearmament, the Győr Program required substantial investments in heavy industry, in transportation, and in telecommunications. The larger percentage of revenues (600 million Pengő) raised to finance the plan were to come from a one-time property tax. This was levied on private individuals and enterprises with assets exceeding 50,000 Pengő. "In the country there were 28,569 private individuals and legal entities who were required to pay 5–14 percent of their wealth in a one time property tax as an investment subsidy, to be paid in 5 yearly installments."[39] The other 400 million would come from banks and large enterprises. Rather than draw on existing stores of capital, however, both wealthy individuals and enterprises turned to the National Bank for credit. In June of 1938, the basic charter of the National Bank was modified to permit it to extend more credit to the state by issuing unsecured banknotes. The potentially inflationary consequences of issuing money were not apparent in the first two years of the plan, during which the state placed orders of nearly 200 million Pengő with industry, of which 70–75 percent was ordered from the iron, metals, machine, and electronics industries.[40] The proportion of national income devoted to military expenses would grow exponentially. Pressures to monetize the national debt were exacerbated by Germany's indebtedness to Hungary throughout the war, a result of its policy of maintaining a balance of trade (an exchange clearing account system) rather than paying for goods outright; the debt grew from 326 million Pengő in 1941 to 2,918 million Pengő in 1944.[41] Germany's role in the Hungarian economy grew at the end of the 1930s. Approximately 50 percent of Hungary's exports ended up in Germany, while German capital interests held 12 percent of Hungarian industrial shares.[42] At the same time, thousands of Hungarian workers were employed in Germany. The number of agrarian migrant workers alone reached nearly 45,000 between 1937 and 1943, the peak years being 1938 and 1939.[43]

In Hungary, as elsewhere, World War II played a significant role in furthering state intervention in economic affairs.

> Beyond the direct equipping and provisioning of the army, [state intervention] spread to every area of economic life. The state became the largest consumer of industrial production, and it became necessary for the state to intervene in questions of production, such as establishing enterprises, the supply of raw materials and energy, regulating prices, the credit system, and the supply of labor power. [44]

By 1941, the state had officially declared a fully planned economy (*tervgazdálkodás*). This affected citizens as well as industries. Problems of provisioning sparked the introduction of measures to ensure that supplies would make their way to soldiers. Local inspections by police were initiated to ensure that sufficient stores of wheat would be made available for bread. By the summer of 1942, guards were stationed at threshing machines "to determine the producers' surplus."[45] This procedure was replaced by the Jurcsek plan, which taxed produce on the basis of the productive capacity of the land.

In the last year of the war, extensive damage was wrought as the Germans' rearguard action against the Russians moved across Hungarian soil. Budapest was under siege for 102 days.[46] The ability of the state to regulate affairs diminished radically; the machinations of encroaching powers, and their proxies, were virtually impossible to prevent. "All in all, 40% of Hungary's national wealth was destroyed: 90% of the industrial plants were damaged, 40% of the rail network and 70% of the rolling stock were lost."[47] At the end of the war, approximately 10 percent of Hungary's population had died in battle—as soldiers or as citizens—or, in the case of nearly half a million Jews, had been slaughtered on the streets or sent to Auschwitz.[48] Moreover, nearly a million able-bodied men were incarcerated as prisoners of war, which deprived the country of much-needed manpower for the postwar recovery. In one year, Hungary's economy experienced the worst monetary inflation in history. In the last five months of 1945, the average daily increases in prices were 2 percent, 4 percent, 18 percent, 15 percent, and 6 percent. In the first seven months of 1946, the average daily increase skyrocketed. From the middle of June 1946 to the end of July, the average daily increase went from 8,504 percent to 158,486 percent.[49]

Plans for postwar reconstruction were being discussed in a variety of quarters as early as 1943. Unfortunately, Hungarian experts who had experienced the economic and social cataclysm that had followed World War I could anticipate the kinds of difficulties they now faced. "The greatest trouble was caused by disorganization and lack of co-ordination. . . . We have two tasks after the war: right away after the war in the first and second year to ensure a peaceful transition to a peace economy, and to prepare productive work in subsequent years on the basis of a comprehensive plan."[50] It is not surprising, then, that in 1947, on the eve of the Third Year Plan, leading figures in the industrial sector spoke earnestly about planning in the language of necessity—moral as well as historical. Example:

If the state recognizes the rights of its citizens to life, then it must help them so that they can live. Whether one likes it or not, agrees or not with our design—theoretically, politically, economically, morally or philosophically—the conditions force us to have a target and planned economy. *The road ordained by economic and social necessity is the one we must travel.*[51]

While concerns about the relative balance between state intervention and individual initiative still hung in the air, few questioned the need for state management. Funding of the 1947 plan rested on hopes that internal economic resources would be sufficient, and that Hungary would not have to apply for foreign credit (although that possibility was contemplated). The financial reserves designated for the plan were to come primarily from the state, various public institutions, Hungarian-Soviet joint ventures, cooperative enterprises, and private capital.[52] Nearly 85 percent of these funds were to be raised with various taxes, including a one-time levy on property and a one-time levy on the increase of wealth due to profits acquired from the war and inflation. It was also expected that, in addition to issuing plan bonds to make use of the population's savings, the surplus from state-owned factories would contribute to the pot.[53] Expectations of a rapid increase in production nationwide and a willingness on the part of villagers to replenish their livestock herds and finance their own building costs would ensure the plan's success.[54] In the cautious optimism of the times, modifications in the plan appeared feasible. "No question but the investment portion in the Three Year Plan constitutes the greatest burden on the Hungarian economy. It is reassuring that, if necessary, the investment program can be reduced without substantially having to change the production plan."[55]

After the war, the dominance of foreign interests—once German, now Soviet—continued. In 1944, Germans dismantled factories in Hungary and shipped them west; as of 1945, Russians were shipping manufacturing assets east. Pillaging and looting accounted for some of the loss of economic assets after the cessation of fighting, but war reparations—paid in kind as well as in cash—were a greater encumbrance on Hungary's industrial base. In addition to shipping existing equipment eastward, Hungarians were also required to manufacture machines and equipment to Soviet specification.[56] As part of the peace treaty, assets formerly owned by Germans were appropriated by the Soviets. Once-active markets for Hungarian goods in Italy and Germany were now trained eastward, strengthening in trade the already substantial interests the Soviets had in the Hungarian economy. After the signing of a treaty for economic cooperation on September 23, 1945, Soviet officials established a number of joint ventures with Hungarians. Clearly, by 1947 the Soviet Union—as shareholder, factory owner, landlord, and occupying army—controlled significant assets in the Hungarian economy.

The process of rebuilding the economy after the war intensified an already strongly centralized and managed economy bequeathed by the previous regime. Ironically, the demands for reparations increased the Hungarian state's participation in the economy. Before the stabilization of Hungarian currency in August 1946, all of the productive capacity of the five most important industrial concerns was devoted to manufacturing to fulfill the conditions of reparations, dropping to only 60 percent as of September.[57] With time, what had been oversight of production by the state became outright ownership. The five most important heavy industry factories were nationalized, much to the dismay of their owners. The state's encroachment on heavy industry grew.

At the end of 1946 roughly 150 thousand workers, or about 43.2% of the employees in manufacturing and mining, were working in state firms. Among them 75,000 worked in the five largest factories taken over by the state. The state role prevailing in energy, raw and basic materials exercised a significant influence over every branch of industry, indeed the entire economy of the country.[58]

The state now held a significant portion of the assets of the Hungarian economy. Finally, banks were nationalized in 1947, transferring capital into the state's hands. Another important milestone in state control of the economy was reached.

Following demands raised by the Communists during the 1947 elections, the Council of Ministers . . ., and the parliament . . . ratified the nationalization of Hungarian owned shares of big banks and banks in which it had financial interests, as well as industrial and commercial concerns. The jurisdiction of the measure indicates that before the war nationalized big banks kept 72% of the capital of all the banks under their control, and more than three-fifths of industrial capital stock, and in 1947 they had properties equivalent to this.[59]

So, as Pető and Szakács explain, "nationalizations in March (1948) did not represent a basic change with respect to substantial growth in the role of the state in the economy, but rather created the conditions for the transformation of the institutional system, and a change in the relationship between the state and enterprises."[60]

As the Cold War intensified in 1947 and 1948, the vision of the Three-Year Plan once supported by the Communist Party was abandoned. Having swallowed the Social Democratic Party, and having forced significant political enemies into exile, the Communist Party could now redirect its efforts to jump start the economy from a more extensively state-led, state controlled industrial base, while turning its back on promises to guarantee private property. The Marshall Plan was rejected, and in February of 1948 new agreements with the Soviet Union were signed. The once-bright vision of a rejuvenated economy founded on investing in modern productive technologies was traded for an emphasis on increasing brute quantities. "Instead of reconstruction, there was only renovation, which displaced potential renewal. Therefore instead of introducing the quick technical and technological changes achieved during the war—of which only a small amount was perceptible in Hungary—they generally restored the earlier, obsolete equipment."[61] The rush to the finish made it possible for the new party/state to declare a glorious and shortened end to the Three-Year Plan. Not surprisingly, a larger proportion of investments were made in industry than initially planned, leaving rural communities to bear the lion's share of the costs and burdens of reconstruction.

The typical picture of the transition to Stalinism in Hungary has been one of the thoroughgoing imposition of a Soviet-style planned economy. "[T]he Procrustean imposition of Stalin's version of proletarian dictatorship resulted in a radical transformation of the entire political, social, and economic system."[62] This view ignores state economic planning during the capitalist period, the growth of state intervention in the economy during the war years, and the Three-Year Plan ratified by the Hungarian Parliament in 1947. Kornai's depiction comes closer to capturing the moment:

The economy that emerged [after the war] was a curiously mixed one, with a "regular" capitalist sector on the one hand and socialist elements on the other. A steady process of nationalization took place. Land reform was carried out on a large scale. This period came to an end around 1948–49 with the amalgamation of the Communist and Social Democratic parties and the elimination of the multiparty system. From then on, construction of a socialist system began with full force, starting straight away with classical socialism.[63]

Kornai's appellation "classical socialism" is an analytic term suited to his systematic approach to economic forms, but tells us little about the particulars of the transitional process. Though no doubt the final goal of the Communist Party was to facilitate the transition to a socialist economy modeled on that of the Soviet Union, it is a very different matter to represent the transitional process as an economic about-face. Viewed in a comparative framework, Hungarian industrialization was much closer to that of developmental states such as Korea and Taiwan than to the Soviet Union.

Economic Expertise

A mixed economy of capitalist and socialist elements was not the only foundation on which to build socialism; economic theories propounded in the interwar period were also ready to hand for the new party/state. Academic economists and government officials active in policy making during the war brought a familiarity with state planning, in practice as well as theory. The standard tale of Stalinist transformation in Hungary is one of Russian experts and Muscovite communists[64] building the new institutions of the Marxist-Leninist state. There is no question that the Communist Party worked hard to exert its influence over many institutions within the state as early as 1945. Political domination, however, did not necessary translate into the wholesale rejection of capitalist knowledge or expertise.

Many features of economic policy were shared by the new disciples of Marxism-Leninism and the so-called bourgeois economists, not least because they shared an allegiance to rationalization and scientific management.[65] A strong compatibility existed between the anti-market principles of Marxism-Leninism and the vision of a well-managed planned economy business economists advocated at the time. Marxist-Leninists were committed to the progressive improvement of society, sharing the utopian dreams of bourgeois economists, and using many of the same techniques to achieve those goals. Bourgeois economists at the time also supported policies that based policy on privileged knowledge and expertise. In other words, bourgeois economists' vision of a modern economy—the importance of economies of scale, modernized forms of work organization and mechanization, and carefully designed wage structures—was fully compatible with the goals promoted by the new regime. Economic growth was seen to issue from the proper institutionalization of these factors, as much as from extensive investment in infrastructure, physical stock, and education. Of course, bourgeois economists did not necessarily embrace a one-party state, welcome democratic centralism, or easily relinquish private property,

State Planning in Hungary

but they certainly were far more sympathetic to expertise-driven policy regulation than to parliamentary adjudication.

Just how the party/state would treat bourgeois experts was an open question. A strong anti-intellectual bias characterized Communist Party workers; suspicions also plagued some new bureaucrats that advisors schooled in the capitalist era were unreliable. There were others, however, who valued expertise in statecraft and economic development, and who were willing to overlook the pedigree of knowledgeable persons. Two countervailing forces were at odds: paving the way for a new elite, loyal to the party/state and versed in Marxism-Leninism, and the keen need for qualified staff to implement party/state policies. The actual process of balancing party allegiance with valuable expertise was a continual problem. This battle was fought over and over again at all levels of government: within ministries, between agencies, and across departments. The clash of these tendencies made the fate of experts unpredictable. Some experts were marginalized, sequestered off in minor jobs, or even imprisoned, while others found themselves promoted up the ranks. The fate of specialists ebbed and waned with the fortunes of hardliners and reformers within the government. The consequence of this often capricious turn of affairs was not only a loss of qualified personnel to jail cells, the usual story told about the Stalinist era, but also the continued presence of a number of seasoned economists and government officials from the previous regime. In fact, the construction of new socialist institutions was carried out by a large number of people whose expertise was firmly grounded in the capitalist era.

Sweeping aside a crucial number of department heads, especially those with a committed political and social conservatism, laid the way for the Communist Party to staff ministries with its own appointees. Some officials were purged from government agencies in early de-Nazification projects (the so-called B listings of 1946), and with time even more were let go as government agencies and educational institutions were reorganized. Yet public pronouncements proudly declaring reactionary bureaucrats had been removed clashed with private deliberations at the highest levels of the party acknowledging the need for expertise. In a meeting of the Central Committee's Agricultural and Cooperative Department held in August 1949, this policy was explicit: "Since we have such a deficit of agricultural specialists, our principle in relation to dismissal should be only to discharge direct adversaries. In the case of experts one must examine whether they are honest, whether they work well, and whether or not they oppose the Party line."[66] Freshly minted university graduates were brought in to head departments, but their lack of experience, coupled with an understandable respect for expertise, led them to rely heavily on more seasoned bureaucrats. Although high-level managers recruited from the working class may have been less sympathetic to specialists in their midst, that did not necessarily translate into obstructionism, since the manager's position relied at least in part on results. No doubt it pained the experts to have to be subordinated to less qualified superiors, but that is not a condition peculiar to socialist bureaucracy. [67] A 1954 chart listing the qualifications of higher echelons of management personnel in the Ministry of Agriculture shows the pattern of senior officials coming from the ranks of proletariat comrades, while lower down on the bureaucratic hierarchy, well-qualified officials remained in place (table 7.1). At the Ministry of Agriculture, this included officials trained in business science and manorial state management, not just the more traditional branches of animal husbandry and crop breeding. As the table illustrates, it was also more common for credentialed officials to have joined leftist parties later than their superiors, and in several cases they had not joined at all. In other words, much of the hands-on policy work—drafting regulations, recording information, tracking administrative debates, overseeing policy implementation in the countryside—was conducted by civil servants who had been trained and worked in agrarian economics and management long before the Communist Party took over.

Academics and researchers were also subjected to extensive review by the Communist Party, and many formerly respected scholars lost their teaching posts or research positions.[68] Recurring campaigns were waged against idealism (i.e., Western genetics) in biology, and against the aberrations of social science. A report reviewing the personnel office of the Agricultural Experimental Center issued on August 22, 1950 identified serious problems:

> The majority of researchers in the field of scientific work in agriculture are petit-bourgeois intellectuals, a significant percentage of whom are older (especially the independent researchers), whose family connections and previous milieu hampers their development. This is apparent equally in their work, behavior, and continuing education of both a specialist and political character. They stick to their old familiar methods, don't recognize the results and observations of Soviet agricultural science, or use this to support their own views. When considering their results they base them on Western results, and that shows that they are satisfied with their work.[69]

Replacing problematic intellectuals with scientists pursuing the new revolutionary avant-garde was not a straightforward affair. Old habits of respect on the part of staff in research centers blinded them to the dangers in their ranks, leading them to be less vigilant in background checks than party officials would have liked. "They don't recognize enemies who are enshrouded in the haze of expertise."[70] Moreover, judging the sincerity of

Table 7.1

Information on leading and mid-level cadres in the Ministry of Agriculture, May 14, 1954. Source: MDP-MSZMP 278 f., 74. csop., 31 ö.e., pp. 4–7. MKP: Hungarian Communist Party (Magyar Kommunista Párt), 1944–1948. MDP: Hungarian Workers' Party (Magyar Dolgozók Pártja), 1948–1956. SzDP: Social Democratic Party (Szociáldemokrata Párt), 1944–1948; as of 1948 allied with MDP.

	Age	Original occupation	Party membership, year joined	Professional or vocational training	School	Communist Party school attendance
Leading cadres						
Ministerial deputy director	30	Agriculturalist	52 MDP	Stock breeder	Postgraduate studies at Agricultural college	2 years in correspondence school
Ministerial deputy	40	Worker	45 MKP	Auto mechanic	3 years middle school	1 year
Ministerial deputy	38	Worker	43 SzDP, 45 MKP	Construction worker		1 year
Ministerial deputy	43	Worker	45 MKP	Steel pourer	Agricultural college correspondence	1 year
Ministerial deputy	32	Gardener	46 MKP	Gardener	2 years gardening courses	1 year
Ministerial deputy	34	Mechanical engineer	44 MKP	Mechanical engineer	Polytechnical university	1 year
Director general	45	Upholsterer	28 MKP	Upholsterer	Commercial high school	
Director general	36	Teacher	46 SzDP	Teacher at agricultural school	Agricultural university	2 years
Mid-level cadres						
Assistant under-secretary	38	Worker	36 MKP	Licensed farmer	Agricultural college	—
Chief of department	40	Lawyer	45 SzDP	Lawyer	University of Law	—
Chief of department	28	Locksmith	45 MDP	Locksmith	6 years elementary school	1 year

Table 7.1
(continued)

	Age	Original occupation	Party membership, year joined	Professional or vocational training	School	Communist Party school attendance
Chief of department	42	Industrial worker	45 MKP	Industrial worker	6 years elementary school	5 months
Chief of department	30	Civil servant	46 SzDP	Economist	Economics university	—
Assistant under-secretary	37	Blacksmith-locksmith	36 MKP	Blacksmith-locksmith	6 years elementary school	1 year
Chief financial officer	34	Industrial worker	45 MKP	Tool and die maker	4 years middle school	—
Department head	40	Civil servant	45 MKP	Practiced farmer	Law school	—
Department head	28	Agronomist	51 MDP	Agrarian economist	Agricultural university	—
Director	41	Plant breeder	46 SzDP	Licensed farmer	Agricultural university	—
Vice-director	53	Estate steward	48 MKP	Licensed farmer	Agricultural college	—
Director	48	Stock breeder inspector	47 MKP	Stock breeder	2 years vocational school	—
Director	34	Veterinarian	—	Veterinarian	Polytechnical university	—
Director	53	Agricultural inspector	—	Licensed farmer	Agricultural college	—
Director	50	Agrarian worker	45 MKP	Agrarian worker	4 years elementary school	3 months
Director	37	Veterinarian	45 SzDP	Veterinarian	Agricultural university	—
Director	31	Poultry breeder	45 SzDP	Poultry breeder	3 years vocational school	3 months
Vice-director	69	Teacher	—	Veterinarian	Agricultural university	—
Vice-director	53	Teacher	—	Licensed agriculturalist	Agricultural university	—
Vice-director	32	Agronomist	45 MKP	Licensed agriculturalist	Agricultural university	—

State Planning in Hungary

Table 7.1
(continued)

	Age	Original occupation	Party membership, year joined	Professional or vocational training	School	Communist Party school attendance
delegate director	52	Agrarian inspector	–	Licensed agriculturalist	Economics university	–
Director	30	Civil servant	50 MDP	Licensed farmer	Agricultural college	–
Director	40	Teacher	–	Licensed gardener	Horticultural college	–
delegate director	51	Estate steward	–	Licensed farmer	Agricultural college	–
Director	42	Agricultural worker	45 MKP	Agricultural worker	6 years elementary school	5 months
Department head	41	Civil servant	48 MKP	Accountant	Commercial high school degree	–
Director	35	Forestry engineer	47 SzDP	Licensed forestry engineer	Polytechnical university	–
Director	30	Agriculturalist	46 MKP	Licensed agriculturalist	Agricultural high school certificate	1 year
Director	63	Civil servant	19 MKP	machine technician	High school graduation certificate	–
Director	32	Gardening inspector	51 MDP	Licensed gardener	Horticultural college	–
Vice-director	32	Agronomist	45 MKP	Licensed agriculturalist	Agricultural university	3 months
Vice-director	41	Mechanical engineer	45 MKP	Licensed mechanical engineer	Polytechnical university	–
Director	27	Engineer	–	Licensed mechanical engineer	Polytechnical university	–
Department head	34	Agricultural worker	45 MKP	Agricultural worker	7 years elementary school	6 months
Vice-director	39	Estate steward	–	Licensed farmer	Agricultural university	–
Vice-director	39	Estate steward	–	Licensed farmer	Agricultural university	–
Director	41	Estate steward	42 MKP	Licensed farmer	Agricultural university	–

Table 7.1
(continued)

	Age	Original occupation	Party membership, year joined	Professional or vocational training	School	Communist Party school attendance
Director	42	Estate steward	51 MDP	Licensed farmer	Agricultural university	—
Director	37	Agrarian teacher	—	Licensed farmer	Agricultural college	—
Director	48	Estate steward	—	Licensed farmer	Economics university	—
Director	34	Estate steward	45 MKP	Licensed farmer	Agricultural high school certificate	—
Director	44	Agricultural worker	39 SzDP	Agricultural worker	6 years elementary school	—
Director	43	Estate steward	—	Licensed farmer	Agricultural college	—
Director	34	Agronomist	46 MKP	Licensed farmer	Agricultural university	—
Director	30	Estate steward	45 SzDP	Licensed farmer	2 years vocational school	—
Director	56	Industrial worker	15 SzDP	Industrial worker	5 years elementary school	5 months
Director	32	Gardener	51 MDP	Gardener	Horticultural high school	—
Director	27	Gardener	52 MDP	Licensed gardener	Agricultural university	—
Vice-director	32	Forestry engineer	45 MKP	Forestry engineer	Polytechnical university	—
Director	38	Forestry engineer	48 MKP	Forestry engineer	Polytechnical university	—
Director	42	Cabinet maker	31 MKP	Cabinet maker	4 years high school	—
Director	42	Cabinet maker	45 MKP	Cabinet maker	6 years elementary school	1 year
Director	36	Forester	46 MKP	Forester	4 years high school	—

epistemological conversions was a gamble, as specialists became good at quoting the right passages from the enshrined literature, all the while holding on to their old ideas.[71]

The intense efforts the party/state devoted to creating a new scientific ethos of Marxism-Leninism are well documented in Péteri's fascinating history of the Hungarian Economics Research Institute (Magyar Gazdaságkutató Intézet, or MGI) between 1947 and 1956.[72] His account situates the transition in a world of new Stalinist epistemologies, where positivism was ridiculed as a pseudoscience and "glorifying facts" as the "pathological symptom" of its "methodological dead end."[73] A thorough analysis of the institute's struggles within larger battles taking place across the party hierarchy demonstrates how specific political agendas dictated institutional reforms. Péteri also illustrates how these efforts floundered, defeated by simple problems of staffing and expertise.

Established in 1927 by the Budapest Chambers of Commerce and Industry to adapt the new insights of the study of business cycles to Hungary, MGI rose to a prominent position as the most respected site for the study of economics in the country. "To this day their studies in monetary policy, investigations of the Hungarian national income, analysis of industrial investment and capital accumulation, examination of all sorts of economic problems in Hungarian agriculture, as well as periodical economic reports are important sources of data and ideas for research in economic history."[74] MGI was both a research institute keeping current with contemporary economic theorizing and a valued but always independent partner in the creation of national policy. After the war, the institute sustained its autonomy from political control, even when several members were appointed to high level positions in the coalition government. The ability of the institute to fulfill its mandate was increasingly circumscribed, however, when previously available data were withdrawn. MGI was dismantled in August 1949, to be replaced with the Institute for Economic Science (Közgazdaság-tudományi Intézet, or KTI). As a result of vetting former members for their party affiliation, Péteri calculates that approximately 30 of the 34 researchers employed at the institute were let go.[75] Only two of the researchers were members of the Communist Party; most were not allied with any party whatsoever, which constituted a major sin in the eyes of the party/state, drawing charges of passivity and indolence.[76] After all, in the world of "socialist science" one could never be detached from the party/state and the working class.[77]

In its haste to revolutionize economics, the party/state had neglected to train personnel, a problem that was never solved during the entire history of KTI (1949–1953). KTI never found a full-time director, and was permanently understaffed. Very few economists in Hungary were well versed in Marxist-Leninist political economy. Those who were occupied leading positions in the Communist Party, in state administration, or in party education. Training a new generation of Marxist-Leninist economists fell to the university, but that presented its own problems, as those teaching the courses were not luminaries in the field.[78] This forced the institute to offer its own courses. As Péteri wryly notes about the students selected to work at the KTI from the graduating class in 1950, "the poor professional training of the staff is attested to by the fact that for the first half year of the KTI a course on the political economy of capitalism was organized for the institute's members."[79] Moreover, classroom materials were inadequate or nonexistent. In 1947, Imre Nagy had to appeal to Jenö (Eugene) Varga in Moscow to provide him with foreign-language editions of Soviet political economy textbooks, preferably in German, French, or English. Péter Erdös, one of the temporarily assigned co-directors of the institute, complained in 1951 that he could not find people qualified to do independent research; the coming generation was going to require many years of experience before reaching that status.[80] Indeed, not until 1955 could István Friss, a committed Marxist-Leninist, make this declaration: "The fact is . . . that for the first time organized Marxist economic research is being conducted in our country, which portends well."[81] The consequences of not having properly trained experts extended beyond staffing research institutes. In 1955 a report was issued by the Academy of Sciences criticizing the quality of economics training for upper-level cadres and calling for reform in higher education. Economists trained in capitalism were the mainstay of government bureaucracies.

Implementing a proletarian dictatorship, the ensuing changes in the Hungarian people's economy and the building of socialism has created a great demand for economics cadres with strong professional preparation in the basics of Marxism-Leninism. This need has appeared in every branch of the people's economy. There has been a particularly strong need for economist cadres in higher government agencies—the National Planning Office, the Central Statistical Office, economic departments—which in the beginning, in the absence of a new economic intelligentsia, were forced to work almost exclusively with old experts.[82]

Science and Sovereignty, or the Question of Imperialism

In the course of several decades, Hungary fell under two different spheres of influence, into the orbits of two different imperialist powers: Germany and the Soviet Union. Germany had long exercised influence over Hungarian intellectual trends and political affairs; Soviet domination was new. Reorienting one's intellectual compass and political allegiances would not be easy, even if the broad contours of economic modernization were agreed upon by both parties. To this point, I have been arguing that the scientific pedigree of business administration guaranteed its neutrality, abetted its "political promiscuity." Of course, the neutrality of technique was illusory. Scientific management enshrined hierarchies of privilege, justifying new standards of inequality in the name of science. Accommodating different political ambitions and a new set of moral priorities—refashioning technopolitical practices—was a tall order.

Countries in the Soviet Bloc have long been portrayed as lackeys or reluctant puppets of Soviet politics, instantiating Lenin's analysis of imperialism and its consequences. Unfortunately, this simplistic representation does little justice to the complex dynamics the transition to Soviet-era rule entailed. Since the opening of archives in Russia and Eastern Europe in the early 1990s, we have more resources with which to rethink our perspective on Soviet control of Eastern European regimes. Few doubt the ability of occupying army to limit political debates and narrowing accepted political solutions (a point also relevant to the western zones occupied by English, American, and French armies). Everything else is under renewed scrutiny, most notably Stalin's postwar plans, the United States' "behind the scenes" actions toward the region, and the balance of power between Soviet and Eastern European regimes.[83] Documents from the Hungarian party/state reveal that the relationship with the Soviets was more complicated and often less congenial than was publicly acknowledged at the time. Interviews conducted with economists and former civil servants about their experiences in decision making and bureaucratic procedures during the early 1950s underscore the impression found in written documents about the strains in Soviet-Hungarian relations.[84] Party officials did their best to please or placate the Russians, yet how that sensibility (whether founded on pragmatic political ambition or on sincere hope for social transformation) translated into policy is a different matter altogether.

In the present volume we have an excellent example of this new scholarship for a later moment in the Cold War: Sonja Schmid's analysis of atomic energy and technology transfer. Schmid is able to demonstrate convincingly that Eastern European states exercised significant influence over the specific character of technological transfer, and moreover, that countries in the region took substantially different paths toward securing atomic energy. Her analysis is a fine critique of technological determinism, but she also undermines the determinism implicit in the "Communist party/state monolith" notion which has been characteristic of much past work on the Soviet Union and Eastern European regimes. With party and government archives at our disposal, we are able to open up the black box of Marxist-Leninist party/states. Battles over ideology, policy, and practice within the party and across government can now be fully documented. The particular role of expertise—in all its complex diversity—can also be better accounted for, providing, as Schmid illustrates, a very different picture of politics and decision making.

During the 1950s, the superiority of the Soviet model was taught in Hungarian universities and party schools, and continuously parroted on the radio. Comments in the documents I read repeatedly described the difficulties of making these policies work in Hungary. The Soviets were keenly aware of the contrast between public pronouncements and private actions, and they were never shy in conveying their displeasure. In a 1948 memo addressed to the Foreign Relations Department of the Soviet Communist Party, Korotkevics and Zavolzsszkij complained about the tendency of Hungarian Communists to prefer their own interests over those of the Soviet Union:

> In public Rákosi, Farkas and Révai emphasize the importance of the friendship between the Hungarian Republic and the Soviet Union. They speak highly of the historical mission of the Soviet state and of the role of Comrade Stalin. In the course of everyday affairs, however, the majority of the leadership of the Hungarian Communist Party—presumably so that they not be seen as the agents of Moscow—ignore and keep quiet about the Soviet Union, and at the same time attempt to demonstrate behavior appearing to be indifferent to the Soviet Union.[85]

Soviet leaders found the same hesitation in the principles concerning agriculture in the Hungarian Communist Party's political platform written in 1947–48. The worries Hungarians felt about collectivization were shared in many capitals of the Eastern Bloc, where party leaders were trying to figure out how the Soviet experience with collectivization would travel. As the showdown between Yugoslavia and the Soviet Union was coming to a head in 1948, Soviet officials made it very clear that they would no longer tolerate Eastern European regimes experimenting with alternative paths to socialism[86]:

> The statement of the party platform properly emphasizes that it follows the ideology of Marxism-Leninism. It deploys and develops the teachings of Marx, Engels, Lenin and Stalin further, but in addition it stipulates "in compliance with Hungarian conditions." ... This stipulation—on the pretext of accommodating Hungarian conditions—legalizes the opportunistic revision of Marxism-Leninism.[87]

Soviet advisors were also free with their advice on the inadequacies of teaching and research being conducted in Hungary. Visiting Soviet delegations made their criticisms known. Research institutes were poorly managed, requiring closer supervision to provide the ministry with analysis based on a thorough review of the Soviet and Hungarian literature directly related to the planning process.[88] The Soviets found the separation of abstract thinking from practical experience especially worrying. The Business Economics Department at the Agrarian University came under serious criticism for the totally abstract nature of lectures, and the absence of any analysis of the practical experience of progressive farms in the curriculum.[89]

Hungarian officials actively solicited the aid of their Soviet colleagues, inviting them for shorter or longer stays as their assistance warranted. Soviet advisors walked the halls of universities and ministries, dispensing advice in policy deliberations as liberally as in the classroom. In 1951, the need to consult reams of Soviet research and government documentation was deemed to require the People's Economic Council to establish a separate bureau dedicated to translating texts, though that bureau's efforts were handicapped by the paucity of qualified translators.

All the efforts to retool research, teaching, and policy development came to naught, however, when the Soviets' vision of collective production proved difficult to implement.[90] Eighty percent of Hungarian farmers refused to turn their land over to the state and join cooperatives, even when intense pressures were brought to bear: high taxes and requisition orders, imprisonment for minor infractions, and dwindling resources to sustain families on their own land. Cooperative farms faced enormous challenges working land with insufficient tools and draft power. Hundreds of thousands of hectares were left fallow, abandoned by villagers who preferred to seek their fortunes in industry. Government offices at the local level were inundated with a flood of regulations. Consistently understaffed, they were poorly equipped to respond to higher authorities' demands. Party committees in villages were rarely effective, if they existed at all. Party agitators roamed the countryside, but their often halfhearted efforts bore little fruit. Bureaucrats in Budapest also found themselves chafing at the expectations Soviets had of their work. As one agrarian economist, who as a recent university graduate in 1952 had been named head of the Department of Labor Relations at the Ministry of Agriculture recalled, "We struggled with the Soviet advisors mercilessly because their primitive conditions shocked us. They wanted us to adopt their advice entirely.... We debated a lot ... we tried to prevail. Occasionally we were able to get our way, sometimes not."[91] In some cases, the value of a Soviet perspective was unclear. One expert who worked at the Ministry of State Farms reminisced about a meeting he attended in which the Soviet advisor seemed completely uninformed about the issues being discussed. When the Hungarians asked their Russian comrade a question, he flipped through the pages of a book in front of him. Exasperated by having to wait for an answer each and every time, the expert finally turned to Hegedüs, who was chairing the meeting, and asked pointedly whether they couldn't just send the Soviet advisor home and translate the book for themselves. His disrespectful comment got him sent from the room, but that was the end of his punishment. The really smart Soviet advisors, I was told, quickly put together a package of policy proposals, then spent the rest of their time in Hungary fishing.

These anecdotes paint a picture of a working relationship fraught with disagreements, misunderstandings, and frustrations—in other words, the everyday struggles of a bureaucracy developing new policies. They took place, however, under severe material constraints. In the first few years of the socialist party/state, the economy deteriorated rapidly. Stepping back from disaster in 1953, the new regime sanctioned by Moscow attempted to rectify conditions by liberalizing agricultural policy and reining in the errant secret police apparatus. The subsequent battles within the party over reform policies fostered confusion in the government, and frustrations among the populace escalated. Accommodations made in the past were no longer forthcoming from Moscow. The Soviet invasion in 1956—what Béla Király has called the first war between socialist states—demonstrated the ends to which Soviets would go to assert their control.

Conclusion

This has been a story about the shifting fortunes of experts, bureaucrats, and businessmen. The history of planning economies in the mid twentieth century has been analyzed within the framework of emerging disciplines—economics, administrative science, and scientific engineering. It is a story about attempts to transform national bureaucracies from the domain of civil lawyers to the property of university-trained economists and engineers. As such, it recounts the emergence of a new "economy of expertise" that transforms economies and state bureaucracies. The value of particular forms of knowledge and specific kinds of authorities are constituted through appeals to objective techniques of scientific investigation and certainty, promising innovative solutions to pressing problems. While the political goals of these policies had changed from rejuvenating a capitalist economy to building a socialist polity, the means by which these goals were to be reached were virtually identical.

In the course of our discussions, our research group came to categorize this and several others papers as contributing to a view of "Long Cold War," i.e., revealing significant features of political and economic practice that preceded the Cold War but had important influence on its early development. I have used the notion of planning as technopolitics, and recounted its interwar history in Hungary, in order to undermine the idea of the socialist state as an originary institution constituting a rupture in Central European economic history we date to the beginnings of the Cold War. In itself, this is an important analytic move, as it forces us to examine more carefully the ways that states were built and economies engineered in a period usually broken in the middle, unnecessarily. The misreading of Soviet imperialism in the course of the transition to socialism also has to be discarded, since new archival materials and the fresh perspective offered by postcolonial theories make it possible to think more rigorously about how power is exercised and with what effects. In the context of the other papers in this volume, we are able to draw the history of economic modernization in Eastern Europe into dialog with the history of development economics and political modernization in other regions of the world. The historiography of Eastern European socialism, especially in its earlier phases, has rarely looked farther afield. By treating Eastern Europe outside the histories of economic development projects and anti-imperialist struggles of the 1950s—by neglecting the analytic insights of postcolonial theory and science studies—we do an injustice to the lived experiences of those caught up in the battles of the Cold War.

In past histories, the ends of state policies—entrenched capitalist interests versus a new politics of redistribution—obscured the ways particular elites established their authority and ensured their positions through appeals to scientific certainty, objectivity, and universal truth. In both contexts, the role of a wider community of citizens contributing to managing the society, and in particular to deciding how wealth would be shared, was rejected. In the end, the utopia envisioned, and so frequently heralded by work scientists on both sides of the Cold War divide, had little room for those outside the economy of experts.

Notes

1. The broader project on which this article builds is a study of scientific management, productivity, and wages in the agricultural sector of the Hungarian economy

(1920–1956). This explains the frequent references to publications focusing on agriculture. Agricultural work science and scientific management are not well known in the literature on rationalization and modernization in the twentieth century, even though these fields were actively pursued across the globe.

2. This misconception has far greater consequences than a mere historical oversight. This image of radical institutional transformation has implicitly, if not explicitly, informed the discussions and policies of the post-1989 transition to post-socialism. Unfortunately, this view has contributed to policy designs which have fostered pernicious social inequalities—the concentration of wealth and widespread impoverishment of the region—and serious problems with political corruption. Many observers assure us that life is improving in Eastern Europe, but the important question, as always, is "For whom? And why for some, and not others?"

3. See, in this volume, Mehos; van Oosterhout.

4. Johnson 1982.

5. I am indebted to David W. Cohen for the notion of an economy of experts.

6. This insight also pertains to the relationship between modernizing economies in Europe and in the colonies, be they in Africa, Southeast Asia, or the Middle East. See, e.g., Cooper 2005; Maat 2001; Mehos, this volume; Mitchell 2002; van Oosterhout, this volume.

7. Hecht 2003: 1–18.

8. Gourevitch 1986.

9. Amsden 1989; Garon 1987; Gorden 1985; Johnson 1982; Wade 1990.

10. Bendix 1956; Guillén 1994; Maier 1975; Merkle 1980.

11. Johnson 1982.

12. Johnson 1982; Shearer 1996; Nolan 1994. The value of personally examining scientific evidence, and the difficulty of reproducing scientific results without hands-on experience, is an important theme in Science Studies. For the classic formulation of this problem, see Collins 1974. See also Bockman 2002: 310–352.

13. Magyary 1930: 2.

14. Ibid.: 3.

15. Daniel Ritschell points out that recent historical scholarship on debates over economic policy and planning in Britain are distinctly Whiggish, exaggerating the similarity among plans over time, and reducing them to variations on Keynesianism. Ritschell 1997.

16. Nolan 1994: 6.

State Planning in Hungary

17. Rabinbach 1990: 272.
18. Maier 1975: 583.
19. Bereznai 1943; Lovász 1942: 41–44; Magyary 1930; Reitzer 1941: 996–1013.
20. Kovács 1940; Nagy 1941; Rézler 1940.
21. Badics 1934: 153–167.
22. Bojkó 1997: 138.
23. Bojkó 1997: 8.
24. Berend 1958.
25. Bojkó 1997: 10.
26. Berend 1958: 26.
27. Berend 1958: 8.
28. Berend 1958: 29.
29. Berend 1958.
30. Berend 1958: 30.
31. Berend 1958: 108, 54.
32. Ihrig 1935: 131. Ihrig goes on to note that in comparison to the United States, England, Holland, and Germany, Hungarian agricultural production was far less regulated.
33. Berend 1958: 109.
34. Berend 1958.
35. Berend 1958: 109–110.
36. Berend 1958: 112.
37. Berend 1958: 28; see also Steuer 1938: 601–611.
38. Reitzer 1941: 996–997.
39. Berend 1958: 300.
40. Berend 1958: 301.
41. Siklos 1991: 51.
42. Siklos 1991: 292.
43. Lencsés 1982: 177.
44. Bereznai 1943: 7.

45. Matolcsy 1943: 18.

46. Ungváry 1998: 9.

47. Borhi 2004: 53–54.

48. Borhi 2004: 54.

49. Pető 1985: 61.

50. Bárányos 1944: 6.

51. Tonelli 1947b, p. 193.

52. Tonelli 1947a, p. 184.

53. Tonelli 1947a.

54. Tonelli 1947a: 178–179.

55. Tonelli 1947a:178.

56. Borhi 2004: 146.

57. Pető 1985: 79.

58. Pető 1985: 81.

59. Pető 1985: 95.

60. Pető 1985: 99.

61. Pető 1985: 122.

62. Kovrig 1979: 236.

63. Kornai 1992: 30.

64. "Muscovite communists" were those Hungarian Communist Party members who spent extensive periods of time in the Soviet Union.

65. It is easy to ridicule claims of scientificity for the kind of historical materialism practiced by Marxist-Leninist states, but this should not blind us to the centrality of scientific aspirations in policy making.

66. Magyar Országos Levéltár (MOL), Magyar Dolgozók Pártja Mezőgazdasági és Szövetkezeti Osztály 276 f., 85/31 ő.e., p. 4–5; 1949.aug.1.

67. I am fairly confident that my conclusions about bureaucratic continuity are well founded, even though they are based solely on the history of the Ministry of Agriculture. Far more empirical research work is required to investigate the history of other ministries in Hungary, in particular those devoted to education and culture, where party ideology may have played a larger role. Yet insofar as the Communist Party considered the Ministry of Agriculture the most intransigent and the most

politically conservative institution within the government, it provides us with a useful case study of bureaucratic elites and the dynamics of the transition.

68. Péteri has written an excellent account of the complex disciplinary and party politics involved in the transformation of the Hungarian Academy of Sciences from an essentially honorific body before 1945 into a Soviet-like ministry of science (in Péteri 1991: 281–299). For a comparative history of the academic transitions in three socialist states, see Connelly 2000.

69. MOL MDP 276 f., 93/348 ő.e.: 89, 90.

70. Ibid.: 92.

71. Ibid.85/55 ő.e., p. 7; 1950.júl.10.

72. Péteri 1998: 185–201.

73. Péteri 1998: 190.

74. Péteri 1998: 191–192.

75. Péteri 1998: 196. As Péteri points out, precisely at the time when the economy was undergoing a grand transformation, highly skilled statisticians and sophisticated economists were excluded from the playing field of national policy.

76. Péteri 1998.

77. Péteri 1998: 189.

78. Péteri 1998: 200.

79. Péteri 1998: 199.

80. Péteri 1998: 200.

81. Magyar Tudományos Akadémia Levéltar, II. osztály (Közgazdaságtudomány), 183. doboz, 6. dosszié, p. 4.

82. Ibid., 2. dosszié, p. 1.

83. For examples of archivally grounded post-Cold War research, see Borhi; Roman 1996; Harrison 2003; Stone 1996.

84. I wish to acknowledge that during interviews, informants may have exaggerated their bravado or emphasized how clever the Hungarian nation was in the face of foreign control. Even taking that into consideration, the number of comments supporting the view that Soviets were far less influential than has been assumed, and the empirical evidence from Soviet sources reinforcing this view, give me confidence that the overall tone of the memories is accurate.

85. Korotkevics and Sz. Zavolzsszkij, "A Magyar Kommunista Párt vezetésének nacionalista hibáiról és a Magyar Kommunista sajtóban érvényesülő burzsoá befolyásról" (Standeisky et al.: 205–208).

86. Rainer 1996: 393.

87. Izsák 1994: 267.

88. MOL XIX-K-1-ah 4. doboz, N/204, 1952.

89. Ibid., 10. doboz, N/4; 1952.jan.3.

90. Collectivization was only achieved in the late 1950s, after 1956, and under intensified pressure from the Soviets and Chinese.

91. Mrs. György Lonti, interview by author, June 18, 1997.

8 Fifty Years' Progress in Five: Brasilia—Modernization, Globalism, and the Geopolitics of Flight

Lars Denicke

The airplane approaches Brasilia, after flying for some minutes over a ondulating tableland [sic], cut by low groves of trees. Then, at a distance, the reflection of a lake surprises us for its vast dimensions; the city, shaped like an airplane, rests near the shore. . . . The plane lands and the passengers are exposed to their first terrestrial contact with the city. The peninsula of our airport acts as a promontory, a terrace from which to admire Brasilia.[1]

This was one of many enthusiastic voices commenting on the inauguration of the new Brazilian capital in 1960. The shift of the political center from Rio de Janeiro to the new city in the country's less developed interior, more than 600 miles from the coastline, was central to Juscelino Kubitschek's presidency from 1956 to 1961. The construction of Brasilia, accomplished in only three years, marked the technopolitics of development at high speed, titled, in Kubitschek's government program, "Fifty Years' Progress in Five." As I will argue, Kubitschek's program reveals the spatial logic of two central concepts of Cold War technopolitics: modernization and globalism.

I use the term *modernization* to indicate the attempt of a nation, in this case Brazil, to upgrade its infrastructure to achieve a status that would put it on equal terms with the great powers. In one of the most prominent theories of modernization during the Cold War, W. W. Rostow propagated a model of linear progress, the "Stages of Economic Growth." Here, the turning point is that of a "take-off into self-sustaining growth." As I will argue, this metaphor is bound to the spatial logic of the connections made by an airplane as it moves from one airport to another. In examining this logic in the context of Brasilia, I will introduce the concept of *reversed development*.[2]

The concept of modernization is inseparable from that of globalism. Rostow's political goal was to diffuse the conditions of modernity throughout the world, and he was influential in doing so as advisor to Presidents Kennedy and Johnson.[3] The two concepts are also interwoven from an analytical perspective. Bruno Latour ascribes lengthened technical networks to a modernity that has been projected into "global totalities." According to Latour, these networks are "composed of particular places" and "are by no means comprehensive, global or systematic, even though they embrace surfaces without covering them, and extend a very long way."[4] This is striking in terms of the technology of flight. With its bird's-eye perspective and the speed that it takes us to far-off places, flight revitalized the fantasy of a global sphere and the coherence of the whole earth. Airplanes transcend the division of sea and land that governed geopolitical thinking for centuries.[5] At the same time, their routes are not continuous but are connected by isolated nodes on the ground—airports.

The construction of Brasilia was intended to represent a center of the imagined global sphere of modernization. This technopolitical strategy was configured within the context of the legacy of World War II, in which Brazil provided essential nodes in the network of allied logistics for air transport. Before examining this genealogy, I focus on the relevance the airplane had for the conception of modern architecture and urbanism, following a thread from Le Corbusier to the very architects who created both Brasilia's urban plan and its most iconic buildings.

Brasilia—Doomed to Modernity

Brasilia is "doomed to modernity," as one author has recently put it.[6] No attempt to tell the story of the city can avoid contextualizing it within the discourse of an architecture and urbanism that originates with Le Corbusier. In the 1920s, the Swiss-born architect emerged on the European scene with his radical visions of a new era. His utopian city, detailed in his 1929 publication *The City of To-Morrow and Its Planning*, was a clear break from the history of the city as a continuous development starting with the medieval village. Le Corbusier drew a regular grid in which 24 skyscrapers were placed in open spaces with parks, museums, residential units, and commercial areas. Each skyscraper was linked to the city's underground system and served as the center of a district. In this way, Le Corbusier wanted to free the city from its fate of continuous density and overlaying. "We must build in the open" was his basic assumption, and accordingly the city was designed not for a specific but an ideal site. The modern age of machinery was to create its own space, a combination of vertical compactness and horizontal openness. The construction of this city would pile up the human habitat within a limited space, leaving the area between these concentrated buildings open. In the words of Le Corbusier, "A sky-scraper is, in fact, a whole district, but verticalized."[7]

The airplane was the privileged medium for connecting this isolated city to the outside world. Le Corbusier envisioned the roof of the central station, surrounded by skyscrapers, as a platform from which "taxi-planes" would connect the city to a big airport located outside of it. This would have been only an intermediate solution: "Who knows whether soon it will not be equally possible for [airplanes] to land on the roofs of the skyscrapers, from thence without loss of time to link up with the provinces and other countries."[8] Rather than showing how the isolated city was to be integrated into a region, Le Corbusier's plan was connected to the fantastic space of the airplane. And, as if to welcome and guide the airplanes entering and leaving the city, the skyscrapers were built so that they appeared to rise up and greet them. As the literary scholar David Pascoe has pointed out, "the flat space at the centre of the urban space must be seen as the generator of its kinetic energy, a machine which sends out physical communications through the body of the metropolis and out."[9]

In 1935 Le Corbusier took another step to put the airplane at the forefront of the new age of urbanism. His book *Aircraft* was an overall rejection of historical continuity and the legacy of the nineteenth century. In its opening scene, Le Corbusier describes how he was part of a 1909 riot that destroyed the train station of a Paris suburb in revenge for the train's delay at transporting the crowd to a flight show. Throughout the book, he put the airplane's aerial perspective to good use: "With its eagle eye the airplane looks at the city. It looks at London, Paris, Berlin, New York, Barcelona, Algiers, Buenos Aires, São Paulo. Alas, what a sorry account! . . . Cities, with their misery, must be torn down. They must be largely destroyed and fresh cities built."[10] Thus the publication's motto, *L'avion accuse*, targeted the city itself to make space for Le Corbusier's own vision.

Twenty years later, Brasilia seemed to make these visions come true. Like Le Corbusier's ideal city, the new Brazilian capital was set on a level area and built in the open. It lay in the less developed interior of the country, with no roads or railways leading to the site. The design of the city did not have to make compromises for an existing structure. In a Le Corbusian manner, the plan divided the city into districts according to their functions. These references are not coincidental. There was a personal bond between Le Corbusier and the protagonists of modern Brazilian architecture. During his second stay in Brazil in 1936, Le Corbusier worked with Lúcio Costa and Oscar Niemeyer—the former would become Brasilia's city planner, the latter its main architect—on a new building for the Ministry of Education and Health in Rio de Janeiro. The structure is seen as the beginning of modern architecture in Brazil and is also the first constructed skyscraper that Le

Corbusier was actively involved in designing.[11] In the 1950s, Lúcio Costa celebrated Le Corbusier as the author of the "Holy Bible of Architecture." But according to Costa, Brazil's contribution to modern architecture was to bring out its sculptural dimension—a dimension inherent to the language of Le Corbusier but generally overlooked.[12]

This sculptural understanding is visible not only in the iconic buildings of Oscar Niemeyer but also in the ground plan of Brasilia as a whole. The plan forms a cross that has often been read as being shaped like an airplane, as if it were not only paying homage to the fascination Le Corbusier had for the airplane, but also referring to the fact that the city lay in isolation in the country's interior. In 1957, when construction began, the site could be reached only by airplane or by oxcart paths. Whereas aircraft were to land within Le Corbusier's ideal city, the airport of Brasilia was built outside the city. The city's inner structure was not designed for pedestrians moving between districts on underground trains, but for the automobile. As Costa noted in his plan, "it was decided to apply the free principles of highway engineering, together with the elimination of road junctions, to the technique of town planning. The curved axis, which corresponds to the natural approach road, was given the function of a through radial artery, with fast traffic lanes in the center and side lanes for local traffic."[13] It was as if, in driving through Brasilia, the cars should follow flight paths projected onto the ground in a continuous flow or along curved lanes. This marks an important difference from the concept of Le Corbusier. Whereas Le Corbusier had drawn up a geometric plan in which the city was to reach toward the sky, Costa sculpted a city corpus in which the axes were shifted from vertical to horizontal. The French Minister of Culture, André Malraux, stated during his visit to the Brasilia construction site in 1959: "For here the first great landscapes of modern architecture which our era is to know are going to appear. This means that this vertical architecture is going to take on a new direction. . . . It is the re-conquest of the skyscraper by the ground."[14] The air age was no longer to be welcomed. Rather, it had arrived and landed on the ground as a spacious city in which the movement of cars would be as unhindered as that of planes in the air.

This brief genealogy helps to explain the fascination architects and urbanists had for the airplane. But the discourse of architecture covers a fundamental spatial configuration of modernization that was the incentive for the project: Brasilia was meant to be the space from which the development of the country's interior would emerge. This integration from within follows the spatial configuration of air traffic and leads us to US operations for military logistics in World War II.

The Legacy of World War II

In World War II, Brazil was a node of major importance in the Allies' logistical network of air transport. With their restricted ranges, airplanes could fly to Africa only from the northeast part of the country across the narrowest portion of the Atlantic. Using this route, the United States transported supplies not only for the war effort in North Africa but also for the invasion of Sicily. On an extension of this route, 5,000 planes were delivered to the Red Army by way of Iran. Later, transport planes flew from North India over the Himalayas to support the isolated Chinese troops fighting the Japanese. This route across the South Atlantic, Africa, and Asia was integrated into a logistical network that covered the North Atlantic, Alaska, and the Pacific. The network enabled the US to deliver strategic materials, airplanes, and troops to the distant theaters of war. No less important, hundreds of airports were constructed, the world's airways were extended, the production of transport airplanes was immensely stimulated, and the US civil airlines, which played a major role in the wartime endeavor, became leaders in the global market.[15] These achievements were popularized as the arrival of a new global age. In a famous fireside chat broadcast on the radio in February 1942, President Roosevelt proclaimed: "It is warfare in terms of every continent, every island, every sea, every air-lane in the world."[16] Similarly, Wendell Willkie's 1943 best-seller *One World* propagated a global vision derived from a plane trip the author had taken around the world: "Continents and oceans are plainly only parts of a whole, seen, as I have seen them, from the air. . . . There are no distant points in the world any longer. . . . Our thinking in the future must be world-wide."[17] And Vice President Henry Wallace foresaw the world being drawn together by "airway[s] so that every man in truth will be the brother of every other man."[18]

A modernized cartography further emphasized a new global age. The common Mercator code projected the sphere of the earth upon the inner walls of a cylinder. Unrolled to a map, the projection distorted the poles and the Pacific. This made the map good for Atlantic shipping along the equator but useless for the new age of flight. Instead, maps using the azimuthal equidistant projection were drawn from a bird's-eye view—a surface touching the sphere at only one point. All places were marked in true distance to this center point where the map was thought to touch the sphere of the earth. But these maps could show only one hemisphere. To create a world map, the second hemisphere was drawn as a strongly distorted ring around the first. These maps sacrificed a form pleasing to the eye for accurate representations of distance and direction—the criteria most critical for flight along aerial routes. The new maps, which became widespread in advertising and education, told exactly how far an airplane had to fly to reach a destined point in a chosen direction. New "air maps" and "air globes" ignored the topography of continents and oceans entirely, marking only the big cities and airports. This new cartography made visible the fact that the airplane had introduced a specific era of globalism, just as the nautical discovery of the New World had done around 1500. This was all before missiles and spaceships left the stratosphere to bring back photographs of the earth as a globe.[19]

The projections of a global sphere necessitated by air travel replaced the geopolitical discourse decisive for positioning the United States in the world. The older discourse had been based upon the opposition of land and sea. In terms of the Monroe Doctrine of 1823, it defined security as the defense of the American coastlines. During its heyday in the 1920s and the 1930s, the politics of isolation was not only a reaction to the US intervention in World War I but, more profoundly, a sentiment originating from this geopolitical perception formulated one hundred years before.[20] This changed at the end of the 1930s as military advisors imagined the scenario of hostile airplanes attacking the US. Strategic maps indicated points in Europe and Africa from which German and Italian airplanes, with an anticipated increase in range, could attack points in the Western Hemisphere. American security was redefined to require the control of points beyond the American continent, such as Greenland, Iceland, the Azores, or even West Africa. As one author put it in an article in *Foreign Affairs* in 1941: "The Monroe Doctrine was concerned with continents. . . . This three-hundred-year period in American life has come to an end. . . . Our coastline is no longer the line of American defense."[21]

In this picture, Brazil was the Achilles' heel. Military airplanes from Germany and Italy could invade the Western Hemisphere by way of Brazil. The civilian airways of the Axis powers, already in operation, were understood as a prerequisite for an invasion of the American continents. In fact, German and Italian airlines had been successful in setting up networks for civil aviation in South America during the 1930s. Another 1941 article in *Foreign Affairs* put it this way: "The drone of German and Italian airplanes over South America is not a new sound. It has been heard, at least in the case of German aircraft, in steadily increasing volume for the past twenty years."[22] However, Pan American Airways had been even more successful in its Latin American engagement. In the strategic interests of the US, it intensified its efforts before the attack on Pearl Harbor. The US administration expropriated shares of German airlines in Colombia and Brazil to eliminate commercial competition for Pan Am and launched the Airport Development Program in 1940 with a focus on Brazil. Camouflaged as an expansion of Pan Am's network for civil aviation, the program provided secret funds to construct airports and upgrade bases for military use. Upon entering the war, the US officially made this network part of the military.[23]

Thus the airplane introduced a new kind of global thinking along two lines. As an aesthetic configuration, the bird's-eye view made the coherence of the earth visible, mapped connections that were becoming available at ever-increasing speeds and, more decisively, ignored the topography of land and sea. This viewpoint encouraged an emphasis on globalism as a basis for a new political order. Within a geopolitical frameset, it put an end to the era in which the US felt separated from the rest of the world, secured by vast stretches of ocean. Parts of the world came closer, and it was imperative to control the vectors of transmission before hostile powers could turn them against the country. Although not the exclusive concern in this new world view, Brazil was one of the most scrutinized strategic areas.

How did the logistics of air transport configure the discourse of modernization? It is important to note that the positions of airports were chosen not with local or regional considerations in mind but according to their locations along respective routes. The system was dictated by the range of the airplanes (mainly that of the C-47 cargo version of the DC-3, which was 1,000 miles). The logic of airports was that of isolated islands. Airports did not primarily belong to their environment but were connected through invisible flight paths to the network of all airports. The personnel lived under special conditions, with supplies being flown in from the US.[24] These isolated points processed airplanes and radio signals for communication, guiding the airplanes along the airways and generating a global connection. As the official historiography of the US Air Force tells it, "wire facilities had literally to circle the globe and at the same time to provide point-to-point, air-to-ground, or ground-to-air communication."[25] In the thinking of Gilles Deleuze and Félix Guattari, the emergence of a "global war machine" produced the ideology of "a smooth space that now claims to control, to surround the entire earth."[26] The paradox of this logic was that it was grounded on a striated space that was projected from the ground to the air, with radio signals defining the paths for airplanes and airports marking the distances to be flown.

I contend that flight is an earthbound technology. As airplanes became more powerful, they became heavier and more dependent on airports with runways long and strong enough to support them. The literature commemorating aviation in World War II has celebrated this fact. The airports constructed have been described as "spots in a wilderness; no railroads, no roads, just forests and then suddenly a huge airport. . . . It was up to America to build the bases and not merely to build them but to bring to them by air most of the equipment for that purpose."[27] For their construction, a veritable invasion of engineers, material, and machinery were flown in from the US and local workers were recruited.[28] Due to the main function of the airports for refueling planes in transit, the airports would be connected only partially to an existing infrastructure. In most cases, such an infrastructure was nonexistent. Rather, the direction was reversed: pipelines were laid from the refueling stations on the aprons to storage tanks newly built within the airport area or to depots farther away. The airport was linked to the other airports by the airplanes and was built as a functional city in itself, fulfilling the demands of airplanes in transit and the needs of the personnel stationed there. It was only then connected to its region by roads, water pipes, electricity lines, and so on.

These multiple connections could not be created without affecting their vicinities. Regional suppliers and workers profited from the invasion of US agencies. For Brazil, the wartime state of emergency opened US markets to an extent not possible under normal competition. The country exported Amazon rubber for airplane and car wheels along with other raw materials of strategic value. A major steel plant was funded and built in the southeast with US assistance.[29] In a message to President Vargas in 1942, the Brazilian Secretary of State, Oswaldo Aranha, commented that the American activities in the northeast of Brazil should prove to be of "incalculable value" to the country.[30]

Brazilian authorities hoped that industrialization would emerge from the airport construction, and forecasts reaffirmed these hopes. As part of the military alliance between the two states, the American Technical Mission to Brazil landed with teams for constructing and operating the airports. It was to come up with ideas for stimulating Brazil's economy into an enduring process of development after the war. The Cooke Mission, named after its head, Morris L. Cooke, was obsessed with cargo planes. From its perspective, an airplane with attached gliders was the "freight car of the air." In a 1942 study, it predicted: "The economies of all civilized countries will be immediately and favorably affected after the war by the introduction of cargo carrying planes."[31] "From Oxcart to Glider" was a central theme for modernizing Brazil by extending the infrastructure of air transport: "[T]he key to the prompt opening up and development of the Brazilian hinterland, without which the dream of extensive industrialization is impossible, is in the air."[32] Explicitly, the US mission put the airplane's chance for success above that of the railway: "The future appears to belong to electricity rather than to steam, to aluminum rather than to steel, and to air transport rather than to railroads."[33] Thus the Cooke Mission introduced a distinctively new perspective on the industrial development and integration of the Brazilian hinterland. For catching up with industrialized countries, the option of development at high speed—the velocity of the airplane itself—opened up an alternative to linear development on the ground.

But after World War II, Brazil waited in vain for the promised alternative along with the economic and industrial development that would, as Brazilian Secretary of State Oswaldo Aranha had predicted, emerge from the bases in the northeast. Its economy grew mainly due to the export of coffee and raw materials, but this did not lead to a coordinated industrialization. Brazil continued to have high hopes for US commissions that came to the country in 1948 and 1951 to formulate suggestions for economic development. In deference to the wartime Cooke Mission, the commissions emphasized the conventional route: extending the railway network, upgrading harbors to increase coastal shipping, and connecting the north and south of the country by way of maritime routes traversing the Atlantic Ocean. The innovative programs of industrialization from above, integrating the vast interior of Brazil by means of aviation, seemed to be forgotten.[34]

The Construction of Brasilia as Reversed Development

The utopia of Brasilia rested on a desire to change the course of Brazil's economic stagnation. Shifting the capital to the interior represented the conscious emancipation of Brazil from a colonial past that had restricted development to the coastal settlements and resulted in industrial concentration in the southeast. Intending to explain retrospectively why and to what end Brasilia had been built, the widely published 1960 lecture "Brasilia and National Development" postulated: "This Portuguese tropical America, this seaboard civilization built up upon exports of raw materials and imports of manufactured goods, this decentralized area of European capitalism, was not strictly speaking a nation at all, but a disconnected scattering of mills, plantations, trading posts and commercial warehouses."[35] Brazil's national myth held that the expeditions of pioneers, the so-called *bandeirantes*, had defined two-thirds of the country's territory, the inner part up to the Andes. Ever since the treaty between Portugal and Spain had legitimized this extension in 1750, the heart of Brazil had been waiting for its integration and development, finalizing the process of emancipation from the colonial legacy.[36]

In the construction of Brasilia, the pioneering *bandeirantes* now came by airplane. According to an essay that was published to coincide with the inauguration of Brasilia in 1960: "The modern 'bandeirante' bring civilization . . . by airplane. . . . We are following the track [of the bandeirante] but at a 20th century method: Telegraph, radio, asphalt, airplane, Brasilia."[37] I contend that the airplane was not understood as a mere instrument for making Brasilia more easily accessible once the city was built, but that it was the very technological condition for the construction of Brasilia. This is the principle I call *reversed development*, which emerged from isolated zones, initiated by airports connected to one another during World War II. This construction had taken place not in the "seaboard civilization" of the south but in sparsely populated and disconnected areas in the northeast. Natal, Recife, and Belém were coastal cities with little extension into the interior, except for Belém, which was located on the Amazon river.

Brasilia initiated the reversed development promised by the construction of airports in World War II. The intersection of the runways symbolized this reversal. During World War II, to accommodate C-47s, airports provisioned two runways 5,000–6,000 feet in length, intersecting at an acute angle. As at every airport, the runways ended abruptly. Each intersection of two runways symbolically connected to the next intersection at another airport 500 miles or more away. Such an intersection of highlighted lines ending abruptly had been marked at the site where Kubitschek came in October 1956 for his first visit to the area designated for Brasilia.[38] No concrete runway was there to welcome him however, and the DC-3 "did not glide but jolted upon the uneven territory," as he later remembered.[39] His rhetoric highlighted the area's isolation and the emptiness of the steppe as seen from the airplane's viewpoint: "All was flat and ample."[40] According to Kubitschek, the transformation of this wasteland would be the "trampoline for the Brazilian civilization" and would "give birth to the nation." Transgressing this "last frontier" would represent the "true roots" of Brazil.[41] (See figure 8.1.)

True to Kubitschek's rhetoric and that of the Cooke Mission, Brasilia was built according to the ideal of the airplane even before Lúcio Costa handed in his plan in 1957. The crossing highlighted on the ground to welcome Kubitschek made a reference to the crossing of runways. It marked the intention to enact a reversed development, revising the spatial logic of continuous development on the ground. What followed was even more distinctively linked to the airplane. Choosing the exact location with aerial photography, flying the first expeditions, and constructing an airport at the end of 1956 marked the aerial beginning of Brasilia. The airport was the

Figure 8.1
Photo of two runways crossing at the airport of Bélem during World War II. Source: Pratt Committee, Investigation of Construction Activities in Latin America, Construction of Certain Latin American and Caribbean Air Bases built by the United States, National Archives, Records of Headquarters Army Service Forces [ASF], Records of the Control Division, Records of the Pratt Committee investigation of airport construction in Latin America and the West Indies, 1945–46, volume I, box 2.

gateway for transporting machinery and materials to the site. In the first year, no paved road led to Brasilia.[42] As the Cooke Mission had proposed, development at high speed should not be achieved through an extension from the coastal areas, but by the development of an interior zone connected to others by air. This marked the claim of Brasilia, to reshape history using the airport rather than to repeat the history of extending terrestrial traffic lanes.

This replacement was a reversal of urban genealogy, as Paul Virilio has imagined: "The ancient city was a space of transit for human and animal movements. Built upon the intersection of tracks, it absorbed the caravan, the moving in of the peasants, the arrival of traders and messengers, as the village on the road junction had done. . . . Way up and staircase, city and harbor, footpath and gangway were what determined an infrastructure of slowness."[43] In Brasilia, airport runways were to replace caravan tracks. The intersection of runways was enthusiastically conceived as a starting point for an *infrastructure of speed*—something Virilio would describe as a dystopia. But Kubitschek determined that the very construction of Brasilia start at the intersection of the airport's runways. First an airport and a residential building for the president should be built, followed by buildings for parliament, court, government, and the National Guard.[44] Upon Brasilia's inauguration in 1960, a cynical perspective could have held that what had been built was not a city but an airport with adjacent conference centers.[45] Later, roads were built from this airport city out to other regions in the country, including the "road of the century" through the interior to Belém in the northeast and a line back to the "seaboard civilization" in the southeast. The airplane was once again at hand to configure construction. Work started in different areas simultaneously, supplied by airplanes landing on specially built airfields. Supplies to farther removed construction teams were parachuted in.[46]

The airplane lay at the heart of Kubitschek-era technopolitics. Speed can be understood as a determining factor for Cold War technopolitics in the threatening race between the two world powers. Although Brasilia's construction didn't occur within this race per se, the speed of the age of aviation was inherent in its genealogy. Following the proposition of the Cooke Mission in World War II, Brasilia was projected as an intersection of runways, stretching up to the global flight paths in the sky and back to the ground. It disrupted the genealogy of the city, replacing it with a reversed development or, as was said in 1960, a process "of converting space into time and geography into history."[47] As a vivid scene for this conversion of space and time, an essay on the inauguration of Brasilia stated: "There goes a story of an Indian, in one of those far off places, who worked at the airport. A pilot invited him to a little excursion, in his airplane. After landing at another place, and seeing for the first time an automobile, he exclaimed: 'Look, look, an airplane without wings.' This our countryman jumped directly from the stone-age, in which his tribe still lives today, to the age of machinery."[48]

The concept of a reversed development serves to analyze the discourse of modernization in both its spatial and temporal dimensions. It makes reference to the discontinuous connections on the ground as configured by aviation. But it also serves to emphasize a temporal quality that cannot be grasped merely as acceleration. It means to install a futuristic structure from which the protagonists pull along the areas left behind. In his governmental work, Rostow dealt with these issues on a concrete level. One of his assistants, based in Vietnam, suggested that villages, in order to be modern, should not only have a school, market place, and bus terminal, but also a landing strip and helicopter pad. "The agro-center would be completely modern—it would 'futurize' village life without killing the old village."[49] It was as if Rostow had written the script for the Kubitschek era and the construction of Brasilia.

Geopolitics as *Aeropolíticia*

Understanding the technopolitics in the Kubitschek era as a reversed development "converting history into geography" is one thing. But one also has to ask which geography this conversion was attempting to establish. From the early 1950s on, the most prominent figure of Brazilian geopolitics, General Golbery do Couto e Silva, tried to enunciate an *Aeropolítica*[50] that would harmonize the space (*espaço*) and position (*posição*) of Brazil. Golbery derived the geopolitical influence that Brazil should wield by recalling its position in air transportation within the Western Hemisphere. He recalled the route from Natal in Brazil's northeast to West Africa as essential to the security of the Western Hemisphere and its major power, the United States. According to Golbery, the northeast of Brazil was the "natural aircraft carrier" for continental security.[51] But instead of concentrating development exclusively in that region, Brazil should move inwards, Golbery believed. "Of decisive importance is a central point, highly sensitive to attacks from the ocean . . . against which it would be secure due to its elevated position above the coast."[52] From this commanding post Brazil would control South America and the South Atlantic. According to Golbery, this central region, together with the northeast, was an asset that Brazil could use to claim a privileged partnership with the United States.[53]

Golbery tried to alter the doctrine of security that the United States had formulated at the end of World War II. His action was nothing less than an attempt to reinterpret the geopolitical reality of the Cold War, which focused on the Northern Hemisphere as the most strategic terrain. The airplane had brought the great powers into easier reach of one another. In 1904, Sir Halford John Mackinder, the founding father of twentieth-century geopolitics, enunciated that North America was potentially enclosed by sea power from the two shores of Eurasia. Forty years later, the airplane opened a third flank over the North Pole. The evolving geopolitical discourse was reinforced by the new cartography based on the azimuthal projection, which focused on one hemisphere and doomed the other half to distortion. A series of publications in 1944 and 1945 propagated the "Principal Hemisphere," which, from its focal point in the north, contained only North America, Europe, and North Asia. Thus the new global age of flight so enthusiastically spread by Roosevelt, Willkie, and Wallace had a bias: It turned out to embrace only half the world.[54] (See figure 8.2.)

The strategy of containment, officially launched in 1947, propagated a perspective of the United States surrounded by the Soviet Union from three

Figure 8.2
Map according to the 'Azimuthal Equidistant Projection,' centered on the North Pole. Source: Spykman 1944.

directions—east, west, and north—and easily reached by the technologies of flight. Translated into an aerial cartography, it defined security in terms of distance to be covered by airplanes. As Hans W. Weigert put it in 1946: "In the Far North the situation is obscure. Here the USSR is our immediate neighbor over the top of the world, across the 'Arctic Mediterranean.' 'If there is a Third World War,' said General Arnold, 'its strategic center will be the North Pole.' . . . Dangerous errors of judgment can follow a habit of looking at Asia across the vast expanse of the Pacific. Look north!"[55] Starting in 1947, the Strategic Air Command built up its potential for delivering nuclear weapons to strategic areas in the USSR with bombers based in Europe, the Middle East, and the Pacific. But in the formative years, only the bombers based on the 509th Air Wing in Missouri were equipped with nuclear weapons, and the scenario for their deployment involved one-way missions to the USSR over the North Pole.[56]

The strategy of containment and the increase of airplane ranges eliminated the importance of the air transport lines in the South Atlantic. Instead of going through Brazil, overseas transit lines used the Azores as their main post in the Atlantic. Following the lead of the combat network, logistical operations also required fewer airports, as the range of airplanes increased. In the early 1950s, the US Air Force transferred its logistical operations from Brazil to a chain of stations running from Puerto Rico to the Azores and on to Europe. Brazil was a subordinated base of an alternate network, and its status as a "natural aircraft carrier" was soon outlived.[57] Brazil fell out of the containment strategy's "Principal Hemisphere." Strategic flying objects no longer transited its sphere. However, Brazilian geopolitics tried to deny these technological realities and their effects. Instead, Brazil would be the second center of the globe, controlling the Southern Hemisphere. In a 1959 essay, updated because of the increased range of airplanes, Golbery drew a circle of 6,000 miles around Brasilia on an azimuthal projection of the world. This circle was meant to define the area Brazil was to control. Reaching far into Africa, this zone was in danger of Soviet invasion, according to Golbery. Brasilia, the elevated point protected from ocean attacks, would be the commanding hill for safeguarding the Southern Hemisphere. By presenting Brazil as a predominant strategic terrain, Golbery again extended geopolitical thinking globally while still keeping it in alignment with the Cold War strategy of the United States.[58] (See figure 8.3.)

To a degree, Kubitschek made this logic his, though not in predominantly military terms. His 1955 plan for the development and modernization of the country set a specific goal for air transport. The airports would be upgraded with concrete runways suitable for the heavier, bigger, and

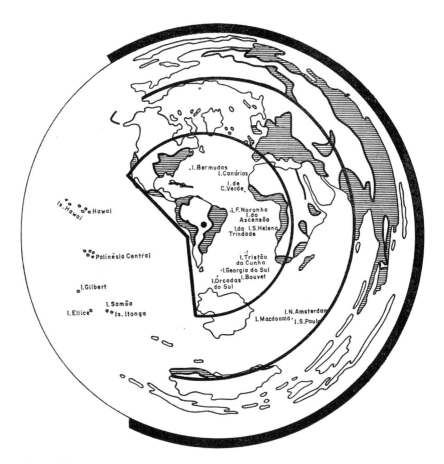

Figure 8.3
The map Golbery used to identify Brazil as a centre of the globe. Source: Couto e Silva 1967.

faster airplanes that had become the standard in international air travel. The fleet of DC-3s that Brazil had acquired at the end of World War II was to be replaced by new technology.[59] Upon his first visit to the site of Brasilia, Kubitschek not only decided that an airport should be constructed, but specified that it should have concrete runways 10,000 feet in length. Brasilia should be ready for the new jet age that was about to arrive. The Boeing 707, which depended on these stronger and longer concrete runways, would connect Brasilia to the global sphere, just as earlier airplanes had done in World War II. But instead of the wartime chain of airports, jet-age connections were direct. The range of the Boeing 707 was 5,000 miles, whereas the DC-3's range had been only 1,000 miles. Brasilia should be a center of the globe, allowing non-stop connections to other major hubs.[60]

Politically, this direct connection was coupled with a discourse for the inversion of distance. To obtain US support for upgrading Brazil from its status as a developing country, Kubitschek established the idea that communism was finding its feet in South America and spreading as a revolutionary movement. Thus he turned the logic of containment and its fixed lines of division in on itself, before this actually happened in Cuba. According to Kubitschek, only economic development would be suitable to counter this danger. "In a dramatic letter to President Eisenhower, the Brazilian president proposed an ambitious new program, 'Operation Pan-America,' which would bring together the United States and Latin America in a long-term multilateral program of economic development."[61]

The Eisenhower administration was unwilling to follow this viewpoint, however, and it kept Brazil in the queue. Upon his visit in August 1958, US Secretary of State John Foster Dulles said that he "recognized that Brazil is a country which offers excellent prospects for economic growth and said that the United States would like to cooperate in 'partnership within limits . . . in this great development.'" But at the same time, he limited the prospects. "He noticed while *flying over Brazil* the vast size of the territory and the low state of development. He compared Brazil's development now with the United States one hundred years ago."[62] From his vantage point in the sky, Dulles had reduced Brazil to the status of the US in the age of railway construction. Rejecting this technological disparagement and trying to achieve proper recognition for his technopolitical regime, Kubitschek intensified contact with President Eisenhower, eventually cajoling him into visiting Brazil in February 1960. The Kubitschek administration proposed meeting first in the northeast, then in Brasilia. It was an attempt to establish a historical continuity. In 1943 President Roosevelt had to stop in Natal on his flight back from the Casablanca Conference, the first international flight of a US president. This was the occasion of his only meeting with Brazil's President Vargas. The symbolic value of renewed association in Natal and Brasilia would have been significant. Both nodes of a global network, in World War II and in the jet age, would have been linked, the project of modernization ratified. But this meeting could not take place because in 1958 Air Force One was a Boeing 707 and the airports in the Brazilian northeast were not equipped for its landing. That Brazil's "natural aircraft carrier" (as Golbery had called the northeast) was not part of the primary infrastructure for US security could not have been made more obvious.[63]

The fact that Brasilia's airport had concrete runways 10,000 feet long proved to be prescient for the rituals of politics. Eisenhower came to Brasilia two months ahead of the capital's inauguration, the first US president to travel to Brazil since Roosevelt. One result of this meeting was that Brazil secured a credit plan with the Export-Import Bank of Washington to purchase two Boeing 707s.[64] Brazil had its direct connection to the US—as celebrated in a song by the Brazilian bossa nova star Astrud Gilberto: "Silver jet / Makes this trip Non-stop / Like my heart Non-stop / So fly me to Brazil."[65]

The Legacy of Brasilia

Astrud Gilberto recorded her song in 1965 while she was living in New York. There is a reason to the melancholy note in her celebration—in 1964 a military coup had put an end to democracy in Brazil. As Kubitschek's successor, President João Goulart increasingly favored communist ideas, resulting in the Brazilian military overthrowing the government. When the relevant records in the US National Archives were declassified in 2004, documents revealed that the US military was ready to support the Brazilian military in "Operation Brother Sam." In accordance with the wishes of the Brazilian generals, the US Department of State initiated the logistical mobilization. When the coup d'état began, freighters were already out to sea and six freight planes loaded with materiel were ready for takeoff. However, their deployment proved unnecessary. Goulart left the country without resistance, and the military took over within a day.[66] The troops did not encounter the "Cuban-trained, Russian-equipped hordes of sergeants, peasants, workers, and students" that had been forecast. In the cold light of day, the threat of Brazilian communism was a ghost.[67]

Still, the military remained in power for 21 years, installing an authoritarian state that allowed US interventions in Uruguay and the Dominican Republic in the 1960s and served as a model for the dictatorships in Chile and Argentina in the 1970s.[68] In the beginning, the regime followed a "diplomacy of geopolitical inspiration," which meant aligning themselves more closely with US politics.[69] The rapid industrialization at the cost of inflation that both Kubitschek and Goulart had pursued was interrupted. Brazilian economic policy returned to a conventional path.[70] Opposed to such a policy, the well-known economist Celso Furtado, who had been in charge of a program to develop the northeast in the beginning of the decade, accused the government of reducing Brazil to the status of a "satellite" in the Cold War global order. Because the "stability of its orbit" was more important than its "proper development," Brazil had been "disconnected from technological progress," its "resources and immense territory turned against itself."[71] In the technopolitical setting of Brasilia, Furtado's argument was that the extension of the runway intersections on the ground of Brazil's interior had been traded for a reduction to the status of a satellite *outside* a globe defined by the lines of the Cold War. His was a rhetoric that identified the national economy with aeronautics at yet another technological stage—but still in relation to the earth imagined as a global sphere.

In accordance with this renunciation of modernization, the utopia of Brasilia was questioned. As the first military president of the authoritarian state is reputed to have said, the government was "forced to choose between the folly of continuing Brasilia and the crime of abandoning it."[72] They chose the folly and maintained the capital in the interior. As in the Kubitschek era, areas the airplane already reached were to be opened up further and continuously integrated. In the 1970s the area for this development turned out to be the Amazon, where 50 airports and landing strips provided the only connections. As for the central Brazilian hinterland, a plan for an integration of that area had been written during World War II by the US Board of Economic Warfare. The author was the acclaimed inventor Richard Buckminster Fuller, a protagonist of aerial cartography in the 1940s and the designer of "biosphere domes" constructed after World War II. Fuller's propositions were even more radical than those made by the Cooke Mission:

> Almost so simple that it will be shunned by those who prefer to plan the hard way, in order to take advantage of their hard-earned specialized experience of the past, is the technique now provided by modern warfare that would approach this whole Brazilian jungleland from above, bombing it open, then parachuting in with well-planned hand equipment and personal protective devices to carve out a complete polka-dot pattern of island airports over the whole country, into which pattern mechanical devices would be fed progressively as parachute deliveries graduate to plane-landed deliveries, etc. Each area would receive its quota of machine tools, drafting equipment, air conditioning, etc., and then its engineering and designing personnel would amplify the hold on the jungle. This "island" network of "tropical research and development stations" should form the nuclear structure of the new Brazil.[73]

In the words of Fuller, the Brazilian government preferred to plan the hard way. As with the Belém-Brasilia highway, their program foresaw that the construction of roads crossing the jungle would enhance the country's economic integration. However, the project did not turn out to be a success, and many of the roads were washed away in heavy rains.[74]

Not anticipating this failure, the military used the Amazon project to shift attention from the development of the country's neglected northeast.

In the words of President Emilio Médici in 1972, Brazil had "not yet reached the era of a finite world, so we have the privilege of incorporating into our economy, step by step, new, immense, and practically empty regions."[75] This rejection of a finite world was mirrored in the external policy rejecting a subordinate status to the "Principal Hemisphere." Brazil took political steps that did not conform to Cold War alignment. It intensified its relations with communist states and African colonies in revolt against Portugal. In the words of President Ernesto Geisel in 1974, these countries were "sister nations in the neighborhood on . . . the other side of the Atlantic."[76] Although Brazil failed to establish a strategically valuable military base on the Cape Verde Islands, it succeeded in acquiring military jets from France and nuclear power technology from West Germany. Eventually Brazil ended the military agreement it had started with the United States in World War II—an agreement that had provided only a very limited access to technology during the Cold War.[77]

Conclusion

Brasilia unravels a fundamental spatial logic of Cold War technopolitics. Both the historical process of its construction and the geopolitical discourse that justified moving the capital to the interior were inseparable from the system of aviation. The discourse of modernization central to that era not only promoted an intensified process of technical development, it aimed to establish a technologically advanced space that would diffuse and integrate into the country. As I have argued, this understanding was based on the airport, a limited zone serving the highest technology and spreading out functional connections in its vicinity. My analytical concept of a reversed development aims at denoting both this spatial dimension and the temporal idea of the airport as an advanced zone of the future.

The logistical network supporting flight in World War II revitalized the fantasies of a coherent global sphere unified by air transport, a unity that became the ultimate goal of Cold War technopolitics. The implementation took place along clear lines of division, not only of east and west but also of north and south. A new cartography located the strategic terrain of the "Principle Hemisphere" in the north, condemning the rest of the world to insignificance, as exemplified in the distortion of the map. The geopolitical discourse in Brazil rebelled against this reality and tried to define another strategic center in the south.

The reality of Brasilia proved to be very different from its blueprint, with traffic jams instead of flights along curved lanes. Uncontrolled satellite towns mushroomed outside the planned boundaries, subverting the ideal of the modern, functional city. But Brasilia's very existence kept its utopian ideal alive. The remote capital of Brazil is a manifestation of the modern era that anyone can visit—preferably by airplane.

Notes

1. Wilheim 1960: 121–122, quote on p. 121.

2. Rostow 1960.

3. Frederick Cooper has recently commented that Rostow "seemed unreflexively to assume that American society— as understood in the 1950s— represented the telos toward which all the world would converge" (2005: 117).

4. Latour 1993.

5. As Gearóid Ó Tuathail has pointed out (1996: 15, 24f.), "The term 'geopolitics' is a convenient fiction, an imperfect name for a set of practices within the civil societies of the Great Powers that sought to explain the meaning of the new global conditions of space, power, and technology." Though its emergence can be traced back to the late nineteenth century, the term still functions to denote the politics of power in relation to space and geography.

6. Wisnik 2004: 20–55.

7. Le Corbusier 1998: 164, 182.

8. Ibid.: 188f.

9. Pascoe 2001: 117.

10. Le Corbusier 1935: 11f.

11. Mindlin 1965: 25.

12. Costa 1959: 63–71, quotation on p. 70.

13. Costa 1960: unpaginated.

14. Malraux 1959: 27.

15. Craven 1958: chapter 15.

16. Roosevelt 1970/1946: 312–322, quotation on p. 313.

17. Willkie 1966/1943: 203, 2.

18. Wallace 1944: 30.

19. "The globes symbolized in tangible form the new world which Americans believed the airplane was about to create, a world of peace where national boundar-

ies and topographical features were no longer pertinent." Corn 2002/1983: 129; see Henrikson 1975: 19–53; Vleck 2007.

20. On the opposition of land and sea, see Schmitt 2003. On the specific function of islands as nodes in this network, many of them leased in World War II, see Ruth Oldenziel's contribution to this volume.

21. Miller 1941: 727f; Cate 1948: 101–150.

22. Hall 1941: 349–369, quotation on p. 349; Davies 1964: chapters 9 and 10.

23. Carter 1958: 46–62; Ray 1973; Kraus 1986. On the economic and political relevance of the alliance, see Hilton 1975: 201–231.

24. Ray 1973: 135.

25. Jonasson 1958: 339–364, quotation on p. 339.

26. Deleuze 1987: 421.

27. Cleveland 1946: 24f.

28. Kraus 1986.: 156ff.

29. Ferraz 2005: 25ff; Hilton 1975.

30. Aranha in a letter to Vargas, November 9, 1942, quoted on p. 212 of Hilton 1975.

31. James M. Boyle and Marcio de Mello Franco Alves, To: Morris Llewelyn Cooke. Re: Cargo Plane Utilization, October 10, 1942, Roosevelt Presidential Library, Morris L. Cooke Papers (1936–1945), Box 283.

32. Cooke 1944: 159.

33. Cooke and Alberto 1942.

34. US Department of State 1949: 361–373; Institute for Inter-American Affairs 1954, 1955.

35. Corbisier 1960: unpaginated.

36. Kneese 1960: 5–16, quotation on p. 7f; Branco 1983: 120ff.

37. Kneese 1960: 16.

38. Buchmann 2004: 12; Wisnik 2004: 40. Costa was to tell his story of the cross in another way: "It was born of that initial gesture which anyone would make when pointing to a given place, of taking possession of it: the drawing of two axes crossing each other at right angles, in the sign of the Cross." (Costa 1960) The most obvious explanation is that of a Christian tradition. In fact, such a holy cross was erected at the most elevated point of the site when Kubitschek first visited the site.

39. Kubitschek 1975: 46.

40. Ibid.: 45f.

41. Oliveira 1998: 225–240, quotation on p. 231f. For a detailed description of Brazil's interior before the construction of Brasilia, see Lévi-Strauss 1973: chapter 13.

42. Ludwig 1980: 7f.

43. Virilio 1978: 19–50, quotation on p. 32.

44. Kubitschek 1978.

45. Ludwig 1980.: 24; Oliveira 1998: 233.

46. Melo 1962: 1–6.

47. Corbisier 1960.

48. Kneese de Mello 1960.: 16.

49. Memorandum from Kenneth Young to Walt W. Rostow, February 17, 1961, quoted in Westad 2005: 399.

50. Couto e Silva 1967: 31.

51. Couto e Silva 1967: 49ff.

52. Couto e Silva 1967: 58f.

53. Couto e Silva 1967: 51. See Child 1979: 89–111.

54. van Zandt 1944; Weigert 1944; Spykman 1944; Henrikson 1975; Schulten 1998: 174–188.

55. Weigert 1946: 250–262, quotation on p. 260f.

56. Converse 2005/1984; Leffler 1992; Moody 1995.

57. This was before intercontinental ballistic missiles were installed in fortified bases, from which the US and the USSR were threatening each other directly (Blaker 1990: 31ff; Schake 1998: 189ff.).

58. Couto e Silva 1967.: 64–95.

59. Kubitschek 1955.

60. Kubitschek 1978: 46; see Kubitschek 1975; Mesquita 1958: 198–199.

61. Skidmore 1967: 173; *FRUS* 1958–60, volume V: 682.

62. *FRUS* 1958–60, volume V: 694. Ellipsis in source text, emphasis added.

63. Ibid.: 743ff.

64. Ibid.: 755.

65. Gilberto 1965.

66. *FRUS* 1964–1968, volume XXXI; Kornbluh 2004.

67. Leacock 1990: 216ff.

68. Leacock 1990.: 216ff; Westad 2005: 151.

69. Bandeira 2000.

70. Skidmore 1988: 12f and 29ff.

71. Furtado 1971: 1–19, quotation on p. 12f; Skidmore 1988: 68f.

72. Ludwig 1980.

73. Fuller 1983: 306f.

74. Skidmore 1988: 144ff.

75. Médici on the third anniversary of his assuming the presidency in 1972, quoted in Skidmore 1988.: 147.

76. Geisel quoted in Selcher 1976: 25–58, quotation on p. 45.

77. Skidmore 1988: 192ff.

9 Crude Ecology: Technology and the Politics of Dissent in Saudi Arabia

Toby C. Jones

In November 1979, in an unprecedented act of dissent, thousands of Shi'i men and women revolted against Saudi Arabian authority. Over the course of several days, rioters attacked local symbols of sate power in the kingdom's Eastern Province, burning the offices of the Saudi Arabian National Airline, demolishing government vehicles, and clashing with heavily armed police forces.[1] The state responded quickly and harshly, dispatching as many as 20,000 members of the National Guard to beat back the protests.[2] Guardsmen fired openly at those in the street, killing dozens and crushing the attempt to throw off the yoke of Saudi repression.[3]

The rebellion was precipitated by a number of factors. After 1913, when forces loyal to the al-Sa'ud (the ruling family of today's Saudi Arabia) conquered eastern Arabia, the Shi'is living there had been forced to endure multiple hardships. In the late 1920s, the religious warriors who helped the al-Sa'ud forge their empire called repeatedly for the Shi'is to convert to Wahhabism or face annihilation. The al-Sa'ud stopped short of genocide. But over the course of six decades the state ensured that its Shi'i minority was constantly subjected to discrimination and violence, including the destruction of mosques, seminaries, and cultural centers, individual persecution, deliberate impoverishment, and environmental devastation.[4] Though the uprising was animated mostly by anti-Saudi outrage, it was also a response to a set of frustrations and conditions connected to Cold War relations and politics.

The demonstrators made clear that they considered the United States directly complicit in the excesses of Saudi power. Since World War II the US had been the kingdom's longest and staunchest ally, helping not only to develop its massive oil infrastructure but also to ensure that its supplies could never be seized by Soviet expansion into the Persian Gulf. Cold War anxieties and the perceived stakes of the struggle made safeguarding Saudi Arabia's security paramount. Saudi Arabian oil, critical to Western industry and capitalism, assumed heightened importance in the context of the global struggle for superpower supremacy. Even though the Soviet Union never seriously threatened American hegemony in Saudi Arabia or the Persian Gulf, the US government and US oil companies took no chances, prioritizing the security of the ruling al-Sa'ud. As was the case elsewhere during the Cold War, this meant that American officials and corporate executives with interests in the kingdom preferred the stability of authoritarianism to any alternative that might threaten access to crude oil.[5] But the US was not merely a witness to the rise of authoritarianism on the Arabian Peninsula. American advisors, both from government and from business, were instrumental in shaping Saudi Arabian governance, including its effects on the Shi'i minority. From the 1950s into the 1980s, US policy makers also courted, and struggled to shape, Saudi Arabia as an Islamic counterweight to the various leftist political movements, from Arab nationalism to Nasserism to communism, that the US believed might push the Middle East into the Soviet camp. In 1979, the year of the Shi'i rebellion, at the height of the Cold War, Saudi-American ties also reached their pinnacle. After the USSR invaded Afghanistan, the US and Saudi Arabia openly supported the anti-Soviet jihad to roll back the communist threat.[6]

Shi'i dissidents, fed up with the effects of the US-Saudi arrangement, took direct aim at it. The rebels justified their dissent by pointing out the ways that the country's treasure had been diverted for dubious purposes. American employees of the Arabian American Oil Company (Aramco) had been exploring for and extracting oil from areas immediately surrounding Shi'i communities since the 1930s. Aramco was a giant in the kingdom's Eastern Province. It spent millions of dollars on development projects, on infrastructure, and on the well-being of its American employees as well as those Saudis looked upon favorably by the state, building hospitals, roads, schools, and even swimming pools to serve them. Shi'i communities saw little of this largesse. As oil revenues mounted from about 1950 on, most Shi'is knew only poverty. The state mostly directed its spending elsewhere, particularly toward the procurement of expensive American-made weapons systems that its military was not trained to operate. When the state did spend money on projects in Shi'i communities, often with American oversight, it had the effect of intensifying the Shi'i marginalization and radicalization. In the days leading up to the 1979 uprising, the Organization for the Islamic Revolution in the Arabian Peninsula (OIR), the group that coordinated the revolt, made plain its frustration with US influence in the kingdom, warning Americans resident in the Eastern Province: "We realize the dubious role that you play in our country and toward our national resources. God willing, we will settle our accounts with you in the near future."[7]

The Cold War context, in which Americans played an important role in shoring up Saudi security and authoritarianism, sheds some light on Shi'i unrest. To make sense of this, it is important to understand the specific terms of and the legacy of US influence. In this article, I argue that it was the effects of the combined American, European, and Saudi efforts to use technology to conquer nature in the al-Hasa oasis, the largest oasis in Saudi Arabia and a unique community home to large numbers of both Shi'is and Sunnis, that in part precipitated sectarian radicalization. While the Eastern Province's oil was the lifeblood of Saudi power, the region gushed with another natural bounty: water, a commodity as precious as oil in the desert kingdom. Rich water and agricultural resources had attracted Saudi interest in eastern Arabia well before the discovery of oil. The primacy of petroleum only intensified Saudi interest in capturing and controlling the region's water. Aramco's role fundamentally shaped the Saudi state's approach to managing nature. Though the oil company did not provide the original push behind the state's desire to conquer nature in al-Hasa, it did shape the terms and the means by which the state would attempt to do so. In the 1950s, Aramco social scientists argued that the Eastern Province was a threatened environment, one that faced an interminable decline without some sort of technological intervention. Aramco's emphasis on ecology and the threats that humans constituted to nature established the foundation for the Saudi state's attempt to reengineer nature there over the next three decades. Over this period, the state and its proxies, including Aramco and other Western engineering firms that were engaged in developing Saudi Arabia's natural resources, would sharpen a way of knowing parts of the Eastern Province that justified their work there, but that also helped produce the dislocations that would lead to violence in 1979.[8] Specifically, it was the construction of a large irrigation network designed to save al-Hasa's water resources that helped produce sectarian politicization.

The impetus behind Saudi efforts to develop an irrigation network in al-Hasa was multifaceted. Part of the motivation had to do with the weak state presence there. Until late in the twentieth century, state institutions across the kingdom remained inchoate and ineffective. Large technological projects offered opportunities to correct this, and to expand the state's territorial reach by establishing physical and administrative presences—networks from which state authority could be projected and subjects enrolled. Given al-Hasa's rich natural resources, both oil and water, the Saudi state considered it a priority to secure its power there.[9] The presence of a large number of Shi'is reinforced the Saudi will to power in al-Hasa. Although the Shi'is there remained politically quiescent through the mid 1970s, the Saudis considered their presence potentially dangerous to state security. Inspired by the increase in oil revenues since the 1950s, Saudi officials also hoped to put nature to work in the Eastern Province. Their aim was to capture the region's resources, intensify agricultural efforts there, and use the bounty to help feed the nation, to wean it from its dependence on foreign sources of food. The state's attempt to address this anxiety only helped set the stage for confrontation.

Into al-Hasa

Dense with vegetation, al-Hasa's lush gardens sprawl across the eastern desert of the Arabian Peninsula. Historically, water flowed from deep aquifers, pooling in cool reservoirs around the oasis. Farmers diverted water year-round into the gardens. So plentiful was al-Hasa's natural bounty, and so unlike the rest of the Arabian interior, that excess water streamed into sandy catchments on the oasis's periphery, where it either evaporated or formed boggy marshes. At mid-century the oasis was home to between 25,000 and 40,000 hectares of date trees, whose harvests provided the most important source of sustenance for local residents.[10] Al-Hasa's green stood in stark contrast to the rocky and sandy landscapes that surrounded it.[11] Between 150,000 and 250,00 people made their homes in al-Hasa's gardens in the 1950s.[12] That number doubled by 1979.[13] Just over half of all Hasawis lived in the oasis's two main cities, with around 60,000 residents in al-Hofuf and 28,000 in al-Mubarraz.[14] The remaining 70,000 or more were spread around the oasis, living in 52 villages located mostly along the perimeter of al-Hasa's date gardens. The populations of the villages ranged from 30 (al-Sidawiyyah) to 3,700 (Taraf).[15]

Oil company executives were drawn to al-Hasa, the largest population center near their operations, early on. In 1940, two years after oil was discovered, what was then California Arabian Standard Oil Company (CASOC) employed 2,668 Saudi Arabian workers and 382 non–Saudi Arabians (including Americans). The total number of workers dipped during World War II, but the proportion remained heavily slanted toward Saudi Arabians. The number of workers swelled in the last year of the war and immediately thereafter. In 1945, three-fourths of the company's workers (8,099 out of 10,683) hailed from Saudi Arabia.[16] Much later, the company would become less dependent on labor. But through the late 1950s, oil-field work was still labor intensive, especially the construction of oil facilities and the supporting infrastructure. Most of the company's labor was from regional towns and villages, especially the nearby oases. Between 1945 and 1959, Aramco hired more than 56,000 local residents.[17]

Aramco's interest in understanding the oasis's society intensified in the 1950s as the company's Saudi Arab workers grew militant. In 1945, Aramco workers kicked off a restive labor movement that would periodically interrupt oil operations. Arab workers expressed a number of grievances. They were especially agitated by the company's racist housing and wage policies and for recreating Jim Crow–style relations in the Arabian Peninsula.[18] Rather than confront the core of labor grievances, Aramco executives manipulated Cold War anxieties to undermine the workers, using the specter of communism and Arab nationalism as pretexts for getting the central government to crush outbreaks of labor unrest. Aramco responded to the first outbreak of labor unrest in 1946 by forming its own intelligence-gathering organization, the Arabian Affairs Division (AAD). Robert Vitalis has written that "the first reports by US consular officials emphasized the company's concern that it lacked such an organization and thus had little capacity to understand what workers wanted." [19] The AAD was literally modeled on the OSS' (the precursor to the CIA) Cairo branch and served as a home for several CIA operatives in later years. Social scientists in the employ of the AAD in the 1950s helped set the stage for violence in al-Hasa two decades later.

The Saudi state was also concerned about the militarization of local labor, especially since, in addition to blasting racism at Aramco, labor leaders directly criticized the state. The state's concerns about local threats to its authority coincided with a growing hope that the oasis could be put to work beyond the oil industry. Representatives of the al-Sa'ud had maintained an official presence in al-Hasa since 1913, when the Saudis conquered the region, using the main city of Hofuf as its capital. But through the middle of the twentieth century the government presence consisted mostly of the local governor, his advisers, police forces, and judges imported from central Arabia. Local residents were forced to endure the power of the governor, often violently, but the central government's authority was limited by its small size, by the quasi-autonomy of the local governor, and by the sheer size of the oasis.[20] State control of the region's non-petroleum natural resources had yet to pay significant dividends. The government had interfered with the operation of local and regional markets, but it had not taken full control of al-Hasa's natural resources. Government officials harbored grand dreams of turning al-Hasa into the kingdom's breadbasket, using its water reserves to expand its productive capacity and to help Saudi Arabia achieve agricultural self-sufficiency. This proved a fanciful dream. But it was hardly unusual in a country where oil riches have often fueled visions of grandeur.

Aramco, and in particular an anthropologist working for the Arabian Affairs Department, would not only prove instrumental in helping the Saudi state in its effort to achieve its political-economic objectives, but also in shaping *how* it attempted to do so. In 1955 Federico S. Vidal, who worked directly for the AAD, published a 216-page report for the oil company titled *The Oasis of Al-Hasa*. That report, the result of several years' research while Vidal was field supervisor for Aramco's Malaria Control Program, remains among the most comprehensive ethnographic and sociological studies of the oasis, and is the single most exhaustive source on the oasis at mid-century. It probably was read with great interest by those working in Aramco's intelligence services. Researchers in the AAD were certainly aware of the oasis's social complexity, but Vidal's ethnographic accounting revealed these in greater detail than the oil executives had seen previously.

The overwhelming majority of the inhabitants of the oasis's villages tended the date groves and farmed a few additional crops. Vidal wrote in 1952 that "first in importance among all the items of al-Hasa's agricultural complex is the date grove, called *nakhil*, which here occupies as prominent a place in life as does the camel among the desert Bedouins."[21] Farming in al-Hasa demanded considerable energy. Because arable land was at a premium in the crowded oasis, agriculture was intensive. Date groves occupied as much as 27,000 of the 30,000 acres under cultivation, with farmers harvesting as many as forty varieties of dates.[22] Date gardening was seasonal. Farmers pollinated the palm trees in early spring. The harvest season occurred later in the summer and during the early fall. After harvesting, farmers heaped the dates in piles on top of reed mats, which they then enclosed around the dates and sewed shut, transporting them to merchants in al-Hofuf and al-Mubarraz, who sold them at market.[23] As we will see, the local date market suffered a severe depression after World War II.

The oasis was home to large numbers of both Sunnis and Shi'is, the only community of its kind and size in Saudi Arabia, with the Shi'is probably enjoying a slight overall advantage.[24] There are few reports of open sectarian conflict before the arrival of the al-Sa'ud and their proxies. But that does not mean al-Hasa was free from communal tension or that there were not imbalances between the Shi'i and Sunni communities. Social hierarchies and social power overlapped with religious differences. Both al-Hofuf and al-Mubarraz were mixed cities, and their demographic makeup, as well as social power in both, was evenly balanced between Sunnis and Shi'is.[25]

Differences between Shi'is and Sunnis assumed a more hierarchical nature outside the two towns, most notably when it came to land tenure and who did the work inside the date gardens.

While many Hasawis owned land, thousands of others worked as day laborers, tending the private date gardens of large landowners. The most successful gardens belonged to absentee landowners who resided in al-Hofuf or al-Mubarraz.[26] Overwhelmingly, landowners were Sunni. Vidal commented that "only a few of the most important Shiite families of al-Hasa can be considered landowners. . . . The bulk of the Shiites, although perhaps owning small garden plots, are either craftsmen or laborers working the gardens for wages. In many of the smaller villages and hamlets this is true of the entire population, including the village headman."[27] For the most part, then, Shi'is living in the gardens constituted the oasis's working class, filling the ranks of agricultural laborers and dependent on their Sunni neighbors for their daily wages. The system was free from open conflict, but it hardly seemed harmonious.

It is unclear what impact Vidal's report had on Aramco's labor policies, if any. The report's impact was more profound on the state and its approach to the oasis. While *The Oasis of al-Hasa* unveiled the dynamism of Hasawi society, its more important political legacy was that it ultimately framed the environmental terms by which the state would know the oasis as well as the means by which the state would assert itself there. Through an accounting no less detailed than his sociological study of the oasis, Vidal documented the oasis's hydrological resources. Vidal showed that al-Hasa could be, and should have been, known by its water resources. And when seen through the lens of ecology, Vidal diagnosed a disturbing trend: the oasis was undergoing an environmental decline. One effect of Vidal's work was to construct in al-Hasa an environmental object, one in need of rescue from those who lived within it, and an environment that could only be saved by the technological powers of an outside power, primarily the state.

Diagnosing "Decline" in al-Hasa

Since early in the twentieth century, al-Hasa's date market, its most important, had undergone a steady decline. Historically, Hasawi farmers and merchants produced and marketed dates and date products for local, regional, and global consumption, although intra-peninsular trade was paramount. Until mid-century, dates constituted the single most important component of the diet for not only the average oasis resident but also for neighboring Bedouin tribes. Until the 1940s the date trade was profitable, dates enjoying high demand and fetching high prices. After their conquest of the east in 1913, Saudi leaders strove to rein in local tribes and their raiding practices, partially by settling them in agricultural settlements as well as in the two main cities in al-Hasa. In addition to sedentarization efforts, the al-Sa'ud took economic measures to accommodate the Bedouins. Most importantly, the government took steps to drive down the price of dates, making them more affordable for Bedouin consumption. The drop in prices proved devastating. From 1948 to 1951 the price of 140 pounds of dates dropped from 48 Saudi Riyals to 10.[28] The government lifted the export embargo in the spring 1952, raising prices back to an improved but still poor 17 Saudi Riyals.[29]

The price decrease hit small farmers particularly hard. Larger farmers and merchants who were not entirely dependent on dates for their income did not suffer as much.[30] As the value of dates and date gardens dropped, small farmers who owned their land were often forced to either join the ranks of day laborers or to look for work in local industries or, more likely, outside the oasis. Those who depended exclusively on labor for their wages were pushed to the brink, thousands leaving al-Hasa to find work with Aramco and in the Eastern Province's rapidly growing metropolis Dammam, located about 100 kilometers to the north. A second blow to small farmers and landless laborers was the rise in real estate prices, locking the lower and working classes into their subordinate status.[31]

Vidal was the first observer to publish on the decline of al-Hasa's date market. He saw little hope that the economic situation in the oasis was set to improve, even going so far as to suggest that "the date agriculture and the farmers of al-Hasa are now at a crossroads."[32] He sounded an ominous warning about the future of agriculture in the region. But his—and what would become the state's—primary concern was not with deleterious effects of the government's drive to suppress the price of dates. Rather, he emphasized that the most pressing set of problems had to do with the slow but steady decline of al-Hasa's cultivable area; the oasis was, according to the Aramco ethnographer, "slowly drying out."[33] It was clear that desert was gradually overtaking some agricultural land, burying parts of the oasis perimeter in the suffocating and desiccating sands that rode the backs of migrating dunes.

It was unclear to Vidal exactly how much of the oasis had given way to the encroaching desert, and he did not hazard a guess. Several years later, the Saudi Arabian Ministry of Agriculture sent in a team of scientists to investigate the scope of the problem. They determined, and thus justified their presence and work in the oasis, that along the entire eastern edge of the oasis "a great mass of sand was advancing on the oasis at an estimated average rate of 30 feet annually. Each year some 230,000 cubic yards of sand were ebbing into the oasis, and near 14 villages [in the southeastern corner of al-Hasa], the dunes were looming over the very roofs of the houses . . . the dunes that were moving—measured five and half miles by 100 miles [were] advancing so fast that it would bury the [closest] village . . . within seven years if immediate action was not taken."[34] In addition to the invasion of the massive desert dunes, the al-Hasa water table appeared to be in decline. By the middle of the century, underground springs that once had needed no assistance required mechanical pumps to bring their water to the surface.

Vidal noted that the underground water that had fed al-Hasa's gardens for centuries was partially renewable but not indefinitely so. Al-Hasa's water resources would not last forever, and the Hasawis were to blame. Vidal argued that the decline in the oasis's water table had "been helped along by agricultural malpractice that produced a rise in the ground water table."[35] It was the use and management of that water that most concerned Vidal. There were between 50 and 60 underground springs spaced throughout the oasis. In 1951 they still provided bountiful, if declining, amounts of water for the area under cultivation. In spite of the large number of springs, however, the use of water for irrigation purposes was tightly controlled and unequal. It was also wasteful. Vidal singled out the dominant mode of irrigation in use in al-Hasa as late as the 1950s, known as *saih* irrigation, as particularly problematic.

Water in the *saih* system flowed from large springs and passed through an elaborate network of irrigation canals called *masqas* and then through secondary ditches, from which it was diverted from farm to neighboring farm. The physical network was complex, with canals and ditches crisscrossing and traversing the length and breadth of al-Hasa. Its complexity, however, was not the product of need, but of social power. The system depended on the timed distribution of water from spring to canal to farm. Farmers abided by a rigid system that determined who enjoyed early access to water fresh from the spring—access that made all the difference in the quality of water received and, not surprisingly, in the quality of the crops eventually harvested.[36] Typically, the right of first access went to land owners and farmers who owned the largest plots of land. Distance from the spring did not affect the system of privilege. Water that flowed directly from the spring (called *hurr* [pure] water) passed through a *masqa* and was then diverted directly into the plots of those who had first access. These farmers lifted flood gates located along the mason or stone lined *masqas* adjacent to their land for specified periods of time, often measured by the changing lengths of the shadows of the palm trees or by the location of the evening stars, allowing the water to flow over and saturate their farms.[37] In almost all instances, farmers that relied on *hurr* water took more than they needed from the *masqa*. That resulted in a surplus of unused water, which pooled on the ground. Because al-Hasa sloped gradually from west to east, the left-over water (called *tawayih*, meaning forfeited or twice-used water) was then redirected through a second gate called a *munajja*, passing "into a common channel, usually referred to as *thabr*, which conducts [it] over longer distances and from which smaller canals branch out into other gardens."[38] *Tawayih* water passed in this way repeatedly—from farm to *thabr* to farm and so on—until it arrived at the perimeter of the gardens and either ran off into the desert or pooled in bogs that lined the oasis.

Cultivators and landowners who held rights to *hurr* water took great pride in that fact. Conversely, those who relied on *tawayih* were held in some disdain. Vidal wrote that it was common for local farmers to malign those lowest in the irrigation hierarchy: "[G]ardens at the very end of this redistributing and regathering system are spoken of with disdain as 'drinking' tawayih al-tawayih, twice-used or twice-forfeited water."[39] Because of the symbolic significance attributed to the superiority of *hurr* water, the local system took measures to guarantee that tawayih water was never mixed with the pure. "Thus," Vidal wrote, "long and complicated channels must be built and maintained, once a farmer is assigned a place in the irrigation order and schedule, even though his garden may be closer to a second source of water."[40]

Twice used water was of poor quality, and its nourishing power diminished as it passed through farms. As water passed from field to field, it rapidly accumulated salt through the leaching of salts from the soil, and its quality diminished correspondingly. Vidal ran a series of tests to determine the salinity of water as it ran from specified points along one *saih* network originating from the al-Haql and al-Khudud springs just east of Hofuf.[41] At its origin, the water collected from the two springs had an average salinity of approximately 1,275.5 parts per million, good enough for drinking and irrigation. But as it passed through the large palm gardens and the predominantly Sunni and mixed Shi'i-Sunni villages of the eastern oasis, it became more saline. At the end of its route, at the edge of the oasis in the Shi'i village of Abu Thawr, it contained more than 4,000 parts per million, which made it "almost useless for anything but date agriculture of mediocre results."[42] The small Shi'i farmers were not the only ones hard hit by the irrigation system. Small Sunni farmers also suffered by being at the end of irrigation order, but there is little doubt that al-Hasa's Shi'is bore the brunt of the system's imbalances.

Vidal's discussion of the deficiencies of the irrigation system directed attention to the human source of the problem, but in his recommendations on how to rescue the oasis from environmental disaster Vidal set aside his own work on the various ways in which social power, land tenure, and (to a lesser extent) sectarian difference constituted the heart of threat to the oasis's natural resources. Rather than arguing for reform to land tenure and social relations, Vidal emphasized that the oasis could only be rescued through technological change, the implementation of scientific techniques, and better environmental management. In the end, for Vidal, Hasawi society only mattered with respect to the various kinds of agricultural and irrigation technologies the oasis's residents employed. His encouragement that observers look away from social complexity represented an important act of erasure and replacement. Rather than worry about social power, Vidal ultimately encouraged his readers to think of al-Hasa not as ecological system in which humans and nature exerted mutual influences, but only as an environment in need of rescue. This move was important for future politics in two ways. First, it suggested that the priority in al-Hasa should be the environment rather than society. Humans faded into the background, subordinated to the interests of nature and the markets that nature could serve. Second, it diminished the significance of the tensions inherent in the social hierarchies that predominated in the oasis, leaving them unaddressed, irrelevant, and simmering under increasingly difficult pressures.

The State to the Rescue

The Saudi Arabian government embraced Vidal's concerns about al-Hasa as an environment under threat. The state also embraced his suggestion that preferred solutions lay in the technological and the managerial. The government took Aramco's warnings about the declining health of the oasis seriously, although it did not immediately move to confront the threats facing the area.[43] It took six years for the state to finally mobilize its resources in al-Hasa, launching its initial investigations there in 1961.[44] From 1961 to 1971 the Ministry of Agriculture (and Water) spent tens of millions of Saudi Riyals addressing the environmental threats first outlined in detail by Vidal, most notably the building of a multi-billion-Riyal irrigation network that aimed to remake nature. The state would make little effort to address the underlying social relations that helped produce al-Hasa's environmental crisis. In fact, it became clear over time that the government had little interest in overturning the social elements of the crisis. Insofar as it was al-Hasa's Shi'i community that was locked into subordinate status, it is likely that the government had no desire to empower them.

In 1962 the Ministry hired the Swiss engineering firm Wakuti A.G. to follow up Vidal's investigation of al-Hasa's cultivation practices and its water and soil resources and to propose a solution to the challenge of improving the oasis's irrigation and drainage system.[45] Wakuti operated in the oasis from 1963 to 1971. On the basis of the findings and proposals submitted by Wakuti to the government, the Ministry of Agriculture hired the West German construction company Philipp Holzmann A.G. to build what would become the al-Hasa Irrigation and Drainage Project (IDP).[46] Philipp Holzmann's construction operations spanned the period from 1967 to December 1971, when the IDP was officially opened. In 1968, Italconsult, an Italian engineering firm that also undertook studies in other parts of the kingdom, included an analysis of al-Hasa in its examination of the water and agricultural resources of the entire Eastern Province. The technical and scientific data recorded by Wakuti and Italconsult not only confirmed Federico Vidal's findings that the oasis faced an uncertain future if nothing was done to rescue it, but also leaned heavily on Vidal's conceptual framework.

The construction of the al-Hasa Irrigation and Drainage Project transformed the oasis. By 1971, when Philipp Holzmann completed its construction work, that firm had built and laid more than 1,500 kilometers of concrete canals that carried fresh water from 34 local springs to al-Hasa's farms, and an additional 1,300 kilometers of canals that carried it away.[47] The new irrigation area exceeded 16,000 hectares. Most of the area relied on gravity to transport water from the springs connected to the network of main, sub, and lateral canals. But 4,000 hectares rested above the main springs, requiring the construction of three pump stations and elevated reservoirs, each with a capacity of 15,000 cubic meters. The main 155-kilometer irrigation and 140-kilometer drainage canals were either rectangular or trapezoidal in shape, varying in width from 1.6 to 9.26 meters and a length of 8–10 meters for a weight of around 20 tons each.[48] All of the 3,185 kilometers of irrigation and drainage canals were parabolic in shape, ranging in width from just under half a meter to a full meter.[49] In the course of building the network, Holzmann excavated more than 9, million square meters of earth and used more than 450,000 square meters of reinforced concrete.[50] In order to carry through the construction of the project, Holzmann engineers had to build cement plants; it also built a separate 750,000-square-meter complex to house various workshops, laboratories, storage areas, offices, a power station, and housing for employees.[51]

The movement of millions of tons of earth and the construction of thousands of kilometers of canals was an impressive engineering accomplishment. In the years after the project opened, it became a source of media and public interest, the subject of a variety of professional and popular magazines and journals in Saudi Arabia and around the Gulf.[52] During the 1970s, the al-Hasa Irrigation and Drainage Project became a technological spectacle and a source of national and regional pride. Its completion was depicted as evidence of the kingdom's will to conquer nature, its technical acumen, and its commitment to developing the necessary resources to escape from its dependency on foreign providers of food. In the early 1970s the completion of the IDP lent substance to those claims being made by Saudi Arabia's leaders that it saw the country's future in the application of science in everyday life and especially the building of large technological systems. For its part, the al-Hasa oasis assumed even greater significance nationally because it was home to the IDP. In various literature published after 1971 the IDP and al-Hasa were cast as one and the same, the technical achievement subsuming the oasis itself. At least in state records and academic studies, al-Hasa's geology, agriculture, soil characteristics, and water resources came to represent how it would be known.

The successful collapse of place and society with environment and technology, first set in motion by Vidal, was completed by the research and work carried out by Wakuti and Italconsult in the 1960s. Wakuti addressed the most basic social data, preferring, when it talked about people at all, to focus on the techniques employed by local farmers in the old irrigation system.[53] As for the existing irrigation system, Wakuti lifted most of its details directly from Vidal's report. Similarly, The Italian engineering firm Italconsult, which worked in the oasis only in 1968, helped in this process. There was no relationship between Italconsult and the construction of the IDP, although the firm carried out its survey of the oasis in 1967–68, when major engineering works were well under way. But although its work proceeded independently of the building of the IDP, it was no less important to the overall remaking of the oasis.

The two companies did note and write about society, although they mostly relied on negative assumptions based on race and culture familiar to observers of Western colonialism. Italconsult's final report on al-Hasa noted the presence of a large Shi'i community. Completely missing, however, was any discussion of Shi'i-Sunni relations or the scope of their interaction, let alone the social relations that bound and shaped interaction between them. The report's authors did offer up a bit of analysis regarding intra-Shi'i relations that, though not inaccurate on a limited scale, overlooked that the same set of relations even better characterized Shi'i-Sunni relations: "[I]nstead of lineage, village and class relationships are the rule [in the Shi'a community]. . . . Class stratification is related to wealth—the city merchants, land-lords and jurists representing the upper layer and to alleged [descent] from the caliph Hussein."[54] The firm's reluctance to draw broader conclusions about land tenure patterns and the overlap of class and sect, or at least to state its conclusions directly, did not mean that its analysts didn't understand them. The final report noted that absentee ownership was widespread, that it was rare to find an owner-farmer, and that much of "the property does not belong to the farmers, but to non-agricultural classes." "In sample villages," the report continued, "80 percent of the plots area is rented."[55]

What is perhaps most remarkable about this bit of social analysis is that Italconsult, after offering up class analysis as one way of approaching knowing about al-Hasa, abandoned it as an explanatory framework. Italconsult was more preoccupied with what it perceived to be the conjuncture of the cultural and the technological, arguing that cultural norms manifested in technological backwardness and environmental degradation. The company also made the sweeping claim that technological backwardness was tantamount to being inhuman. The final Italconsult report remarked:

[T]he persistence of the traditional farm, with all its implications, i.e. low technological level, subsistence production, absence of enterprise with regard to stocking and to the market, seems to be due both to the "clinging nature" of the traditions and to the insufficient transformation of the human element prevailing there. The modern structures and infrastructure have so far modified a number of external conditions . . . but they have not radically modified either the human element or the farm structures.[56]

Italconsult pressed for the modification of the "human environment through extensive training of adults, both at the level of personal maturity and of awareness of the *real* conditions of the economic and social environment in which they will work and at that of technological and professional modernization."[57] It was through technology, scientific education, and management that the entirety of al-Hasa's social and cultural makeup would be transformed and that Hasawis would apparently be humanized. The company made clear that technology was more than merely a set of implements and practices, but that it also constituted a set of cultural values that demanded the acculturation of the typical Hasawi farmer.

The work of constructing al-Hasa as an environment threatened by its human residents, rather than a complex socio-ecological system, produced a number of effects. The companies' approach was reflected in the actual construction of the IDP itself, in its operations, and in its impact on local farmers and workers. Not only were the socially informed aspects of life in the oasis ignored nearly entirely; the building of the irrigation system also helped intensify inequality. The result was that the religious and social-hierarchical dimensions of life in al-Hasa were mostly black-boxed, subordinated to thousands of pages of technical data and commentary on methods, hydrology, geography, and geology. Seen in this light, both the act of producing knowledge about al-Hasa and the work of reengineering it were hostile to the residents of the oasis. From the perspective of the state, the companies' surveys and reports contributed important technical information about geology and agriculture to the growing reservoir of knowledge being accumulated by the Saudi government. Armed with the combined works of Vidal and the two European engineering firms, the Saudi government did not exhibit any compunction to undo or even address al-Hasa's social hierarchies. Instead, the government embraced the way of thinking that emphasized al-Hasa as an environment, one threatened by outmoded practices, that needed a more intrusive physical and administrative apparatus to rescue it. The government's political interests, most notably expanding its presence in a place where it had been weak, were served in the short run. Ultimately, however, the government's broader political economic objectives were not met. And its approach to managing al-Hasa helped unleashed the forces that led to violence in 1979.

Toward Discontent

By the end of the 1970s, the Irrigation and Drainage Project had failed to achieve almost all of its technical objectives. Within three years of its completion, the project helped expand the area under cultivation by over 2,000 hectares, a promising start to the project of environmental rescue.[58] But by the end of the decade it had added little more, falling well short of its goal of expanding the total area from 8,000 to 20,000 hectares.[59] Tests carried out after the construction of the irrigation and drainage network determined that al-Hasa's water resources were declining rapidly in spite of its purpose to manage and conserve them. Within the network some of the main springs performed better than others, leading to higher levels of water in some canals and lower levels in others. The water level proved so low along some canals that the area under cultivation declined by a third or more. The total area committed to date cultivation also declined from 4,750 hectares in 1967 to 4,547 in 1980.[60] Even in spite of the expansion of vegetable harvesting, the oasis did not produce any more for domestic consumption than it had previously. In fact, throughout the 1970s the decline of al-Hasa's agricultural and natural resources accelerated. The new irrigation system proved unable to stem the "environmental" decline of the oasis.

In spite of the project's technical shortcomings, it did achieve some of the state's political objectives. It enrolled the entirety of Hasawi society into its orbit, imposing on local residents a massive new bureaucratic and administrative apparatus. Farmers and citizens alike were forced to negotiate with the Irrigation and Drainage Project and its offices in order to get access to irrigation as well as drinking water. Where the state had struggled to assert its power through other kinds of institutions—municipal councils, the governor's offices, courts—it was more successful in establishing itself as an administrative and material presence with the Irrigation and Drainage Project. It was precisely the connection between the IDP as a technological system and the authority of the state that would ultimately lead local residents to revolt.

For not only did the IDP fail to achieve the objective of expanding al-Hasa's productive capacity, it actually accelerated the pace of decline of the oasis's natural resources. Among the most enigmatic outcomes of the Irrigation and Drainage Project was the decline of the oasis's once over-abundant water resources. There are several possible explanations for this decline. Wakuti carried out extensive tests on the oasis's water supplies in 1963–64 and confidently attested to their long-term availability and to the viability of the 34 springs ultimately used to feed the irrigation system. By the end of the 1970s, the oasis did seem to be drying up. The boring of private wells (more than 300 of them by the mid 1960s) by small farmers certainly had some effect, although the flow rate of the wells was too low to have diminished the reserves dramatically. Local residents suggested that Aramco and the techniques it employed in extracting oil were responsible for the disappearing water in the 1960s and the 1970s.

Saudi Arabia's oil reservoirs are not pressurized, so oil has to be pushed to the surface. Aramco pumped water into the subterranean fields in order to force the oil out. One group of Shi'i landowners claimed it was widely believed that the oil company relied heavily on the same sources of fresh water that Hasawi farmers relied upon to pump oil from the Ghawar field, which lay less than 20 kilometers from al-Hasa but over 80 kilometers from the Persian Gulf, the next closest source of water. The IDP, they suggest, was an attempt by the government to cover up Aramco's destructive practices.[61] Even if Aramco did not use water from the underground reservoirs to pump out oil, the company, along with the state, probably put pressure on those reserves in other ways. In addition to al-Hasa, the growing cities of Dammam, Dhahran, and al-Khobar, which were north of the oasis, also relied on the Neogene and Khobar formations. As the cities expanded under the weight of intra-Peninsular migration and the addition of Aramco facilities after mid-century, especially housing and structural facilities that accommodated American workers, they no doubt strained the entire region's water supply. Diminishing water in the region accounted for the expensive efforts to build two massive desalinization plants along the Persian Gulf near Dammam.

Whatever the explanation for the decline in al-Hasa's water resources, residents of the oasis who were dependent on agriculture and nature were under tremendous pressure by the end of the 1970s. They were acutely aware that the IDP had failed to address their social status and their suffering. Thousands had left the oasis looking for work elsewhere. Many remained, however, and the worsening conditions of the oasis, in combination with the continued imbalance of power between rural and urban and between Shi'i and Sunni, contributed to the emergence of violent sectarian politics. When Shi'i activists issued a threatening letter to American employees of Aramco in 1979 citing the US's complicity in abusing the region's "national resources," they were speaking as much about water and the decline of regional agriculture as about Saudi Arabia's wasting oil revenues on expensive weapons systems. The IDP was more than just a symbol of state neglect, however. Alongside other large development projects in the region, it was evidence of the systematic theft of *their* resources, a belief held by Shi'is in al-Hasa as well as in other communities in the Eastern Province.

By the end of the 1970s, Shi'is felt that they had little recourse other than violent rebellion to press for the amelioration of their worsening social conditions. Projects such as the IDP helped crystallize sectarian politics in the Eastern Province. Political Shi'ism emerged in Saudi Arabia only in the mid 1970s and remained a mostly marginal phenomenon until late in the decade. Young religious scholars and activists who had studied in Iraq and Iran during the 1970s had periodically returned home throughout the decade, attempting to organize a grassroots political movement that attempted to formulate a specifically Shi'i political identity in the face of Saudi oppression. Through the 1970s, most Shi'is, at least those who participated in various kinds of political movements, had been attracted to secular nationalist politics. Even in communities such as al-Hasa, where many Shi'is were socially subordinate, there had never been a viable political movement based on sectarian affiliation that cut across social difference or geographic location. Most politics took shape at the village level or were concerned with shared social problems. As social problems went unaddressed or were exacerbated by development work, sectarian politics and identification assumed growing importance. As state power grew, and as it became clear that the state not only didn't intend to address the social problems affecting the Shi'i community but would pursue projects that made them worse, the appeal of political Shi'ism grew considerably, leading ultimately to the outbreak of violence in 1979.

Though the uprising of 1979 was in part a response to the effects of the state's technopolitical approach to "rescuing" local environments, it is important to keep in mind that it was also driven and framed by Cold War relations between Saudi Arabia and the United States. In part Shi'i critiques of Cold War politics had to do with the general nature of US-Saudi relations and with the impact of work done by Aramco on Saudi Arabian authoritarianism. But Saudi Arabia's Shi'is also held the US directly responsible for the devastation of their communities and the damage done to local natural resources. In addition to seeing the Cold War and Cold War relations as an object of Shi'i disillusion, it is important to see the Cold War as set of forces that helped set in motion tremendous political, environmental, and social change, much of which took place beneath the surface of state-to-state relations but that directly flowed from concerns about communism, Soviet aggression, and the potential consequences of pursuing alternatives to Third World authoritarianisms.

Acknowledgments

Material in this article was adopted from my book *Desert Kingdom: How Oil and Water Forged Modern Saudi Arabia* (Harvard University Press, 2010). I am grateful to Harvard for allowing me to reproduce it here.

Notes

1. Organization for the Islamic Revolution in Arabia (OIR) 1979: 42. For a full account of the 1979 rebellion, see Jones 2006: 213–233.

2. Saudi Arabia's Eastern Province is home to the vast majority of the kingdom's Shi'i minority. There are no reliable data for the numbers of Shi'is in the kingdom in the late 1970s or today. Most estimates put the Shi'i population at around 10–15% of the total Saudi population. The Eastern Province is also home to all of Saudi Arabia's vast oil reserves, making the region vital to the lifeblood or the regime and strategically sensitive.

3. The uprising compounded what already constituted a political crisis in the Arabian Peninsula, one that directly threatened the security of the House of Sa'ud. The unrest occurred contemporaneously with the occupation of the Grand Mosque at Mecca by a group of Sunni religious radicals. The occupation of the mosque in Mecca, which lasted several weeks and required the use of Saudi and French Special Forces to end, shook the kingdom to its core.

4. Although the riots were driven by local social and political grievances, revolutionary fervor that year in nearby Iran also influenced the course of events. Shi'is in Saudi Arabia were partly inspired by the toppling of Iran's brutal authoritarian dictator and by political Shi'ism in inspiring mass resistance to the Shah.

5. See Vitalis 2006.

6. See Citino 2002; Yaqub 2006; Mahmoud Mamdani 2004.

7. OIR 1979: 18. After the uprising, the OIR claimed that its letter forced 140 Americans and their families to flee the kingdom.

8. Vidal 1954.

9. In 1948 Aramco geologists discovered the Ghawar oil field, the world's largest, a few kilometers to the West of the al-Hasa oasis.

10. See al-Subayī 1989: 71.

11. Vidal 1955: 13–15. Vidal 1954 is a condensed version of the longer report.

12. In 1951 the Saudi newspaper *al-Bilad Al-Sa'udiyya* put the oasis' population at 500,000, considerably higher than every other estimate. Most estimates vary between 150,000 and 250,000. See Vidal 1955: 17.

13. See Bill 1984: 6; Goldberg 1986: 230.

14. For sizes and populations of some villages, see Vidal 1955: 40–41. Data on al-Mubarraz are from p. 109 of the same source. Figures on al-Hofuf are from Vidal 1954.

15. Vidal 1955: 40–41.

16. See California Arabian Standard Oil Company and Arabian American Oil Company Annual Reports, 1940–1945.

17. Turnover rates were high early on, approaching 91 percent in 1945, but they stabilized within 15 years, reaching 3 percent by 1959. See al-Elawy 1976: 375.

18. Vitalis 2006.

19. Ibid.: chapter 6.

20. In 1953 the capital was moved from Hofuf to Dammam in order to put it closer to the headquarters of Aramco and the commercial capital of the region.

21. Vidal 1955: 149.

22. Vidal 1955: 162–163.

23. Vidal 1955: 166–167.

24. According to Vidal (1955: 34), the probable ratio was 60 percent Shi'i and 40 percent Sunni.

25. Vidal 1955: 96.

26. Absentee landownership dates back at least as far as the sixteenth century, when the Ottoman Empire awarded fiefdoms in al-Hasa to its janissary military leaders. See Mandaville 1970: 504–506. The Ottomans also ruled over al-Hasa between 1871 and 1913. For two accounts of Ottoman administrative rule in the late nineteenth and early twentieth century, see al-Subay'ī 1999; al-'Idrūs 1992. It is worth noting that nineteenth-century Ottoman administrators, who imposed taxes on al-Hasa's merchants and farmers, also sought to expand the oasis' agricultural area. The basic rules governing property rights were also established in this period, with the Ottomans setting up four kinds of ownership: private property, government owned land, land overseen by endowments (*awqāf*), and common land. See al-Subay'ī 1999: 34–38.

27. Vidal 1955: 37. Vidal went on to note that "from the point of view of over-all social structure, the al-Hasa garden villages are fairly simple. The population of the hamlets and the smaller villages consists of only one class, that of the agricultural workers, since the land owners live in the big towns. In the larger villages, a rudimentary upper class is found, consisting of a few families of landowners who dominate the village economically and socially. The rest of the population is also made up of garden laborers. In recent years, some villagers who had left to do contracting work have returned to build permanent residences in the place of origin, though still working outside. It is possible that these people may start competing for social prominence with the older leading families, but a more likely prospect is that the new rich will marry into the village aristocracy." (ibid.: 37–39).

28. Al-Hasa converted to a cash economy in the early twentieth century.

29. Vidal 1955: 193.

30. Vidal 1954: 215.

31. Vidal 1955:196. Debt grew among small farmers and landless workers in the twentieth century as well. With the decline of date prices, borrowing from larger merchants in order to buy foodstuffs increased, leading to a cycle of indebtedness that proved difficult to break free from.

32. Vidal 1954: 215.

33. Vidal 1955: 186.

34. Tracy 1965.

35. Vidal 1955: 187.

36. The irrigation system predated the arrival of the Saudis, but once the family's representatives took control of the oasis it recorded the distribution rights in the Finance Office. Al-Subayʻī 1999: 77.

37. Al-Subayʻī 1999: 77.

38. Vidal 1955: 136.

39. Ibid.

40. Vidal 1955: 137.

41. One of the largest springs in the oasis, it had a discharge rate of approximately 22,500 gallons per minute (Vidal 1955: 120). See also "'Ain," *Aramco World* 11, no. 10, December 1960. There were several women's bathhouses along the length of the course taken by water from al-Haql. Aside from al-Haql and al-Khudud, which also had a discharge of over 20,000 gallons per minute, there were two other large springs in al-Hasa: 'Ayn Umm Saba'a and 'Ayn al-Harra. The four largest springs delivered more than 150,000 gallons of water to al-Hasa per minute.

42. Vidal 1955: 141–142.

43. The sharing of technical and other data was stipulated in the 1933 oil concession. Article 26 of oil concession stated: "The Company shall supply the Government with copies of all topographical maps and geological reports (as finally made and approved by the Company) relating to the exploration and exploitation of the area covered by this contract."

44. Tracy 1965.

45. In the contract hiring Wakuti, the Saudi Arabian government outlined its technical ambition as well as its and the company's legal obligations. It is interesting to note that the Saudi government stipulated that Wakuti must abide the country's boycott of Israel if it wanted to operate in the kingdom. To prove that the Swiss firm was honoring that commitment, Saudi Arabia demanded a manufacturer's certificate stating that no materials were manufactured or made in Israel as well as a statement from the Swiss Chamber of Commerce attesting to the country of origin of all materials (Kingdom of Saudi Arabia 1964).

46. Wakuti engineers remained on board in an advisory capacity, working closely with Philipp Holzmann engineers to ensure that the project was implemented according to specifications.

47. Kingdom of Saudi Arabia 1964: 11–20. Holzmann also put down over 1,500 kilometers of roads to carry out its work. The irrigation and drainage canals were kept separate, effectively doubling the amount of material and space required for the

project. Wakuti claimed in 1971 that "this length corresponds to the distance from London to Rome" (Kingdom of Saudi Arabia 1964: 11).

48. Holzmann 1971: 16.

49. Holzmann 1971: 3–5.

50. Holzmann 1971: 27. Wakuti put the figure at 7 million square meters (Kingdom of Saudi Arabia 1964: 22).

51. Holzmann 1971: 6.

52. Congratulatory articles emphasizing the technological achievement of the IDP appeared in Aramco's monthly Arabic magazine *Qafilat al-Zayt* in March 1972 and in July 1991. *Al-Manhal*, one of Saudi Arabia's longest-running cultural magazines, ran a laudatory piece in its December 1971–January 1972 edition. The good news spread beyond the kingdom. The Kuwaiti monthly *al-'Arabī* also celebrated the IDP in August 1974.

53. Wakuti, Studies Volume 2: 28.

54. Italconsult, Final Agricultural and Water: 24–27.

55. Italconsult, Final Agricultural and Water Report: 38.

56. Italconsult, Final Agricultural and Water Report: 34.

57. Ibid. (emphasis added).

58. Lutfī 1986: 126.

59. Rajab 1980: 122–124.

60. Rajab 1980: 124.

61. Interviews by the author, al-Hasa, Saudi Arabia, September and October 2003.

10 A Plundering Tiger with Its Deadly Cubs? The USSR and China as Weapons in the Engineering of a "Zimbabwean Nation," 1945–2009

Clapperton Chakanetsa Mavhunga

Does it make sense to talk about the "Cold War," let alone "The Global Cold War," in the Global South? What happens to local time when "watershed moments" in the Global North are extended uncritically to mark global time? Are we sure that the materiality and meaning of these "local" events are shared beyond their borders? How do other locals measure their own times?

Like "the First World War" and "the Second World War," "the Cold War" falls within a continuing way of defining what counts as worldly (what is globally significant) from Europe and North America, using war as if it is the only marker of time. The rivalry between two countries—the United States and the Soviet Union—and the trickery they deploy to outwit one another, and using other countries as unobvious weaponry, is transformed into a universal moment in which everybody is living.[1] On occasion Cuba is mentioned, if only as a Soviet surrogate and base-plate position for Moscow's nuclear warheads.[2] China enters the fray as a Soviet ally—until it gets fed up with Moscow's duplicity when striking nocturnal deals with Washington.[3] In the end, whenever scholars insist on "the Cold War" in the Global South, their defense is no more than following the North's footsteps and pathways in the Global South. Of late, even scholars of such oft-omitted "Cold Warriors" as Cuba, China, and the Nordic countries have followed suit.[4] Any apportionment of agency to African players becomes no more than a work of charity in which the Africans can do no more than respond as opposed to initiating events and synchronizing the North to their own time and circumstances.

So what are the modalities of inverting the commonplace synchronization of Southern time to Northern time into a synchronization of Northern time to Southern time? As this essay will proceed to show, it would force us to spin the narrative of American and Soviet users of Southern puppets into one of Southerners as designers (or political engineers) of post-colonial futures using the Union of Soviet Socialist Republics and influential Southerners Cuba and China as unobvious weaponries (armories as well as strategic assets) to achieve their objectives.

Thanks to emerging memoirs of Soviet operatives in Africa during the 1960s, the 1970s, and the 1980s, we now know that the Soviets did not even use the term "Cold War," whether prefixed with "the" or "a." The Soviets considered such vocabulary "the creation of 'war mongers' and 'imperialist propaganda.'" From Moscow, the battle was not between "two 'superpowers' assisted by their 'satellites' and 'proxies'" as depicted from Washington but "a united fight of the world's progressive forces against imperialism."[5] Official America had borrowed the term from the English novelist-journalist George Orwell, who in 1945 had deployed it to deride how atomic power had equipped the US and the USSR with a bully-boy mentality of dividing and ruling the world between themselves.[6] Of course, Orwell's anger toward hegemonic forces and their powers of permeation blinded him to the very same permeation as an avenue for local resistance against or even manipulation of the hegemonic, or, as James Ferguson recently showed, the likelihood of such seemingly universalizing forces to anchor in some while completely steering clear of other places.[7] It could very well be that the North viewed the period as one of a nuclear arms race while the South viewed it as an anti-colonial and postcolonial era.[8]

The question at stake in this chapter is this: Does the Cold War conceptually and analytically belong in the South, and if so, on whose terms? The Norwegian scholar Odd Arne Westad rejects the position that it does not, on two grounds. First, "without the Cold War, Africa, Asia, and possibly also Latin America would have been very different regions today." Second, "Third World elites often framed their own political agendas in conscious *response* to the models of development presented by the two main contenders of the Cold War, the United States and the Soviet Union" [emphasis added].[9] In making his powerful argument, Westad is writing against traditional diplomatic historians for whom the Cold War is only about superpowers and their shenanigans.

Whereas Westad makes a case for the inclusion of the South in the Cold War, I do not see the necessity. The Cold War cannot suffice as an analytic to explain developments in the South, particularly because it misses so much about motivations and agencies. Calling this anti-colonial resistance period a 'Cold War' era would be tantamount to a notion of time I have already rejected: using Northern temporal benchmarks that are very situated and specific to certain countries' foreign relations to envelope benchmarks (e.g., colonialism, anti-colonial struggles, and independence) that are specific to the South. The further effect is to see Southerners as "using the opportunities offered by Cold War logics" for their own purposes, such that the Cold War sneaks right back to belong in the South analytically, even if it is not a dominant explanatory mode. That too is not what this chapter means: rather, it seeks to show that opportunities were not "offered" by anybody but were a result of local initiatives.

The moment one uses the term "response" to describe what the actors discussed in this chapter are doing, their status as initiators is lost. They become "surrogates," "satellites," or "puppets"—exactly what Zambian President Kenneth Kaunda meant in 1976 when, referring to the worrying increase in Soviet presence in Southern Africa (especially Angola), he spoke of "a plundering tiger with its deadly cubs coming through the back door."[10] Yet at the level of practice, those who use others as puppets are unaware that the so-called puppets are using them (in Shona, kushandiswa). They are what one might call puppets of the puppets. Those who see themselves as engineers or designers of artifacts are in fact artifacts of the artifact: the user of the user is, in fact, the used.[11]

The term "using" extends beyond the traditional STS sense of designers as engineers making artifacts for users (consumers).[12] Rather, it is a process of designing through inversion. The Soviets and the Chinese were entitled to think these black politicians were their puppets. Upon closer scrutiny, these politicians were 'playing puppet' as a camouflage to use these communist countries as weaponry for designing themselves into nationalists, create guerrilla movements, and assemble ideological repertoires to engineer colonies into independent nations through warfare and diplomatic trickery. This, I suggest, is how the North became a weapon of the South (not just the Sino-Soviet blocs but also the US and Mobutu Sese Seko and Jonas Savimbi, France and Félix Houphouët-Boigny, and so on). There were moments in such encounters when countries of the North projected themselves as "superpowers" fighting a Cold War. Meanwhile, in the eyes of Africans, they were merely sources of guns, military training, and communist ideology—tools with which to liberate power for themselves. At moments, in local contexts, the "superpowers" were virtually "superpowerless" in the face of the agencies of local actors, for whom the logics and exigencies of "liberation wars" and seizing power were paramount.

They became weapons of local actors. When used as a noun, "weapon" means anything used against an opponent, an adversary, or a victim. In its verb sense, "to weapon" refers to two things. First, it refers to how local politicians in white-minority-ruled Rhodesia (now Zimbabwe) used the Soviets and the Chinese first as chisels to carve their raw civilian men into guerrillas, then as quartermasters to supply or equip them with guns. Second, it refers to how the Soviets and Chinese, on one hand, and the Zimbabwe African People's Union (ZAPU) and the Zimbabwe African National Union (ZANU), on the other, transformed one another into weapons. These senses of weaponry are very destabilizing: the weaponizer is, without being aware of it, being transformed into a weapon. Thinking they are using others, the users are, by using, being used to perform a function by those they are using.

For Mamadou Diouf, such "trickery" or "appropriation" (hence my sense of "appropriate technology") raises two critical questions about the location of locals (Africans) in the narrative of globalization: "Is it a matter of appropriating this process by 'annexing' it? Or, rather, of exploiting this process to lend new strength to local idioms, so as to impose on the global scene the original version in place of its translation and adaptation?"[13] Contrary to Stuart Hall, Diouf is reluctant to accept that localism is "the only point of intervention against the hegemonic, universalizing thrust of globalization." According to him, Africans have constantly remodeled their traditions to create a new memory that differs from that of Western "modernity" in order to "anticipate a future saturated with projects of an indisputable modernity."[14]

The story told in this essay is precisely one of intersections between these globalizing, hegemonic thrusts, on the one hand, and the local imperatives that create a buy-in or rejection of the same, on the other, with guerrillas, guns, ideology, and history as specific weapons serving mutual purposes. They occupy that intermediate space between the North and the South, between communist countries and anti-colonial movements, as boundary objects.[15] Once the topographies are mapped, the essay then discusses the war in Rhodesia as a process of engineering a postcolonial state through violence. The conclusion narrows the discussion to the rise and reign of President Robert Mugabe of Zimbabwe, suggesting an examination of him as an engineer of political survival in power through the weaponization of forces that would ordinarily render him the hegemon's weapon or victim.

Mutual Weaponization

There are two types of genealogies that intersect here that make it difficult to accept a notion of "the global Cold War." The first is one of US-Soviet rivalry, both countries trying to turn the Global South into "topographic weaponry" to outsmart each other from 1955 on. The process involved recruiting Africans from leftist organizations for civilian training on scholarships, and engineering them into hybrid vehicles of communism and anti-colonialism through political indoctrination. By 1965, the Soviets and the Chinese were "sculpting" African guerrillas out of civilians recruited under false pretenses (they were told they were going for civilian courses, only to arrive and be shepherded into barracks).[16]

The second genealogy, which occupies most of my attention, is with regards to Africans' trickery and appropriation of the external to fill spaces in their own crossword puzzles. A proper archeology of this innovative tradition is better located in the colonial (and pre-colonial) moments of African history to head off any misconception that it came either with Europeans or Sino-Soviet "advisers" in the 1960s, the 1970s, and the 1980s. This chapter does not deal with pre-colonial appropriations, preferring instead to focus on developments immediately leading up to how black politicians helped themselves to and used the Soviets and Chinese as weaponry. In the first fifty years of white settler rule in Rhodesia, those blacks that got into the school system subverted missionary education that was supposed to make them meek colonial subjects into master keys opening doors to further studies in the diaspora since Rhodesia had no university until 1957.[17] The idea of "Zimbabwean nationalism" owed much to this intellectual exposure to black thought in Africa, America, France, and Britain. In 1959, Pan-Africanism—particularly negritude—became a glue for binding multiple workers' organizations into Zimbabwe's first black mass party, the National Democratic Party (NDP). By 1961, a second transformation had occurred: the battlefield had shifted from negotiating tables to streets. By 1963, the battlefield shifted to the bush, the weapons no longer books and eruditions, stones, and petrol bombs but guns.

This turn to guns is the stage where the US-Soviet rivalry and the Zimbabwean trajectories meet. The turn to guns after 1961 is located within a long local tradition of spiritual and secular weapons for taking and sparing life for purposes of human security. Its predecessors are poison, witchcraft spells, bows and arrows, spears, axes, and snares.[18] Before European colonization, southern Africa's inhabitants had for centuries innovated upon their metallurgical, pharmacological, ecological, and biotechnological traditions to produce goods exchangeable for overseas products, including guns.[19] The resistance of the Shona and Ndebele people (Africans) to the British colonization led by Cecil Rhodes in 1893 and in 1896–97 can be attributed to their initiative to acquire muskets and Martini Henry rifles and subvert them to the practice of African kingship. They were defeated not because they had failed to adapt their customary fighting technique to guns,[20] but because the equipment was outmoded in comparison with their European enemies' Maxim guns. After all, the foundries were in Liverpool, and what reached the South was mostly decommissioned or trade stock, considered unsuitable for military purposes and therefore fit only for export.[21]

It was not until after 1961 that Africans contemplated using guns to challenge the rule of (descendants of) colonial settlers. Since the 1930s, the emerging African elite educated in universities in South Africa and America had used the "civilized" language of diplomacy without success. When in 1961 the British bowed to Rhodesian pressure and refused to grant independence to Africans, the die had been cast. Power would not be given; it had to be taken—by force if necessary.[22] by any means necessary.

In 1961, ZAPU—under the leadership of Joshua Nkomo, a trade unionist and intellectual educated in South Africa—became the latest African political formation to challenge the state. Banned a year later, it continued "underground" as the People's Caretaker Council (PCC). In August 1963, tired of the politics of entreaty, a few—mostly Shona-speaking radicals led by a US-educated Wesleyan minister, Reverend Ndabaningi Sithole—broke away from the PCC to form the Zimbabwe African National Union. A year later, Rhodesia banned the PCC and ZANU and detained most of the leaders, including Nkomo, Sithole, and Mugabe.

Before ZAPU's banning, however, African politicians had already taken the first practical steps toward acquiring guns. This is the point at which the local genealogy intersected with the national interests of China, the Soviet Union, and Cuba. Until 1955, Africa had remained at the edge of Chinese and Soviet foreign policies.[23] China's revolution had ended only recently (1949). Cuba was still four years from the end of the Batista regime. The Soviets had just buried Joseph Stalin two years earlier and installed Nikita Khrushchev.

China found its feet first: with Indonesia it convened the first Afro-Asian People's Solidarity Organization (AAPSO) conference at Bandung in 1955. This bold move established what Richard Wright called "the Color Curtain" contesting the validity of an "Iron Curtain" in the ordering of the world at the time.[24] From Moscow, the world division of "haves" and "have-nots" was not necessarily a color problem but ideology. In the Soviet imagination, the world had to be cleansed of capitalism, the monster that had given rise to imperialism and the colonization of the South. Subsequently, Khrushchev provided arms and training to equip African leftists with the technical means to weaken Western imperialism from within.[25]

African politicians saw Moscow as strategic political and military capital to stiffen resistance against Rhodesia. The foundations of ZAPU-Soviet networks were built at three successive AAPSO meetings—one in Conakry (April 1960), one in Beirut (November that year), and one in Moscow

(January 1961). The National Democratic Party (NDP) started the construction work before it was banned and ZAPU was formed to replace it. The affable Tarcissius George "TG" Silundika was the builder. The weapons ZAPU sought in 1961 were not guns but a printing press and scholarships; the motive was clearly to use Marxism as "subversive weaponry" to replace Rhodesia's oppressive capitalism with a more equitable political order.[26]

Moscow did not just open its armories to ZAPU. In fact, in September 1962, after the banning of ZAPU, the PCC dispatched Joshua Nkomo to Egypt to buy guns from a contact named Mohammed Faiek and smuggle them in by commercial airline via Tanzania and Zambia.[27] Inevitably, the party's armory began as a motley collection of small arms that the market could supply and the little money available could buy.[28] Later the collection grew to include AK-47 rifles.[29]

From 1967 to 1970, ZAPU had good guns but used the tactics and strategy the Soviets had drilled into its trainees without critical thought and without practical adaptation. After its disastrous joint conventional campaign with the South African National Congress (ANC) military wing, the Umkonto weSizwe (abbreviated MK), in 1967, the Rhodesian Security Forces (RSF) captured AK and SKS rifles, RPG-7 shoulder-operated rocket launchers, RPD machine guns, PPSH submachine guns, and explosive devices.[30]

Until 1975, ZAPU's faith in guns had been restricted to small arms, as heavy weapons were thought to give away troop positions while slowing down the mobility that typifies guerrilla warfare.[31] Beginning in 1976, however, it made the transition to conventional weapons by attaching relatively heavy pieces to its guerrilla units and rear bases. From July 1978 on, convoy after convoy of armored cars rolled out of southern Angola into Zambia.[32] By May 1979, giant Russian Antonov-12 transport planes were landing "tanks, artillery and heavy machine guns" into the capital Lusaka daily to equip mobile battalions.[33] Previously, all ZAPU's weaponry was portable; from 1978 it "grew legs" and "gained weight." T-34 tanks, MTU-55 bridging equipment, BTR-152 armored personnel carriers, BM-14 and BM-21 multiple-rocket launchers, and Soviet-made command cars were brought in to equip the regular force trained in Angola.[34] Some ZAPU operatives say the Soviet Union delivered an unknown number of MiG-21 fighter jets to provide air cover.[35] Soviet operatives disagree only in details.[36] It is clear that the USSR had become a rather generous quartermaster to ZAPU.

ZANU struggled with shortages of guns throughout the war. After a militant beginning involving acts of arson and public violence directed at both the state and at ZAPU, ZANU scaled up its operations to a military strategy in 1966. The main technopolitical structure for this was the Zimbabwe African National Liberation Army (ZANLA), which was composed of trained and armed politicians. Its commander Josiah Tongogara said of its first battle near Sinoia in north-central Rhodesia: "We bought guns, bought dug-out canoes and crossed the Zambezi, landing on Zimbabwean soil as a people's army for the first time. . . . We had only a hundred guns and 50 soldiers, but it was the best army one could dream of. That is what we felt about it." [37] These early operations were miserable failures, prompting reorganization for the next four years.

When it re-launched operations in 1972, ZANLA had negotiated a pact to fight alongside the Mozambican guerrilla movement FRELIMO (Frente de Libertação de Moçambique) and was receiving arms shipped from China to Dar-es-Salaam and, after Mozambican independence, to the port cities of Beira and Maputo. Most of them were AK-47 rifles in crates.[38] This materiel was stored in depots for distribution to the operational headquarters of Tembwe, Chimoio, and Xai Xai. Each guerrilla detachment going into the operational area (Rhodesia) on foot then carried its own supply of guns and ammunition, which was usually "enough . . . to fight for months a hundred kilometers in every direction."[39] Once inside Rhodesia, they cached these supplies within a certain radius of their mobile operations.

Assembling Human Weapons

Guns were useless without the acquisition of the necessary skills to kill or spare life. This section attempts to locate some of the spaces where ZAPU

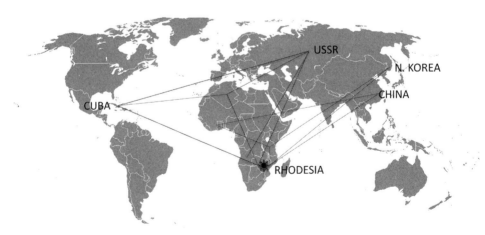

Figure 10.1
The engineering of a Zimbabwean nation through the weaponization of countries.
Source: Clapperton Mavhunga. Copyright 2007.

and ZANU sent raw human material to be engineered into mobile weaponries with which to physically carve Rhodesia into Zimbabwe. One might see the guerrilla as a boundary object, a kind of weaponized body at the intersection of two designers (his own organization and the communist countries training him), a vehicle through which these two designers used each other.[40] I have omitted training in North Korea, eastern European countries, Egypt, Ethiopia, Libya, Uganda, and other countries because I am currently examining the OAU Liberation Committee archives in Dar-es-Salaam.

The Soviet Union

We must remember that while ZAPU and ZANU had designed a strategy to get into power, they did not have total control over the means; nor were they self-sufficient.[41] Soviet training assistance to ZAPU began with students on scholarship attending political training at the Institute of Social Science. They were not meek sponges for communism; in fact, on numerous occasions they embarrassed ZAPU by making withering critiques of Soviet communism. If Moscow detained or expelled them, it risked defeating its purpose of engineering mobile vessels for its ideology. On the other hand, ZAPU forced its cadres to apologize lest they jeopardize Soviet support.[42]

Especially for civilian trainees, racism was a major problem. The testimonies of African students in Georgia in 1962 suggest that local students subjected their black counterparts to "enmity and antagonism." The problem was so serious that the Communist Party of the Soviet Union (CPSU) terminated the admission of blacks and transferred those already in Tbilisi to Moscow.[43] In mid 1962, race riots torched the Bulgarian capital, Sofia. The police intervened "violently" to protect Bulgarian students angered by "a Ghanaian's dancing with a Bulgarian girl." In 1963, violent anti-African riots hit Prague, sparked by "the preferential treatment the government offer[ed] to Africans."[44]

One year after the Prague and Sofia riots, ZAPU formally requested military training in the Soviet Union. On December 24, 1963, the party's vice-president, James Chikerema, delivered a letter to AAPSO representative Latyp Maksudov in Cairo requesting the CPSU to provide four months of training for 30 people "for subversive work [and] for military sabotage." He also requested six months of training for three recruits to manufacture "simple small arms," since bringing arms into Rhodesia was "impossible." In mid 1964, two ZAPU groups were admitted to Northern Training Centre to undergo "a ten-month comprehensive course, which ... included general military subjects and specialization in guerrilla and conventional warfare and even field medicine." The first group included Akim Ndlovu, the second Pelekezela Mpoko. Both would become senior ZAPU commanders.[45]

Subsequent groups underwent 12–18 months of phased instruction involving first some heavy communist indoctrination at the Central Komsomol (Communist Union of Youth) School in Moscow.[46] The idea was to subordinate the anti-colonial projects of these recruits to Moscow's universalistic communist "war" against Western capitalism. Once the crust of parochialism was peeled off, the trainees were then sent to Odessa Military Academy in the Ukraine for officer cadet training under the mentorship of General Alexei Chevchenko.[47] From there they were taken to a training center in Tashkent (Uzbekistan) or to one in Perevalnoye (Crimea),[48] where conventional warfare and command-and-control skills were drilled into them. The training emphasized loyalty to the Communist Party's objectives and structures, the role of decisive force as "the midwife of revolution," mobile warfare with heavy armor, artillery, and airpower, an orderly war theater delineated into "tactical areas of responsibility," speed, and surprise.[49] The final phase involved espionage training at the Higher Intelligence School near Moscow.[50]

Figure 10.2
Ambassador of the Republic of Zimbabwe Pelekezela Mpoko presents his letter of credentials to President Vladimir Putin of Russia, February 3, 2006. Source: http://www.kremlin.ru.

Cuba

In his book *Conflicting Missions*, Piero Gleijeses argues that, whereas the history of "the Cold War" has been written from the perspective of the Global North, there is no reason why it cannot be viewed "from Third World country to Third World country." Cuba was just a small island, yet shaped "the Third World" profoundly. Gleigeses shows that Cuba was no mere puppet or satellite of Moscow. In fact, it was the Cubans that sought a Soviet alliance in 1959, while Moscow found Cuba's location on America's doorstep strategic.[51] Through its direct combat involvement in Africa and Latin America, Cuba solidified South-South solidarity networks that had begun to manifest since Bandung.

Looking at the Cuban Revolution of 1959, African nationalists saw a successful homegrown revolution they could use to inspire their own revolution.[52] Three years later, ZAPU's first recruits arrived at the Minas del Frio training base in Cuba's Sierra Maestra Mountains to commence training. This coincided with the Cuban missile crisis. The subject of the training was not communism but the homegrown ideals of the Cuban Revolution and the tactics that had delivered its success: infiltration, operating in small groups, sabotage, and the training and command of a guerrilla army. Nkomo admitted that "the training [the Cubans] gave our soldiers was better and more realistic than that offered by almost any other country."[53] Including the Soviet Union.

China

For all the thunder about the Soviet Union as a plundering tiger, China was the first nation to train guerrillas to fight the Rhodesian government. In 1955, after nearly five centuries of isolation, China confirmed its return to the international scene when it jointly organized the Bandung Conference with Indonesia. The next year, the government in Beijing recognized Egypt's independence, hoping for Cairo to rally pan-African, pan-Arab, pan-Islamic, and pan-"Third World" support to help it gain recognition as the sole government of China at the expense of Taiwan. As a non-member, Beijing used Africa—and, by extension, solidarity within the Non-Aligned Movement)—as its voice in the United Nations. And, by sponsoring students (including military trainees), China created advocates in Africa. In 1971, the UN General Assembly voted to recognize Beijing as the sole government of China and expelled Taiwan. With that goal accomplished, China cooled its relationship with Africa.[54]

After China's pact with Egypt in 1956, the stream of African students seeking academic, military, and scientific knowledge in China gathered momentum. In 1961 some 118 of them came to Beijing.[55] The purpose of seeking education in China was not merely to acquire tools to dislodge colonial governments but also to govern after they were gone. This training took two forms: civilian and military education. All new arrivals were taken to the Institute of Foreign Languages to learn Chinese as a medium of common instruction.

China's major apparatus for engineering these trainees into ideological weapons was political education, which started at ages 2–3 with kindergarteners being taught how to sing revolutionary songs in praise of Mao and songs deploring American imperialists as "the worst enemies of the Chinese people." Afterwards the children entered the Youth Pioneers and completed their ideological construction in the Communist Youth League. No matter how well one did in one's core academic studies, a continuous streak of poor grades in political education could result in expulsion. "Politics" in Chinese education meant one thing only: "Marxism as expounded, commented on and interpreted by Marx, Engels, Lenin and Mao." Books routinely started with, expanded on, and concluded with praise for the Chinese Communist Party.[56]

Though the "national liberation war" narrative of Zimbabwean history lauds China as a "friend" of ZANU, it is silent about the purely civilian training, which engendered some rather bad feelings from the Chinese public. At least one student from Rhodesia lived through this period of deep Chinese public hostility arising from being "left half-starving on evil-smelling cabbage while the foreigners can eat good food in almost unqualified quantities." Deans received the same pay as their foreign students. African students were instructed to "jump [bus] queues" as one of many gestures designed to make them see how good communism was and report back home to their political parties accordingly. Instead the students ridiculed it and "stole Chinese women."[57] There was also a racist streak: Chinese often gave way to Africans to avoid skin contact or breathing contact, or to avoid being seen in the proximity of a black person.[58]

It is not clear if such prejudices governed the more hierarchical and disciplinarian military camps such as Nanjing Military Academy, where ZANLA's commander Josiah Tongogara was trained.[59] The syllabus made sure that every recruit understood the historical symbolism of the place to be inspired by it. For it was through this very place that in 1949, Mao's Red Army had crossed the Yangtze into southern China to seize power. Tongogara arrived with ten others in 1966, four years after the pioneering group.[60] The training started with two months of ideological indoctrination in the "Chinese Revolution" and its communist ideals. A three-month phase devoted to mass mobilization, military intelligence, political science, mass media, and guerilla strategies and tactics followed. In the last phase, the trainees were taken to another school of military engineering for two months of training in land-mine warfare. The critical difference between Soviet and Chinese training was the emphasis on mass mobilization, guerrilla-oriented tactics, land-mine warfare, and Mao's stature as father of revolutionary warfare.[61]

Engineering Zimbabwe from Africa

The Organization of Africa Unity (OAU) was founded on May 25, 1963, on the precept that no part of Africa was free as long as any particle of its soil was still under colonial (white) rule. Kwame Nkrumah of Ghana, Julius Nyerere of Tanzania), Ahmed Ben Bella of Algeria, and Gamal Abdel Nasser of Egypt) were convinced that military means—not just fiery rhetoric—were necessary to achieve this objective. To these men, Africa was no mere "idea" but a reality choking for air under the heavy weight of colonial rule: it needed liberation. The discourse of Pan-Africanism explained a reality that already existed through the shared experience of living under colonial rule.[62] What was required was a liberational structure to confront what Valentin Mudimbe called a "colonializing structure."[63] Nkrumah initially proposed an African Liberation Army. Idi Amin Dada of Uganda and Olusegun Obasanjo of Nigeria pledged brigades. But only the establishment of the OAU Liberation Committee to coordinate international assistance to anticolonial movements materialized. Though many African countries chipped in with training facilities from 1962 to 1979, the available data permits only a discussion of Tanzania, Mozambique, and Angola. The intention here is to show how Zimbabwe was engineered from outside, by African elites in partnership with the Soviets, the Cubans, and the Chinese, through various forms of support.

Tanzania: The Headquarters of Southern African Liberation

Nyerere's philosophy of Pan-Africanism revolved around his concept of "African Socialism"—pan-Africanism critically using socialism as an instrument to engineer a postcolonial modernity—as outlined in the four main points of his Arusha Declaration of 1967. First, Nyerere anchored Chinese collectivization within the east African philosophy of ujamaa (familyhood)—the belief that the individual existence is subsumed under the communal good—in the hope of achieving an "agriculture-based modernity."[64] Second, he declared a one-party state to steer this policy, borrowing as much from pre-colonial Africa's kingly traditions as from the Chinese Communist Party. Third, in the 1970s, Nyerere entered into a Commodity Agreement under which ujamaa would deliver crop produce to China in return for the construction of "turn-key projects" such as the Tanzania-Zambia railway (TAZARA).[65] ZAPU and ZANU benefited directly from the fourth aspect of the Arusha Declaration: liberating the rest of southern Africa from colonial rule. Tanzania became the headquarters of Southern African liberation not merely by hosting the Liberation Committee (whose head, Brigadier Hashim Mbita, was Tanzanian), but also by acting as the point of arrival and distribution for incoming Sino-Soviet, Western humanitarian, and other support, as well as guerrilla headquarters and bases.

ZAPU graduates from Cuba, the USSR, and Algeria and ZANU graduates from China and Ghana returned to Tanzania to establish training camps and guerrilla armies. Albert Nxele, a graduate of the Intelligence School in Moscow, opened ZAPU's first two training camps, one at Kongwa (1966) and one at Morogoro (1967). The latter became the first OAU-sponsored guerrilla training camp in Africa. The courses took from six to eight months and involved basic infantry training.[66] Meanwhile, Nanjing's graduates established camps at Mgagao and Itumbi with Chinese, Tanzanian, and ZANLA instructors.[67] Nyerere also hosted FRELIMO training at Nachingwea until the guerrilla movement took power in Mozambique. He then turned the base over to ZANU.[68] Nachingwea was "the biggest and best-equipped training base in Africa," covering 16 square miles and including an airfield and barracks for 10,000 trainees.[69] The Chinese, Tanzanian, FRELIMO, and ZANLA instructors blended field experience with Maoist guerrilla warfare.[70]

Mozambique: ZANLA's Eastern Gateway into Rhodesia

FRELIMO and ZANLA had been fellow travelers on the road to freedom since 1970. In 1975, when Mozambique gained political independence, FRELIMO's new president, the Mozambican Samora Machel, granted the "Zimbabweans" freedom to establish training bases, headquarters, refugee camps, and farms from which to feed and fight for their own freedom. Mozambique became ZANLA's gateway into Rhodesia and a new incubator for manufacturing fighters from raw material recruited from the overflowing refugee camps of Chimoio, Nyadzonia, and Doroi. The biggest such training base was Tembwe (established in 1976), which could accommodate 4,000 trainees.[71] Located in the war zone, the camp gave trainees immediate battle experience. As ZANLA turned to its own version of conventional warfare, fighting with small arms (AK-47 rifles), the instructors adjusted their syllabus accordingly. Tembwe's biggest problem was a lack of enough weapons to match the glut of recruits. Trainees made their own "guns"—dummies carved from wood or plain sticks—and used stones as grenades. They got guns only as they deployed to the field.[72]

Angola: ZIPRA's Soviet-Run Training Camp at Luena

Portugal's departure from Angola invited the MPLA (Movimento Popular de Libertação de Angola) to march on Luanda and take power in 1975. One year later, President Agostinho Neto offered ZAPU's armed wing, the Zimbabwe People's Revolutionary Army (ZIPRA), training bases near Luena (formerly Vila de Boma) for the purpose of building up a conventional force capable of seizing power in Rhodesia. The offer also included an alternative route for "the transportation of arms and other supplies." Until that point, ZAPU had experienced problems with the Liberation Committee—and the Tanzanian government—over diversion and occasional resale of guns and other supplies meant for it to Chinese-backed guerrillas.[73]

After the failure of talks in Geneva in 1976, ZIPRA increased its combat personnel from about 70 troops inside the country to about 600. And 1,200 were undergoing training, 1,000 were starting training in Luena, and 3,000 recruits were in transit camps in Zambia and Botswana. That year, Nkomo asked the Soviets to send military instructors to Angola, to provide a transport plane for personnel and equipment airlifts from Angolan camps into Zambia, and to accept 200 men for specialized military training in the USSR. Twenty would be trained as pilots and an unspecified number as artillery gunners; others would learn intelligence work. The Kremlin expedited the request.[74]

ZAPU's objective was to seize power by force. To achieve that, the party would need "a big number of fighters, trained in using small arms and able to act efficiently as combat units." This defined the Soviet instructional mission: to train soldiers and give them guerrilla instruction in case conventional operations suffered temporary setbacks. The first twelve Soviet military instructors arrived in July 1977. Lieutenant-Colonel Vladimir Pekin, the chief instructor, had been deputized by a political commissar, Captain Anatoly Burenko. Pekin's group later stood down for Lieutenant-Colonel Zverev's training team. The Cubans shared barracks with the Soviets and the ZIPRA trainees, but it seems their role was mostly confined to logistics and camp security. The combined mission was to receive and graduate 2,000 recruits every two months. In all, more than 10,000 soldiers and commanders were trained.[75]

Zambia: ZIPRA's Northern Gateway into Rhodesia

Between Zambia and Rhodesia was the crocodile-infested Zambezi River. ZAPU's political headquarters were in Lusaka, capital of Zambia. The troops trained in Angola had to pass through Zambia to get into Rhodesia. ZIPRA ran its own "survival course" at Westlands Farm just outside Lusaka to "grease" the Angola graduates into combat readiness. Reliance on training alone, no matter how good it was, had proved suicidal in the 1967 campaign. One MK guerrilla trained at Westlands described ZIPRA's training as "tough and rough and we were so lean because we went for days without food or food was so little."[76] The training was designed to adapt the guerrillas to the terrain in the operational zone, where the itineraries I have been tracing all led.

Engineering Zimbabwe through War

Through the application of weaponry and trained soldiers, African politicians shifted the nature and venue of combat from talks and roundtables to gunfire and bushes. From 1961 onwards the means and modes of combat changed. Hitherto, the educated and politically active African elites had anointed themselves nationalists and pleaded for Rhodesian recognition as representatives of "every African oppressed by whites." They engineered their status as "the racially discriminated against" into a "Zimbabwean" identity and nation to which the white man came as an invitee, at the mercy of "the natives."[77] How did this happen?

How the Political Elites Became "Nationalists"

When the politics of entreaty turned violent, the instruments changed. Now the politicians—or perhaps their followers—hurled stones, petrol bombs, knobkerries, and spears. They had worn jacket and tie—the emblems of learnedness and "civilization"—to claim that they were modern enough to rule their own people (contrary to Rhodesian Prime Minister Ian Smith's insistence that they were not when declaring UDI, or unilateral declaration of independence from Britain, in 1965). They had talked in the language of negritude and Pan-Africanism, their South African and overseas education making them giddy in the registers of Marx, Engels, and Lenin. No one had heard them. So they climbed down from this lofty perch to speak in local idioms, to define their "nationalist" status through a return to the ancestral traditions they had shunned for the clever trickery of Western intellectualism.

These engineers of a Zimbabwean identity discarded the garb of Western "modernity" and, using intellectual skills acquired in "the white man's academy," carefully selected from their ancestral histories moments that best symbolized and inspired their own mission. From architecture they

A Plundering Tiger with Its Deadly Cubs? 247

got the name of the country they wished to create: Great Zimbabwe, the stonewalled center of Shona power before 1500. Refusing to be called "Rhodesians" (which meant "white settler descendents of Cecil John Rhodes"), they began calling themselves "Zimbabweans" and "Africans." Ancestral clothing—hats made out of wild felines (leopards, cheetahs, and genets)—such as only rulers and spirit mediums had worn before colonization now became the official garb of the leaders at public rallies and meetings. These were not mere cosmetics but the transformation of history into weaponry—what Terence Ranger (who as a nationalist wore his own feline skin hat) called "a usable past."[78]

As pioneering Marxist revisionists who dismantled Eurocentric histories that had stripped Africans of agency, members of the black elite anointed themselves new spokesmen of "their" people at the expense of the chiefs who had "sold their souls for sugar."[79] Understanding the divisive dangers of ethnicity, they turned their Shona and Ndebele ancestors—who had fought each other before colonization—into one unified pillar of anti-colonial struggle. Then they carefully selected the battles their ancestors had won decisively or lost gallantly, and the right heroes and heroines. Then they installed themselves heirs to a rich tradition of resistance.[80]

Adopting the title vana vevhu (sons of the soil) in the fight against vasvetasimba (power-suckers or whites), they inverted the negativities the Rhodesians had inserted into "native as primitive" into "native as rightful rulers of the land." And they went further: their ancestral religion, which missionaries and the state had dismissed as primitive unchanging tradition, now became the head cornerstone in these self-appointed "nationalists'" reinvention of themselves into miracle workers performing masaramusi (indecipherable feats).[81] Nkomo not only adopted the title of the powerful eighteenth-century Shona ruler Mambo (King); he also cast himself as a mystic whose voice could be heard but whose form could not be seen. People made pilgrimages to get party cards from him at Gonakudzingwa, where he was restricted (banished to) in 1964. He said these pieces of paper would automatically imbue them with a magical power to resist the weapons of the state, especially guns.[82]

Through trickery, these "nationalists" changed even the definition of combat. "African nationalism" was cast not as a new struggle but as a continuation of the chimurenga (uprising against colonial rule) the ancestors had "valiantly" lost in 1896–97. Ranger was in their midst, putting his historian's skills to work, turning fictions of continuation between spiritually inspired first and second zvimurenga (singular: chimurenga) into sacred truths. They narrowed the reasons why their ancestors had lost power to

Rhodes down to inferior weaponry. To them, the first chimurenga was the real beginning of "Zimbabwean nationalism."[83]

Using Appropriate Technology to Carve Zimbabwe out of Rhodesia

The vision of Zimbabwe could only become a reality through the erasure of the reality of Rhodesia. Since talks and street protests had gone as far as they could, there was only one appropriate technology to carve Zimbabwe into a reality: the barrel of the gun. Yet the "nationalists" soon discovered that guns and training did not make guerrillas. Nor did the bestowal of a Zimbabwean sovereignty upon their bodies necessarily free them from their quotidian Rhodesian existence. Soon Nkomo, ZANU president Sithole and his secretary-general Mugabe, and other leaders were imprisoned; the rest fled into the Zambian and Tanzanian diaspora to plot an invasion. The first operatives paid the price of failure to appropriate (subvert) the weaponries acquired from abroad into local realities. The political leaders were not militarily trained, and their political strategy was not translatable to military strategy or tactics. Some of the trained commanders were "bloody cowards" who stayed behind in Lusaka enjoying nightlife and sex while sending their men to "commit suicide" before Rhodesian troops.[84] The operatives were under the illusion that "immediately the gun was introduced into the country the masses would rise and join the army" since they "felt oppressed."[85] It didn't happen. The first groups went about armed to the teeth among villagers who had no clue what they were fighting for, and who saw them as a danger. They called the police.[86]

ZAPU did not fare any better. In 1967–1969 it undertook joint operations with the MK, hoping to escort the latter to South Africa's Limpopo border and to return with more recruits.[87] Once the groups had crossed the Zambezi into national parks country, they lay low, going into the villages now and then to mobilize the masses. They scared the civilians, who ran to the police. The casualties were knee-breaking. In Lusaka the hard questions began. Many troops criticized large-scale "invasions" in favor of infiltration, recruitment, and mass mobilization.[88] Damaging ethnic clashes wrecked the top hierarchy, then soon spread to the ranks.[89] Key Shona-speaking commanders defected to ZANLA, carrying with them Soviet-acquired techniques and inside knowledge of ZAPU strategy.[90]

ZAPU now convened a strategy planning conference at which a Revolutionary Council was formed to organize war and political strategy, to source arms and training from communist countries, and to coordinate mass mobilization. The ZAPU "armed wing" was re-branded into a full-fledged army, ZIPRA. Its immediate mission was to infiltrate small and inconspicuous groups, to "move as fast as possible towards villages," and, once there, to "change into civilian clothes and start attempting to recruit and train that population." They would "carry out sabotage without full engagement [involving] a continuous cycle of retreating, planting landmines and hiding."[91]

In ZANLA, the setbacks also triggered a strategic and political review. Tongogara became Director of Operations in charge of all military responsibilities. He operated under a newly established High Command that represented logistics and planning matters to the politicians, while the Dare reChimurenga (Revolutionary Council) concentrated on political affairs.[92] But this clear separation of powers between soldiers and civilians did not translate into immediate success. Recruits were in acute supply, so ZANLA resorted to kidnapping, targeting refugees in Zambia.[93] In addition, the Zambezi was a difficult barrier: guerrilla operations were most effective in summer, but at that time the river was flooded and impassable. The tactics were wrong: the guerrillas "would fight until they exhausted the last bullet and then run."[94]

Any success depended on addressing three factors: the Zambezi, ammunition shortages, and tactics. To solve the first, it was imperative to operate from Mozambique, following, not crossing, all the major rivers. FRELIMO agreed to let ZANLA use its rear bases in 1970. Second, for the next two years, Tongogara sent in small groups (from three to five men) to create subversion cells, cache arms, and read the terrain. They were to avoid military contact, and to go via dependable people who could not readily "sell out." Third, the materiality of the AK-47—their major weapon—was to be adapted to local conditions. That rifle's lightness and portability made it well suited to stealthy infiltration and high mobility. ZANLA's operations in the rainy season (November–April) were specifically designed to take advantage of the AK-47's resistance to dampness, dust, and heat, whereas the Rhodesian FN rifle was susceptible to jamming.[95] Success transformed the AK-47 into the signature of the struggle for freedom.

Even with these remedies, the most important factor was mass mobilization. Here the guerrillas deployed ancestral religion to give potency to skills acquired from training and guns sourced abroad. In reverse, local custodians of ancestral spirituality deployed their immortal powers to use the guerrilla and his AK-47 as instruments to challenge their local colonial tormentor. Chiefs and headmen could not be trusted; ZANLA deemed most of them "puppets" and "sellouts" working for the state. Besides local party officials, ancestral spirit mediums—adult men and women through whom the spirits communicated with the people—were the most ideal contact persons because they had a vested interest in the "nationalist" project. It would restore the ancestors' traditional role of guiding the mortals that Christianity and colonization had diminished, one reason why spirit mediums Nehanda and Kaguvi had orchestrated the 1896–97 chimurenga and paid with their lives.[96]

Nehanda reportedly had declared, before the colonial settlers executed her, "My bones will rise again." Right on time, the guerrillas presented themselves as Nehanda's bones arisen from the dead. In the first chimurenga, spirit mediums had performed rituals before warriors set off to battle to render them impregnable to enemy bullets and sharpen their spears and muskets. Now the guerrillas followed suit: their bodies were cleansed and their AK-47 rifles blessed.[97] And, in Shona spirituality, the forest is the domain of the spirits and masaramusi; it was there that vakomana ("the boys," as the guerrillas were affectionately called) operated, the immortals watching over and guiding them.[98]

By submitting themselves to spirits revered by entire clans and villages, guerrillas secured local mortals to their cause, thereby enabling the AK-47 to be effective as a weapon. What was an AK-47 worth in the hands of a ZANLA person commanding no popular support? The mediums commanded the youths to obey Nehanda's risen bones. ZANLA created in each operational area dependable corps of youth militias called mujibha (male) and chimbwido (female) for gathering intelligence, food, and mobilizing "the masses." Through pungwe (all-night meetings),[99] the masses told the guerrillas government's molestations—taxation, land dispossession, cattle de-stocking, and cruel, racist local farmers and state laws. They had anxieties: "How can we defeat the whites who are well-armed and well-equipped? *Do you have arms?*"[100] With Maoist analysis earned through training, the guerrillas gave convincing and thoughtful answers; punching the air with the AK-47, they chanted "This is your new voice! Pamberi nehondo! (Forward with the Revolution!)" And yet the same voice also silenced—permanently—those who "sold out" to the state: "Pasi nevatengesi! (Down with sellouts!)"[101] The pungwe became an alternative public sphere to that defined for Africans by the state; the guerrillas were behaving like a state.

The mujibhas and chimbwidos provided labor for transporting arms and ammunition that the guerrillas then secretly cached in the mountains and forests to sustain operations for long periods. Upon seeing the "guns, the machine guns, the bazookas, the mines," the masses "danced with excitement and joy." Some were ecstatic to just feel the weapons, others actually wanted to keep them so that they could go and 'sort out' their local white nemeses personally.[102]

The Work of Engineering Zimbabwe out of Rhodesia

With arms cached and masses mobilized, ZANLA waited for the rains to green the landscape into good camouflage. Then, on December 21, 1972, it re-launched its war from northeastern Rhodesia. The guerrillas carefully reconnoitered the area, plotted escape routes and targets of strategic value—military posts, police camps, homes of white police and army reservists, and farms used as command posts. Their tactics involved surprise night attacks, ambushes, and the laying of land mines. The attacks were "so quick and swift that we disappeared before the . . . enemy troops got a chance to fire back." The masses were mystified: "Stories spread around the villages that freedom fighters turn into logs or snakes or bush at the approach of settler [Rhodesian] forces."[103] After all, had their guns and bodies not received ancestral spiritual blessing? The workings of the ancestral spirits and the material capabilities of the AK-47 conflated into the mysterious ways of the guerrilla.[104]

ZANLA's success forced Smith to use peace talks as a weapon to buy time for rebuilding his army, deliberately stretching negotiations to a point where his army was ready to fight again before scuttling them.[105] The strategy worked so well that in 1974 ZAPU and ZANU ordered a complete cease-fire. The chimurenga died. It was left to Machel to organize and make available facilities for a unified ZANLA-ZIPRA army called the Zimbabwe People's Army (ZIPA) in late 1975. ZIPA was doomed from the start. In particular, ZIPRA command elements looked down upon their Chinese-trained ZANLA colleagues as half-baked, their strategy weak.[106] The cleavages between ZAPU and ZANU, between Ndebele and Shona, and between Soviet- and Chinese-trained and armed were too strong. When the political leaders were released from prison in 1974 and joined the guerrillas in exile, ZIPA disintegrated.

In 1976, ZAPU drafted a working document titled Our Path to Liberation, laying out a broad strategy "to conquer state power" through conventional warfare. It recommended the political unification of ZAPU and ZANU into a Patriotic Front (PF) while the armies fought the same enemy separately using conventional means.[107] The exhaustive discussion resulted in the Turning Point Strategy (TPS), which outlined steps toward mobile warfare. First, ZIPRA would create rear bases inside Rhodesia by capacitating guerrilla units to hold ground and to merge into infantry platoons and companies for larger-scale operations. Once field commanders had secured "liberated zones," half the High Command would move into Zimbabwe (meaning areas wrested from Rhodesian control) to direct the war and set up civil administration. ZIPRA veterans deny that there was any Soviet pressure or role in TPS, insisting that it was a logical outcome of the Our Path to Liberation debate. Only when it came to application did ZIPRA request assistance.[108]

TPS was supposed to culminate in a final military push to seize power called Zero Hour—a five-battalion multi-front attack on Rhodesia. The piercing force would consist of three infantry companies, with two more logistics and artillery support companies in reserve. Their mission was to seize bridgeheads—fortifications at bridges within firing range of the strategic garrison towns of Kanyemba, Chirundu, and Kariba—to facilitate resupply, deployment of the main body of armor (tanks and armored cars), and the maneuvering of artillery batteries into base-plate (firing) positions. The armor and artillery would macerate the Kariba, Victoria Falls, and Wankie airfields and any Rhodesian military aircraft in their hangars. Once these airfields were secure, ZIPRA would land MiG-21 jets from Angola to achieve aerial superiority. Using air power, artillery, armor, and infantry, the conventional brigade would then advance toward the Rhodesian capital, Salisbury. ZIPRA now remodeled the earlier strategy of civil disobedience to Zero Hour: its Training Department moved in and cached 50,000 AK-47 and SKS rifles in strategic bush hideouts to hastily arm citizens.[109]

ZIPRA's plans hinged on securing pilots to fly the assault aircraft. The twenty recruits mentioned earlier were already training at the Air Force Centre in Frunze, capital of Kirgizia. Meanwhile, Nkomo resolved to use pilots from "friendly countries" other than the USSR as a stopgap measure. Moscow and Havana were skeptical. The former Soviet operative Vladimir Shubin refutes assertions that the MiGs "arrived in Zambia" but were not "uncrated," that the Soviets withheld the ZIPRA pilots, and that Nkomo refused to replace them with foreign ones.[110] The fact is that ZIPRA pilots completed their training after Zimbabwe was already independent. In December 1979, Smith had agreed to settle with ZANU and ZAPU. Zero Hour was never executed.

Meanwhile, ZANLA was taking its own path to conventional war. The mother parties—ZAPU and ZANU—had agreed on a tentative Patriotic Front, but the two armies pursued a "separate armies, one enemy" policy as a military strategy to encircle the cities from the rural areas, Vietcong-style. ZANLA would storm the eastern garrison city of Umtali. Once it was secured, the political leaders would be moved in to declare an independent republic of Zimbabwe, which would be extendable to the rest of the country through conventional warfare.[111] A boisterous Mugabe declared 1978 Gore reGukurahundi (The Year of the People's Storm). The term is deeply anchored in Shona traditions related to the first torrential rains in October or November, which come after a long, dry winter in which deciduous trees shed leaves, riverbeds dry up, and pools become still and filled with algae. The torrents cleanse the land, washing away the dirt into the floodwaters, leaving the lands clean, green, and full of better life.[112]

Gukurahundi represents ZANLA's final sweep of the dirt (Smith) through the thunder of gunfire and the heavy droplets of bullets. Like TPS, the storm derived its power from merging small guerrilla units into companies and battalions (50–500 each, depending on the strength of the target) capable of washing away the enemy with firepower. Unlike ZIPRA, however, ZANLA did not have aircraft, armor, and heavy artillery, so the storm would come from large "cumulonimbus clouds" of guerrillas armed primarily with AK-47 rifles raining bullets onto their enemy.[113]

This twin strategy was the final nail in Rhodesia's coffin. By 1979, Smith was facing an untenable situation. ZANLA and ZIPRA were picking strategic targets—national fuel depots, railroads, city centers, farms, mines—at will, and were choking life out of the economy. A mass exodus of whites was in motion despite stiff conscription laws barring white men from leaving. The BBC journalist Richard Lindley summed it well on April 20, 1978:

> The whites have now allowed Africans to discover that power can, indeed, grow out of the barrel of a gun. . . . The lesson to the leaders and the men of the Patriotic Front is plain: if you want to keep up the pressure for more radical political change, for a real transfer of power to the black majority, then keep hold of your guns.[114]

There is no better summary of the notion of appropriate technology. Yet whereas the guerrillas amplified the demand for black rule through the barrel of the gun, in 1976 ZANLA found the perfect weapon to amplify the gunfire itself into demands for independence. That rhetorical weapon was Robert Mugabe.

Mugabe: The Itinerary of a Political Engineer

Engineering Power: Rhetoric Amplifier of the Gun Barrel

In a bold declaration in July 1977, ZANU leader Robert Mugabe rejected without ambiguity any illusions the Soviets and the Chinese might have harbored about turning them into mere cubs as Kaunda had feared. Mugabe emphasized that "Zimbabweans" would fight their own war if the communist countries played their part as quartermasters:

> True the socialist countries have shown a greater preparedness to assist the process towards decolonization of Southern Africa. They have given help through the OAU

Liberation Committee and we are grateful for that. They have done so because we have asked for this help, but we reject completely that those who give us help should turn themselves into our masters.... We believe that the war must be fought by us. It is our war, the struggle is our struggle.[115]

Robert Mugabe epitomizes the spirit of this essay: a story not of Sino-Soviet "puppets" but of political engineers whose senses of initiative had not suddenly awakened when "the Cold War" began, but whose trajectory had predated and then outlived such North-North rivalries.

At first sight, his birth at Kutama Mission to an aspiring Catholic nun named Bona Shonhiwa on February 21, 1924 and a life of poverty fits snugly into the narrative of victimhood—until one discovers that the young Robert inverted this same poverty into energy that inspired him to become perhaps Africa's most educated politician,[116] a man who used the colonial education system to acquire a learned vocabulary that would, one day, become rhetorical weaponry (word-turned-weaponry). The education system had been designed to make blacks tea-boys, clerks, and nannies; instead, Mugabe used it as a passport to the University of Fort Hare (in South Africa), a lectureship at Chalimbana Teacher Training College (in Zambia), and a teaching post in Tekoradi (in Ghana), where he imbibed

Figure 10.3
Robert Mugabe, a political engineer. Source: Wikipedia.

lectures from Kwame Nkrumah. When Mugabe returned to Rhodesia, in May 1960, his rise within the nationalist movement was rapid.

At an NDP rally at Stoddart Hall, Mugabe was suddenly catapulted onto the podium to share his Ghanaian experience of independence. The party appointed him secretary for information and publicity the following year. When the NDP was banned, the black politicians simply changed its name to ZAPU. In 1962, ZAPU was banned and its leaders restricted to their rural homes. Mugabe was banished to Kutama. State repression forced the leaders to relocate to Tanganyika in April 1963 to fight from exile. Host president Julius Nyerere told them to leave and fight from within. Shona-speaking elements blamed this bad decision on Nkomo's "inept" Ndebele leadership and formed ZANU in August. Mugabe was not there when ZANU was formed but was elected secretary-general in absentia as he attended the birth of his son, Nhamodzenyika, in Dar-es-Salaam.[117]

Upon his return from Tanganyika, Mugabe was promptly arrested and sent to prison for the next 11 years for plotting terrorism. Books now became weapons for negotiating prison life.[118] Designed by the state to break his spirit, prison became for Mugabe a school, the cell a classroom where he studied for three undergraduate degrees with British universities. He became headmaster over fellow political inmates. Recognizing his intellect, fellow prisoners Edgar Tekere, Enos Nkala, Moton Malianga, and Maurice Nyagumbo would oust Sithole and nominate Mugabe to lead ZANU.[119]

From 1975 on, Mugabe deployed "the book" to amplify the message of the gun barrel. After being released for talks in Zambia in late 1974, he fled with Tekere to Mozambique to avoid re-arrest. He focused on amplifying the lethality of gunfire through the gunfire of words; he was now the voice of the AK-47 and the bazooka. By 1976, the military men in ZANLA badly needed a civilian orator to articulate what they were fighting for at the British-organized Geneva Peace Conference. Mugabe rose to the occasion. On August 31, 1977, he was proclaimed party president. Mugabe's tough rhetoric on liberation through the barrel of the gun outgunned the promise of a negotiated "Internal Settlement," involving Smith and the moderate clerics Ndabaningi Sithole and Abel Muzorewa, that was signed on March 3, 1978. To amplify Mugabe's rhetoric, ZANLA blew up Salisbury's fuel depot into a five-day inferno in December as the guerrillas circled menacingly on the urban areas. Only the intervention of Samora Machel forced Mugabe to negotiate with Smith at Lancaster House in 1979. If he did not sit down and talk, ZANLA would be unwelcome in Mozambique.

Having been dragooned to settle, Mugabe now focused on the impending elections. Nkomo proposed a joint ticket with him as candidate for the premiership; ZANLA chief Tongogara agreed. Mugabe said "No, it must be me" and ordered Tongogara off to the barracks to address the troops. About 100 miles north of the Mozambican capital, Tongogara was "involved in a fatal car crush." In January 1980, Mugabe returned home from five years of exile. At Highfield Grounds he preached peace and reconciliation to a crowd of 200,000. Yet ZANU (PF) fought the election with the sort of violence befitting guerrilla war. A triumphant Prime Minister Robert Mugabe moved swiftly to assure a jittery white public: "The wrongs of the past must be forgiven and forgotten."[120]

Mugabe deployed racial reconciliation as a curtain he drew to mesmerize the West into believing he was a post-racial statesman, meanwhile going after blacks who opposed his vision of a one-party state. Not only did he let Ian Smith live in Zimbabwe unscathed, he also made sure that Smith kept his farm, as a monument to his generosity and fairness.[121] He let whites keep their land not because the Lancaster House peace agreement legally prevented him from forcefully seizing the land (he could simply have used force and worried about costs afterwards), but because the threat to his personal hold on power from black opponents in a black-majority country was far more urgent. Keeping the land question on ice narrowed the circle of any likely conspirators against him as Mugabe went after ZAPU.

Engineering Survival: The Gun (Bullet) Is Mightier Than the Pen (Ballot)
First, Mugabe neutralized the Soviets to deny ZAPU any source of weaponry. The Soviets had "backed the losing horse," China the winning one. That meant there were many pressure points Mugabe could apply. Pressure Point I: to deny Moscow an embassy until it had severed "all contact with the PF-ZAPU." After further talks, Moscow met its conditions and received an invitation to post an ambassador—in February 1981! Pressure Point II: to neutralize ZIPRA's Soviet-made arsenal. Both ZIPRA and ZANLA, fearing that assembly points were a trap and if they completely disarmed their whole fighting forces might be slaughtered by the Rhodesians, kept some of their battle-hardened elite forces in the "rear." In 1982 Mugabe's intelligence suddenly "discovered" arms caches at ex-ZAPU assembly points.[122]

While addressing a rally in the eastern city of Marondera in February that year, Mugabe declared: "ZAPU and its leader, Dr. Joshua Nkomo, are like a cobra in a house. The only way to deal effectively with a snake is to strike and destroy its head."[123] The Western countries were lauding Mugabe for "reconciliation." Economic aid and honorary awards kept coming. The Soviets had been reduced to spectators glad to finally get an embassy in Harare. The bulk of the Shona (Mugabe is Shona) warmed up to, urged

Mugabe on, or stood by as he and his lieutenants used incendiary rhetoric justifying the crackdown against ZAPU (a Ndebele party). Mugabe could unleash Gukurahundi, the North Korean-trained Fifth Brigade, on the cobra and its neonates in peace. Nkomo became a hunted animal, escaping to Botswana disguised as an old woman and then flying to exile in Britain. An estimated 20,000 overwhelmingly Ndebele people remotely suspected of supporting him were slaughtered. A grieving Nkomo chose to let ZANU swallow ZAPU to serve his followers. This Unity Accord was signed in 1987.[124]

With the liquidation of ZAPU, a pattern emerged in Zimbabwe: small parties mushroomed a few weeks before an election, only to vanish soon afterwards. There was only token opposition to Mugabe until 1997, when Secretary-General Morgan Tsvangirai of the Zimbabwe Congress of Trade Unions (ZCTU) led a broad-based coalition that culminated, in 1999, in the Movement for Democratic Change (MDC). In 2000, Mugabe blamed rising dissent (including rejection of his proposed constitution giving him a ten-year term in office) on whites, and sanctioned invasion of their farms. Land had effectively become an incisive political weapon. Mugabe successfully profiled the MDC—which had substantial white membership and sponsors—as a British puppet sent to prevent land redistribution.

We saw earlier how "African political elites" transformed themselves into "nationalists" through the connection of the 1896–97 Shona-Ndebele risings and the 1960s–1970s resistance into First and Second Chimurenga respectively. ZANU (PF) went further: the ancestors of 1896–97 had started the revolution and handed it over to their grandchildren (the "nationalists"), but Lancaster House prevented them from completing it. Hence the necessity for yet a Third Chimurenga, whose mission was crystal-clear: to invade and occupy white farms and secure the land. While Africa celebrated the seizure of land, Mugabe was busy using land as a divide-and-rule weapon, just as he had used reconciliation and ethnicity to hoodwink the West and the Shona as he exterminated ZAPU. The West cried foul when a few white farmers were brutally murdered, even as black opposition activists were being burned alive, tortured, raped, and murdered with little or no whimper of protest from the North. This racial bias played right into the hands of Africa, which stoically supported Mugabe as a pan-Africanist who was a victim of Western neocolonial interference, with the MDC as a front. Africa's Robert Mugabe was a great African liberator giving his people land stolen from them. Zimbabweans' Robert Mugabe was an old despot who had stayed in power for too long and caused their suffering.

Conclusion

Mugabe is an intersection of many narratives of inversion. With his anti-Western rhetoric and his practice of seizing land from whites, he has for a long time mollified and endeared himself to Africa's heads of state and government as a pan-Africanist. Traveling around Africa and overseas, and talking to ordinary Africans, I have discovered how Mugabe's land seizure and acerbic onslaught against the West has become a cause célèbre. To some, he is the best president they never had. The West has become Mugabe's chisel for carving his internal opponents into a front for the British and the American re-colonization of Zimbabwe, specifically during the administrations of Tony Blair and George W. Bush. This has reduced the West's public pronouncements and support for democratic change in Zimbabwe into a massive liability to the struggle for better governance, for example, when these countries have admitted to funding peaceful and contemplating violent "regime change."[125]

Mugabe has also tooled (and fooled) the West in much more audacious, if odious, ways. Consider, for example, what happened after September 11, 2001, when the US Congress passed the USA PATRIOT Act and Bush implored countries throughout the world to join him in "a coalition of the willing" against terrorism. One of the signs of willingness would be to pass a raft of anti-terrorism legislations similar to the PATRIOT Act. Very well, said the Zimbabwean President. Zimbabwe's amendments to the Public Order and Security Act (Chapter 11: 17) on January 22, 2002 used wording similar to that of the PATRIOT Act, but Mugabe's terrorists were not Al-Qaeda but the MDC and other internal enemies.[126] The irony? These pro-democratic forces were the same people whose plight the US Congress had sought to alleviate through "regime change" when passing the Zimbabwe Democracy and Economic Recovery Act (ZIDERA).[127] Here is a case where two US foreign policy weapons turned their muzzles upon each other and pulled the trigger.

If this essay dismisses the notion of "the Cold War"—let alone "the global Cold War"—as not belonging to the South, it is because such extensions of Northern time into universal time have continued to pose existential questions in instances like Zimbabwe. Such labels of time, in the hands of people like Mugabe, are important as sources of raw material to design instruments for their own survival. The idea around the Bush doctrine of "axis of evil" and "coalition of the willing" was in every sense a continuing Washington tradition of defining the world as a geography of "them" and "us." Mahmood Mamdani has already made a compelling case when

A Plundering Tiger with Its Deadly Cubs?

tracing this long genealogy,[128] which rests on a belief that the North's priorities define what matters to the rest of the world. When actors in the South agree with and follow propositions of and sometimes impositions from the North, the ready assumption is that they are "responding" or are "puppets." Little space is left for analyzing the calculations, agencies, and priorities of such "puppets."

Take Mugabe's relationship with the Chinese, for example. After the West declared him a pariah, the Zimbabwean president retreated into history to search for verbs and nouns to profile the West as far worse than "fair-weather" friends. Was it not the West that had hemorrhaged Africa through slavery, colonialism, debt, and structural adjustment programs whose workability has not been proved anywhere, including in the West? When Africans asked for guns to fight colonial rule, had the West—Britain and America specifically—not refused and branded them "terrorists"? Here Mugabe massaged and pampered China with praises as a friend of all seasons. Was it not China that had stood side by side with "blacks" as they fought the "racist white regime of Ian Smith"? And then the coup de grace: "We have turned east where the sun rises, and given our backs to the West where the sun sets."[129]

If I question the Cold War as a category, it is because China's footprint in Africa is already being represented from Beijing as a story of "how China lures Africa." The short history of how Mugabe engineered Rhodesia into Zimbabwe in the 1970s suggests a far more nuanced calculus that goes into decisions to give space to outsiders. Knowing he cannot get a penny from the West after taking land from white descendents of Europe, Mugabe has tried to use China as a weapon against criticism from that quarter in exactly the same way as ZAPU and ZANU did in the 1960s and the 1970s when the West rebuffed them in the fight against Smith. Moreover, China's policy is that it does not interfere in the affairs of independent states, especially those it helped free from "Western imperialism." That means it can deal with rulers who oppress their own citizens: the Chinese respect national sovereignty, even selling guns to regimes on which the West has imposed military sanctions.[130]

Few Zimbabweans take the priorities of China in Zimbabwe seriously. What they know is that China is one of the weapons Mugabe uses to stay in power. When Mugabe introduced anti-terrorism legislation, Zimbabweans dreaded what would happen next: summary arrests, torture, rape, and murder of Mugabe's opponents on trumped-up charges of plotting terrorism with the British and Americans against his government. Suddenly China's non-interference and sales of arms to Zimbabwe and America's "war on terror" found themselves side by side as Mugabe's weapons against the people of Zimbabwe. Similarly, it is these local experiences that defined the priorities and imperatives of the time that might be called "the Cold War" in the North.

Certainly, these hegemonic imperatives can only look "Cold War-ish" from specific Norths, but that is only one way of seeing—and experiencing. It is possible that people can share the same experience, but they may feel it very differently, and represent it according to their own designs, priorities, and expectations. If two people attended a musical show and one was disappointed but the other enjoyed it, neither can claim that his experience of this event was a universal one. It is even worse where one part of the world was fighting against colonization by the North, while the North was locked in rivalry over nukes. It amounts to the trivialization of Southern time and the struggles, initiatives, and triumphs invested in describing the period 1951–1994 as the era of African liberation from colonial rule. The countries that were fighting their "cold" war—of words, artifacts, and troop deployments—became weapons in the hands of the designers of this African independence. If they see themselves as designers of a program of thwarting their Northern rivals and using the South as puppets, that does not mean they were the only ones capable of doing it. Africans were also busy designing them as puppets and weaponry in the liberation of the continent from colonialism. The question for Africa is what to make of this moment when Africa used the North for its own purposes, now that Robert Mugabe and other political engineers are turning such weaponry against the people they said they were liberating in the 1960s and the 1970s.

Notes

1. Gaddis 1999; Walker 1997.

2. For a rebuttal of this tendency and a discussion of Cuba's independent African and Latin American forays, see Gleijeses 2002.

3. Of late, a number of books have emerged that rush to discuss China's current role without a detailed historical role in the decolonization of Africa. Examples: Eisenman, Heginbotham, and Mitchell 2007; Rotberg 2008; Alden, Large, and de Oliveira 2008.

4. Eriksen 2000; Morgenstienne 2003; Sellstrøm 2002.

5. Shubin 2008: 3.

6. Orwell 1945.

7. This framing draws from a recent unsettling of similar universalistic synchronizations of world time by Ferguson (2006).

8. Exactly what it means to live in the postcolony or not is a matter of conjecture. See Mbembe 2001.

9. Westad 2007: 3.

10. "Zambian Paper's Challenge to USSR and Cuba," *Times of Zambia*, February 13, 1976; Moorcraft 1990: 124–131.

11. Although the original idea was inspired by Mamdani's immaculate framing of the agency of the "puppet" to turn against its creator or handler, and for victimization to transform victims into killers, the more I read two of his works on the subject the more I saw also that the victimizing action (using) is simultaneously acting as an ingredient or even conveyor belt for the victim to reclaim their own agency. See Mamdani 2001 and 2004.

12. I am trying to get away from the conventional arguments in Science and Technology Studies (STS) on designers and users, summarized well in Oudshoorn and Pinch 2003.

13. Diouf 2000: 680, citing Appadurai 1996: 4. The notion of "trickery" is inspired by Bayart 2000: 217.

14. Diouf 2000: 682, 684.

15. Star 1989: 37–54.

16. Tungamirai 1995.

17. Ranger 1995.

18. For this view of "indigenous knowledge" as a resource for a native modernity, see Geschiere 1997: 22.

19. Beach 1993; Mudenge 1986; Bhila 1982.

20. Beach 1971: 143–144.

21. On the domestication of guns into hunting cultures, see Mavhunga 2003: 201–231.

22. Sithole 1968. These political elites had gone abroad to get tertiary education because Rhodesia had no university until 1957.

23. Legvold 1970; Mbembe 2000: 259–284; Mbembe and Nuttall 2004.

24. Wright 1956. For the cinematic origins of the notion of "Iron Curtain," see Wright 2007.

25. Bell 1971: 43; Morrison 1964: 102–193.

26. Shubin 2008: 152–153.

27. Nkomo 1984: 102–103.

28. Lessing 1962: 60.

29. *Keesings Contemporary Archives* (June 6–13, 1964): 20110.

30. Shay and Vermaak 1971: 8; *Rhodesian Herald*, August 10, 1968.

31. National Archives of Zimbabwe/MS589, Interview with Joshua Nkomo in the Second Half of 1977 by Australian Film Crew in Lusaka: 7; Nkomo 1984: 165.

32. *Times* (London), July 24, 1978.

33. *Intelligence Digest*, May 16, 1979: 3.

34. Moorcraft 1985: 117.

35. Brickhill 1995: 61; Ellert 1989: 52, 64.

36. Shubin 2008: 177.

37. Tongogara 1978a: 29.

38. *Star* (Johannesburg), February 18, 1978), citing *Tribune de Geneve*, February 10.

39. Interview with Josiah Tungamirai, Zanu PF Headquarters, Harare, May 24, 1999; Tongogara, 1978a: 30; National Archives of Zimbabwe/MS536/11/4 Terrorist Weapons; *Zimbabwe News* 10, 6 (December 1978): 20.

40. For a framing that helped me think toward this critique, see Nash 1979.

41. Some of ZAPU's guerrillas also received military and civilian technical training in—and were trained by instructors from—other Warsaw Pact countries besides the Soviet Union. ZANU sent its recruits to other countries overseas besides China—Romania, Bulgaria, Yugoslavia, and North Korea.

42. Shubin 2008: 156.

43. Shubin 2008: 154.

44. Hevi 1963: 136–137.

45. Shubin 2008: 155.

46. Greig 1977: 165.

47. Bopela and Luthuli 2005: 38–44.

48. Shubin 2008: 174.

49. Scott and Scott 1979.

50. Barron 1974: 57.

51. Gleijeses 2002: 7, 9, 18.

52. Valenta 1975: 23; Hodges and Shanab 1972: 169–170.

53. Nkomo 1984: 177; Pinto 1973: 74.

54. Johnson 2006: 29–30, 34.

55. Johnson 2006: 40.

56. Hevi 1963: 28–29, 35–39.

57. Hevi 1963: 133.

58. Johnson 2006: 77.

59. A Wikipedia entry on Emmerson Mnangagwa mentions his having undergone training in 1964 at Nanjing alongside Felix Santana, Robert Garachani, Lloyd Gundu, Phebion Shonhiwa, and John Chigaba, but it is not referenced.

60. Mubako 2007.

61. Maxey 1975: 80; Greig 1977: 170; Hutchison 1975: 247; Fiennes 1975: 31–35.

62. For a contrary view, which seeks to explain that whites are also Africans just like blacks and that the reduction of "Africa" to "blackness" is a rhetorical and poetical myth dreamed up by the founding fathers of pan-Africanism, see Mbembe 2002; Appiah 1992. There are high emotions and stakes in this debate.

63. Mudimbe 1988: 4.

64. Ibhawoh and Dibua 2003: 59–83.

65. Robinson and Shambaugh 1994: 287.

66. Government of Zimbabwe 1989: 41.

67. Tungamirai 1995: 40; Shay and Vermaak 1971: 27–30; National Archives of Zimbabwe MS 536/7/5 Press Release, Rhodesian Ministry of Information, August 26, 1968; *Zimbabwe News* 10, 2 (May-June 1978): 16–17.

68. Avirgan 1975.

69. *Keesings Contemporary Archives* (October 20–26, 1975): 27402.

70. Evans 1981: 21–25.

71. Government of Zimbabwe 1999: 2; Government of Zimbabwe 1989: 51–52.

72. "Interview with Tungamirai."

73. Shubin 2008: 167.

74. Shubin 2008: 171.

75. Shubin 2008: 172–173.

76. Truth and Reconciliation Commission: Human Rights Violations: Submissions—Questions and Answers, Day 4, Olefile Samuel Mnqibisa, Soweto, July 25 1996, available at http://www.doj.gov.za.

77. Shay and Vermaak 1971: 10–14.

78. Ranger 1976.

79. Lonsdale 1968: 119–146.

80. Ranger 1967.

81. The term "ancestral religion" refers both to the religion of the forefathers and to the specific belief that when an adult person died and proper rituals were performed, the dead person's soul did not die but became a spirit that returned to protect the living. See Shoko 2006.

82. Wright 1972: 358–394.

83. Ranger 1967; *Rhodesian Herald*, October 22–November 17 and December [?], 1961).

84. Bopela and Luthuli 2005: 54–55.

85. Tongogara 1978a: 20; Venter 1974: 132–133.

86. Tongogara 1978b.

87. Bopela and Luthuli 2005: 59–87.

88. *Rhodesian Herald*, January and February 1970; Shay and Vermaak 1971: 74; Wall 1975: 34.

89. Dabengwa 1995: 31.

90. Brickhill 1995: 48.

91. Ibid.; Dabengwa 1995: 32; *Keesings Contemporary Archives* (January 8–14, 1973): 25668; National Archives of Zimbabwe IDAF MS 590/3, Rhodesian Liberation Movements July–December 1973: columns 450–560: "ZAPU War Communiqué," *Zimbabwe Review*, August 28, 1973.

92. Tongogara 1978a: 29.

93. Tungamirai 1995: 40–41.

94. Tongogara 1978a: 29.

95. "Interview with Tungamirai."

96. Ranger 1967, 1999.

97. Conversations with village elders in Mudzingwa and Machangara village, Chihota Communal Lands, May 23, 2001; Lan 1985; Shoko 2006.

98. For a good discussion of this practice, see Daneel 1970.

99. For a video clip of the Rhodesian Security Forces reconnaissance on a pungwe, see "Rhodesia— Internal Operations," http://www.youtube.com/watch?v=tpHVvYAtJj8&mode=related&search=.

100. Tongogara 1978a: 29.

101. Nyarota 2006: chapters 1 and 2; Kriger 1992.

102. Tongogara 1978a: 29.

103. Ibid.

104. Daneel 1970.

105. Moore 1995.

106. Dabengwa 1995: 33.

107. Brickhill 1995: 50.

108. Dabengwa 1995: 33.

109. Brickhill 1995: 62, 64–5.

110. Shubin 2008: 176–7.

111. *Times* (London), April 20, 1978.

112. The Gukurahundi of 1978 is not to be confused with the post-independence Zanu government-orchestrated army genocide in which 20,000 Ndebele people were killed. See Catholic Commission for Justice and Peace and Legal Resources Foundation 1999.

113. Martin 1978.

114. "The Boys in the Bush," *The Listener* (UK) April 20, 1978.

115. "Interview with Robert Mugabe," *New African Development*, July 1977.

116. Holland 2008: 3, 7.

117. Sithole 2006: 32.

118. Norman 2004: 51; Holland 2008: 27. For an academic study of prisons in Rhodesia, see Munochiveyi 2008.

119. Tekere 2007: 68–69.

120. Norman 2004: 72–75.

121. Norman 2004: 75.

122. Shubin 2008: 185–188.

123. "Mugabe Makes Bitter Attack on Nkomo," *The Guardian* (UK) February 15, 1982.

124. Shubin 2008: 189.

125. Peta 2004.

126. Government of Zimbabwe. 2002. Public Order and Security Act (Chapter 11: 17). *Government Gazette* (January 22).

127. US Congress, Zimbabwe Democracy and Economic Recovery Act of 2001.

128. Mamdani 2004.

129. Karumbidza 2006.

130. Beresford 2008.

11 Cleaning Up the Cold War: Global Humanitarianism and the Infrastructure of Crisis Response

Peter Redfield

Sven Lindqvist's book *A History of Bombing* grimly details the brutal fantasies and colonial violence accompanying the advent of aerial warfare. It also incorporates counterpoint themes: growing concern for civilians and fitful claims to common humanity.[1] Sudden attack from the skies produced dramatic destruction, after all, creating a new theater for suffering. Once rendered visible, and matched with the proper structure of sentiment, scenes of broken bodies and panicked refugees could inspire moral response. Between the founding of the United Nations in 1945 and the breakup of the Soviet Union in 1991, the planet saw no shortage of dramatic human tragedy, particularly in the geopolitical margins where most of the Cold War's actual conflicts took place. Innovations in communications technology allowed the international transmission of images in close to real time by the middle 1960s. Among other things, the technical fact of satellite broadcast extended the potential range, speed, and intensity of war journalism. Whereas Guernica had been the exception to a rule of silence about civilian agony, especially in colonial settings, television audiences could now peer through screens to witness suffering at a distance.[2] In the metropolitan centers of former empires, the anguish of physically remote populations now flickered—still intermittently but more graphically—into public view. The contemporary drama known as "humanitarian crisis" emerged as a variation on the theme of charity, often amid the very colonized peoples Lindqvist's more extreme sources once hoped to eradicate.

After World War II, a globally oriented humanitarian infrastructure gradually came into being. Alongside "sharper" instruments (such as the AK-47) that refocused power through the threat of death amid proxy wars, a set of tools emerged to respond to human suffering, together with new international agencies and organizations to wield them.[3] Sharing a longer lineage with military and industrial logistics, as well as common circuitry of international mobility and expertise, this humanitarian apparatus represents a form of technical mutation and reconfiguration as much as innovation. By the final decade of the Cold War, it had coalesced enough to offer stable objects and routines designed for emergency settings. Although hardly a remedy for the upheaval and inhumanity accompanying geopolitical contestation, humanitarian equipment offered a means of temporary relief in particular locales, and thus literal, as well as figurative, sanitization.

In this essay I examine the emergence of this humanitarian infrastructure by focusing on the development of one significant component: mobile medical supplies. I do so primarily through the example of the nongovernmental organization Médecins Sans Frontières (Doctors Without Borders). The case of MSF is particularly germane to this endeavor for three reasons. First, although emergency response is no longer the group's only form of action, it remains its trademark of technical expertise. Second, MSF's ideology of outspoken independence means that it has often played the role of self-appointed humanitarian critic, commenting on humanitarian shortcomings while also seeking to provide aid. Thus MSF can serve as a barometer of sorts for the general state of the field. Third, MSF emerged during a geopolitical period that featured proxy wars between superpowers and significant crises along the borders of the state socialist bloc. The group's focus on civilian suffering and truth telling thereby exemplifies an internationalism of moral engagement amid political disillusionment, in marked contrast to earlier internationalisms committed to utopian politics.

Here I sketch the outlines of MSF's techniques for rapid medical intervention in adverse circumstances, including pre-assembled, standardized "kits" of equipment, guidelines for responding to varieties of public health crises, and minimal forms of health evaluation. Tracking the general story of the kit through a series of examples drawn from specific settings, I illustrate the manner in which the kit emerged amid a broader standardization and professionalization of humanitarian action. The product of modest inventors, and often engagingly simple in design, this technical assemblage nonetheless sustains the expansive ambition of reaching and stabilizing a population almost anywhere in the world within 48 hours. Although geographically mobile it is a temporally restricted set of instruments. Defined for a state of emergency, these tools remain limited by a concentrated concern for an uncertain present rather than an expansive future. In this sense they are, by their very design, not "sustainable," no matter what greater hopes they might absorb or how lengthy their actual use life might prove.

Considered from the perspective of technopolitics, the humanitarian response to suffering suggests a form of strategic engagement practiced in the name of ethics, contested, manipulated, and incorporated as the other side of warfare.⁴ Humanitarian equipment thus represents a distinct variant of global technology, separated from those born of corporate growth or state ambition by its oblique relation to political economy. Directed neither at profit nor at power, these artifacts instead reflect a moralistic ethos and limited goals associated with welfare and basic public health. This is not to suggest that they lie beyond all market logic or ordinary political life, let alone that they only produce intended effects. Rather, I simply wish to echo other critical observers who note that nongovernmental organizations pursue good works primarily in the name of moral values, largely disavowing conventional political ends.⁵ By delineating and enacting a minimal politics of survival, I suggest, MSF's logistical efficiency reveals inherent tensions within any humanitarian project that offers only temporary relief from political and social failure. The organization's relative "success" in achieving a degree of instrumental standardization ultimately highlights the limits of its mobile biomedical project, even as it frustrates its aversion to institutional forms.

The appearance and standardization of mobile tools for aid also serves to highlight the manner in which the Cold War ran hot along its edges. It recalls the extent to which shifting military technology and political strategies produced civilian suffering, laying the groundwork for morally inflected practical action. The recent efflorescence of aid organizations, the prominence of humanitarian justifications for military action, and the mediated morality of emergencies all refer to an established precedent of expectation: by the last decade of the twentieth century, it had become technically possible to both identify distant suffering and mount a rapid response. However limited, contingent, and fragile, the material means for humanitarian action thus carries significant symbolic weight. The fact of its normalization, I suggest, is one of the lasting legacies of the Cold War era. Long overshadowed by nuclear weaponry and the new military-industrial prowess in death, this smaller constellation of artifacts associated with life subsequently emerged to play another defining moral role in geopolitical discourse.

A Prototype: *Materia Medica Minimalis*

Before proceeding to a description and analysis of my specific case, I will first introduce the general problem through an orienting moment of its prehistory. To understand the nature of the equipment MSF eventually set in place, as well as its spatial significance and temporal politics, it is helpful to return to World War II, and the advent of large-scale aerial warfare and the landscape of mass destruction it produced. Humanitarian logistics has many obvious lines of descent, from military supply lines to industrial food distribution, or, amid colonial encounters, the naval surgeon's chest.[6] But just as with bombing itself, the importance of a portable medical infrastructure became critically visible amid the rubble of European cities during the 1940s. An essential antecedent for this Cold War story thus appears in the fading centers of empire, newly pulverized by bombers.

The degree of devastation in World War II presented the newly formed Joint Relief Commission of the International Red Cross with a significant technical problem. Created to coordinate the efforts of the Red Cross's mosaic of national societies with those of the Swiss-based International Commission of the Red Cross, the commission found itself at a loss in the face of massive aerial bombardments that left civilian populations in urban centers medically bereft:

> "There is a total lack of medical supplies here." It was by a summary appeal of this kind that the Joint Relief Commission of the International Red Cross was asked in the beginning of its activities to send medical relief to a capital which had just undergone an air raid. Such a request, put so tersely, left us somewhat nonplussed. What should be sent? What medicaments would be required by a large city which had been devastated by an air attack? What quantity of each medicament would be required? No statistics were there to enlighten us, no document on the problem was available. We had to improvise.[7]

How best to provision a landscape of total devastation? Most urgently, what medications to provide when the entire health infrastructure was knocked out? The commission first surveyed the national Red Cross societies about the medical requirements of their respective countries. The response was, however, "surprisingly diverse, one might almost say, disconcerting." No simple, uniform agreement could be found. Therefore the commission took it upon itself to quickly marshal medical experience and science, in an effort to determine what was "absolutely indispensable to ensure medical care and to meet the emergency needs of a population which has been deprived of food and medical supplies."[8] Newly sensitized to local culture, the commission also took note of the fact that national preferences and therapeutics varied across the European continent. The Red Cross, then, needed a document that would be simultaneously encompassing and precise, allowing for regional differences and yet conducive to medical and pharmacological accuracy.

The condensed result was a booklet titled *Materia Medica Minimalis* (abbreviated *M.M.M.*). First issued in Latin, it subsequently appeared in

French, German, and English editions. For a European embroiled in war, the inherited tongue of Rome served as a convenient means of scientific expression.[9] Balancing this scholarly touch with quartermaster's eye for practical detail, the authors offered estimated quantities necessary to treat a "population unit" of 100,000 persons for six months. They based their estimates on the consumption of medicaments in Switzerland, recognizing that these figures might prove controversial and might require alterations. In view of the urgent need for immediate use, however, they ventured into the messy realm of calculation. Since "circumstances and difficulties" might affect actual delivery, they further divided their listing into two categories, the first of which should receive the greatest priority. The *M.M.M.* itself included only the pharmaceutical end of medical supplies; bandages, cotton, and surgical instruments were to be handled in separate consignments.[10] Nonetheless, its content lived up to its name, defining a baseline state of medical infrastructure.

Materia Medica Minimalis marks a catalytic moment in humanitarian thinking. While the Red Cross's international meetings had addressed a variety of training activities related to medical techniques in the past, it was now constructing a mobile template for crisis response around a principle of flexible standardization. The final report of the commission composed after the war mused that "this work, which was called into existence by the needs of the moment, possessed a usefulness which it seemed would outlast the war period," an assessment that would prove ultimately prophetic.[11] Although *M.M.M.* itself did not directly dictate later relief work, its conceptual descendents would proliferate in the coming decades.[12] As the zone of crisis recognition shifted beyond Europe, the reconstruction of a minimal biomedical infrastructure emerged as a central problem for all manner of disasters in resource-poor settings. Effective medical assistance required basic equipment and guidelines, preferably prepared in advance. Global mobility depended on it.

MSF and the Geopolitics of Suffering

Although the Holocaust shadows contemporary conceptions of human suffering and disaster, cameras never framed Auschwitz as a "humanitarian crisis."[13] The close of World War II may have ushered in a new political configuration, and categories and institutions for its governance, but the era's massive relief works occurred in a time of less instantaneous and less visceral communications. Rather, the conventional watershed moment for televised suffering arrived with the Biafran war in Nigeria at the end of the

1960s, when satellite broadcast first brought starvation into middle-class living rooms worldwide. Whatever its complexities (which were rife with propaganda and manipulation as well as starving children), that moment of anguish provoked reflection and reorganization on the part of a number of existing humanitarian organizations and inspired the formation of others.[14] In the subsequent decade, a generation of aid workers increasingly embraced media exposure for their causes.

By the time of Biafra, the term "humanitarianism" had narrowed, primarily identifying an impulse to alleviate suffering caused by human conflicts and natural disasters.[15] The Red Cross movement's historic efforts to transform military medical practice and international law had rarely addressed colonial settings. Colonial warfare operated by different rules, as Sven Lindqvist makes acutely clear, and suffering beyond Europe was a largely a matter for missionaries and the civil service.[16] The slow dismantling of European empires, however, created a new humanitarian terrain for the second half of the twentieth century. As the United Nations expanded fitfully to a patchwork institutional framework for international governance, it enlarged expectations, if not always results. The development of emergency medicine out of military medicine (institutionalized in the French case in the 1960s with the establishment of a national service known as the SAMU, which stood for *service d'aide médicale urgente*), and the routinization of air travel and rapid transport, extended the scope of potential action.[17] Suffering, whether near or far, could now elicit a prospect of response.

Médecins Sans Frontières itself appeared at the end of 1971 in Paris, when journalists from a medical publication helped bring a small group of doctors who had volunteered for the French Red Cross in Biafra together with others concerned about disaster and conflict in Bangladesh. Troubled by the constraints of established relief organizations, they sought to establish an independent alternative to the Red Cross, unfettered by existing mandates. The veterans of Biafra, impassioned by their experience of what they considered genocide, also yearned to be more outspoken— particularly a telegenic young doctor named Bernard Kouchner. Publicity soon played an increasing role in MSF's operations, if not quite to the extent of later myth. At the outset MSF existed largely on paper, piggybacking on the interventions of others. By the end of the 1970s, however, it had mounted a number of short interventions into areas afflicted by natural disasters and war. It also achieved rapid prominence within France as a result of a *pro bono* publicity campaign by an advertising company that featured the slogan "2 billion people in their waiting room."[18] The ambition of aid was now grandly global.

In keeping with its anti-establishment generational ethos, MSF soon presented itself as an alternative to both anti-colonial and Cold War loyalties. Rejecting all justifications for civilian suffering, it came to oppose the French intellectual romance of "Third Worldism" and denounced leftist regimes that proved inhumane. Amid the disillusionment of post-1968 France and the now apparent excesses of state socialism, MSF offered the prospect of ethical action to defend the life and well-being of ordinary people. A "rebellious" form of humanitarianism would be simultaneously non-aligned *and* thoroughly engaged through the practice of medicine. Thus Bernard Kouchner, a one-time student activist, could, as an outspoken physician noisily practicing "without illusion," discover both himself and the Third World. Rony Brauman, Kouchner's influential successor in the organization, could similarly trade street protests for clinical work in refugee camps.[19] Although the group would experience loud internal squabbles and schisms, the various factions all shared a common ethical aversion to political justifications for human suffering.

Even as MSF found its collective calling, the geopolitics of the Cold War shifted increasingly in the direction of proxy wars after the United States left Vietnam.[20] Alongside the paralyzing shadow of nuclear apocalypse, irregular armies fought savagely in Angola, Mozambique, and Afghanistan during the 1970s and the 1980s. These confrontations only enhanced the international flow of conventional weaponry, fueling ancillary conflicts and alliances, while prompting the displacement of civilian populations. The number of refugees grew exponentially during those decades, providing a surplus of humanitarian need well beyond the capacity of UN agencies. A nongovernmental group with a global vision thus had plenty of opportunity to offer medical assistance.

Three early episodes proved particularly formative for MSF and for the technopolitical concerns of its operations. First, the exodus of "boat people" from Vietnam, combined with mass suffering in Cambodia under the Khmer Rouge, set the ground for an ethics of action that prioritized humanity over political ideology. As the longtime political opponents Raymond Aron and Jean-Paul Sartre marched together in Paris, young MSF volunteers worked in camps on the border of Thailand, becoming radicalized through encounters with suffering rather than revolution. The refugee crisis would emerge as a natural habitat for medical humanitarianism: massed populations in need of urgent, basic clinical care provided a ready stage for moral dramas of suffering. The "boat people" episode also provoked a schism within the organization when a younger contingent, disagreeing what would constitute the best response, overthrew the veterans of Biafra.[21]

In 1980, after this power struggle, Kouchner left MSF to found Médecins du Monde (Doctors of the World); later he would emerge as a significant French politician, championing a humanitarian "right to interfere" on the part of state and interstate actors. Kouchner's successors at MSF would struggle—not always successfully—to distance themselves from this full-throttle version of *sans frontiérisme*, in part by resisting a complete embrace of the human-rights discourse and in part by emphasizing independence and operational efficacy.[22]

Next, the Soviet invasion of Afghanistan only further confirmed the French humanitarian repudiation of state socialism. MSF undertook clandestine missions in the Afghan mountains, experiencing its own romance of Third World solidarity alongside the mujahideen, not to mention the CIA and the future al-Qaeda. Together with Kouchner's new group, they acquired the label "French Doctors" in American press reports of the era, an English phrase that remained iconic in France long after it was forgotten elsewhere. The Afghanistan period constituted a high-water mark in MSF's break with Red Cross discretion, as well as its engagement in Cold War politics. The Afghan adventure also involved a striking degree of local adjustment; in contrast to later practice, mission volunteers immersed themselves among the people with whom they worked, living and traveling in the same manner. Transport involved the torturous navigation of mountain passes with pack animals, medicines carefully packed into small parcels. Teams were isolated and re-supply difficult. In this case solidarity did not ultimately translate into political influence; the organization watched with dismay as its erstwhile partners baldly pursued their respective power games, and the country disintegrated into civil war after the Soviet troops withdrew.[23]

Finally, during the mid-1980s famine in Ethiopia, the original French branch of MSF found itself evicted after denouncing the Derg regime's policy of forced resettlement. The episode, which resonated amid the televised glamour of the Live Aid concert, established the group's reputation as outspoken and willing to oppose all political orders that produced suffering. At the same time, MSF faced criticism suggesting that it was an amateurish organization, long on hot air but short on actual capacity. The charges stung enough that MSF redoubled its efforts to improve its technical abilities and professionalism.[24] By the end of the 1980s, the group had both a new logistics system and an epidemiological subsidiary in place. Once the French section had grudgingly accepted its newer European relatives as equal partners, the larger collective emerged as a truly multinational operation.[25] Increasingly it would be known not only for outspokenness but also for speed and efficiency.

Discovering Preparedness: Uganda, 1980–1986

What would effective humanitarian action entail at a material level? To illustrate the technical problems involved, I will focus briefly on another case from the early 1980s: that of Uganda. Uganda was never a central front in the Cold War; its post-independence turmoil had deeper colonial and regional roots.[26] Nonetheless, the country's crisis occurred at a transitional moment, and offers the comparative advantage of combining a less mythic profile with widely recognized inefficiencies.

At the beginning of the 1980s, the Karamoja and West Nile regions of Uganda experienced extreme famine. The crisis in Karamoja, an arid area bordering Kenya and peopled largely by semi-nomadic, photogenic cattle herders with a fierce reputation, received a good deal of media attention, and a number of aid agencies responded to the images of starvation by rushing teams and materials into the field. Amid the greater aftermath of the fall of Idi Amin, the general situation in Uganda was, in the words of a UNICEF official of the time, "at best chaotic," and the relief operation quickly encountered a host of problems. Subsequent analysis by a group of scholars and humanitarian workers identified a long list of specific setbacks as well as some general issues: lack of coordination and turf struggles between different organizations (and even branches of the same organization), a greater landscape of need extending beyond the targeted recipients of aid, and a "disaster within the disaster" of food supply and the greater infrastructure of logistics required in for its movement and distribution.[27] A former representative of another UN agency observed that many of the people who had worked alongside her in Uganda had participated in major relief operations elsewhere over the previous decade, and that their discussions identified a repeated pattern of failure: "One of the recurring themes was that time and time again the same problem arose in every disaster situation: logistics."[28] She imagined creating, within the UN system, a "strike force" of reservists—a cadre of experienced professionals, with access to stockpiles of equipment, who would be ready to leave at a moment's notice. The UNICEF official similarly concluded that responses should be "quick, rational and experienced" rather than "prolonged, irrational and nonexperienced," but doubted that his own agency, created for long-term activities, would be suitable for the task: "To use a metaphor, such a rapid shift in activities and allocation would amount to demanding a shipping company to turn into an airline overnight."[29]

Among the many organizations briefly present in both the Karamoja and West Nile crises was Médecins Sans Frontières. Not yet ten years old, it was still a relatively minor, if flamboyant, entity in the world of humanitarian affairs. The missions to Uganda were its first in a famine zone, and they were not particularly successful. As a leading participant dryly noted in an interview with me years later, "in that era we improvised; later we'd become more efficient." The group's bulletin report at the time summed up the general situation with the graphic image of a stranded, bullet-ridden bulldozer, its brand new tires stolen by raiders to make sandals.[30] Within ten years, however, MSF had grown into a large and complex organization, fully capable of both technical innovation and logistical efficiency in crisis settings. The professional system of logistics that it had developed enabled it to embody the UN administrator's vision of a global humanitarian strike force.

The Humanitarian Kit

When MSF reoriented its logistics system in the mid 1980s, it focused on creating modular, standardized kits. The concept of the kit itself has a long military and medical lineage. The Oxford English Dictionary suggests that by the late eighteenth century the meaning of the English term had expanded from a wooden vessel or container to the collection of articles in a soldier's bag. An equipment case or chest had long been the steady companion of naval surgeons and other mobile healers, and by the early twentieth century the Red Cross and other groups were assembling all-purpose first-aid kits, combining essential materials in a more modest version of the *M.M.M.* assemblage described above.

MSF's variant would be more comprehensive, recombinant and ambitious: collections of supplies designed for a particular need and preassembled into a matrix of packages. The organization could then stockpile these packages and ship an appropriate set rapidly to any emergency destination in the world. The mature MSF catalog of kits later summarized the conceptual approach it embodied as follows: "A kit contains the whole of the needed equipment for filling a given function. Intended for emergency contexts, it is ready to be delivered within a very short time frame."[31] Thus the diffuse problem of acquisition was effectively translated into a concentrated one of transportation, more easily solved from a central office. Essential materials no longer had to be hastily assembled anew in response to every crisis, or uncertainly negotiated on site amid fluctuating availability, quality, and prices. Moreover, by preassembling materials with a checklist, the kit could function as a form of materialized memory whereby previous experience extends directly into every new setting without having to be actively recalled. For an organization built around both crisis settings and a constantly shifting workforce of volunteers and temporary employees, such continuity would prove especially valuable.[32]

From the perspective of MSF, the kit system was the product of a small number of early masterminds, now receding into organizational legend. Its immediate origin lay in the experience of Jacques Pinel, a French pharmacist posted to Cambodian refugee camps on the border of Thailand in 1980. Previously, MSF team members—a small appendage to a larger, often chaotic operation—had rotated responsibility for logistics; now that they were running their own independent mission, they created a new category of supervisor [*intendant*] positions.[33] Guerilla raids over the border led to periodic Vietnamese bombing runs, whereupon the Thai army would seal the camps, preventing access for several days at a time. In due course the MSF teams learned to assemble essential equipment until they had the process down to a system. As Pinel recalled in 2004, this evolved less from any grand design than from the "banal" practice of packing a bag for a series of weekend trips, and then translating such experience into anticipatory habit:

> The kit, it's nothing more than someone who's leaving for the weekend . . . who needs his backpack with something to drink, something to eat, something to put on his feet if they get sore. He needs all that. So, how does he do it? The first time he imagines what will happen, and assembles his bag with that imagination. And then after that first experience, he sees that there are things that didn't amount to much and others he was missing. And then after the second, third time, he'll finally have a perfect bundle and he prepares it before the weekend, checks it, and then leaves and it works.[34]

In its initial form, the proto-kit was a relatively heavy box made by local carpenters. Carried in the back of a pick-up truck, it quickly earned a nickname of the "semi-mobile endowment [dotation semi-mobile]."[35] MSF's general approach mirrored that of the French SAMU emergency system, which transported medical materials and expertise directly to the site of care. Its originality derived less from its form or content than from the setting in which it was deployed and the extent of its adaptation. Reworking a guide from the UN High Commission for Refugees and a nutritional package from Oxfam, MSF volunteers eventually assembled standard lists of medicine and supplies for a kit to meet the needs of 10,000 people for three months, along with instructional manuals for its use. The project of procedural simplification on site grew into one of standardization between locales. As Pinel later noted in published work, the kit now responded to an "ecology" of humanitarian emergency as exemplified by the refugee camp, where a large number of people lived under temporary, crowded, and often precarious conditions, receiving care from a shifting group of personnel not all of whom might be familiar with the setting or the diseases.[36]

Pinel went on to coordinate MSF-France's new central logistics operation in 1982, and, together with associates, applied the model developed in Thailand to analogous problems elsewhere. The key principle behind the kit approach was to break down a larger predicament (e.g., a flood in Central America), identifying its critical components and developing smaller, specific responses to each in turn. For example, a common health concern for displaced people living in crowded conditions is cholera. Anticipating this problem step by step in detail, the MSF logistics team developed a general kit for the disease:

> We know that we were going to have a cholera epidemic there. OK, we get together people who have already worked on cholera, when we get there there's nothing of what we need to put in place for a cholera epidemic. So, we need a cholera camp, that is to say an isolation tent. . . . If there are thousands of people that's too many, so we'll create a unit to treat 500 patients. . . . What will be necessary? Some tents, OK, how many tents? OK, we'll need a hundred 50 square meter tents. We'll have perfusions because we're going to give infusions and on average there are those who have 2-3 liters and then there are those who have up to 20 liters. So we'll say 10 liters on average, OK. Out of 500 patients there are how many who will receive 10 liters. . . . OK, there will be a hundred. . . . When we finish planning, *voilà*, we have the kit. We try to really make this kit, in order to see how it is, how it fits into boxes, how much it weighs. We physically create this kit, and then we use it in the next cholera epidemic . . . and then an evaluation. And then we revise it. . . . It's like that that the kits advanced, succeeded, not so much because of the notion of the kit, which is really something supremely banal [*archi-banal*], but following many years where we imagined the kits and evaluated them in numerous situations. And then we divided the operations up like sausages [*saucissonné*], we cut, we sliced. That is to say, there's a cholera epidemic, a measles epidemic, put in place in a dispensary of a refugee camp. In doing all that, all the units like that, then when it's necessary to mount an operation we have all our equipment.[37]

Through this combination of organic practice and assembly-line routine, MSF created a more global, component variation of the Red Cross's *M.M.M.* By the latter part of the 1980s, the concept of the kit had grown central to the group's emergency work.

To get a sense of the level of detail involved, let us briefly examine MSF's mature cholera Kit 001, designed for refugee camps and capable of being modified for either rural or urban displaced populations. Designed to provide 625 treatments, it weighs in at just over 6,000 kilograms and includes an array of medicine (e.g., 6,500 sachets of oral rehydration salts and 10,000 tablets of the broad spectrum antibiotic doxycycline) as well as materials for taking patient samples (e.g., dissecting forceps and a permanent black marker) and performing basic medical procedures (e.g., surgical gloves, tunics, trousers, and boots in several sizes, ten 500-gram rolls of cotton wool, 25 arm splints, and catheters and bandages galore). But the kit also features support items—for example, more than 100 buckets and 100 disposable razors, not to mention such logistical articles as notebooks, pens, wire ties, and staplers. The degree of anticipation evident in this collection of trunks and boxes would put most Boy Scouts to shame.[38]

In addition, MSF has accompanied the kits with short, informative instruction books and pamphlets detailing responses to practical problems. These guides are available in international languages common where the group works: English, French, Spanish, and sometimes Russian or Arabic. The subject matter addresses clinical and engineering dilemmas volunteers might encounter in the field, such as how best to conduct minor surgery in a war zone or how to set up a simple water sanitation system. The guideline system acknowledges that even volunteers with established general expertise may possess inadequate technical background for unfamiliar conditions; neither a nurse from Lille nor a logistician from Toronto, for example, is likely to have much training in combating cholera or in building a pit latrine. The guides further seek to address audiences with different levels of knowledge, and to do so in a concentrated way that does not require access to larger libraries. Simple to copy or replace, in a pre-digital era their minimal form guarded against the potential disintegration facing more expensive books in humid climates.[39]

MSF also developed a wider supply chain and communication infrastructure. Upon taking over logistics for the group's operations in 1982, Jacques Pinel became aware that Thailand had been a relatively simple environment in which to operate. There the team had easily acquired vehicles and drugs locally, and the phone system had functioned reliably. Much of Africa proved a different story, as its crisis zones generally lacked phone service, transport, and even basic drugs. Pinel and his associates created a radio communications network, standardized drug lists, and a vehicle pool. They restricted vehicle purchases to Toyota Land Cruisers, already deployed in many Red Cross and UN missions, both to simplify their parts list and to use the garages of these other organizations.[40] Simplification and standardization were their watchwords. MSF-France also established a logistics depot in 1986 in order to provide a standing reserve of equipment for emergency operations. One of several "satellite" organizations created during this period, the depot moved around several locations in France before settling near the Bordeaux airport. From this base the logistics team could—political conditions permitting—quickly launch mission material toward any corner of the world with the assistance of a chartered plane.

While the MSF's different sections have pursued slightly different logistics strategies, the kit system expanded throughout the overall organization to produce a set of relatively stable forms. Kits are now available for all manner of eventualities. The Toyota Land Cruiser, still MSF's workhorse vehicle, comes as a kit (modified for either warm or cold climates); so too does a collection of stickers and flags to mark its affiliation. Members of a mission can order an "Emergency Library Kit" and request items from a field library list that includes such assorted titles as "How to Look After a Refrigerator," "Human Rights in a Nutshell," and "Blood Transfusion in Remote Areas."[41] Governing the overall design are principles of quality, efficiency, and simplicity of maintenance. In some domains a spirit of standardization dictates a particular brand of product (for example, MSF still orders only Toyota vehicles); in others a desire for flexibility of procurement allows substitution of any generic equivalent (most articles are "open" rather than brand specific).

The kit system has never operated in a vacuum. MSF also boasts a long tradition of improvisation and of modifying designs to fit its needs. In an office setting this primarily implies working to simplify systems and reduce their cost.[42] At local mission sites logisticians serve as more general "bricoleurs," tinkering with the means at hand to achieve a desired result. A capacity for improvisation remains essential, though less pronounced in an era of improved communications, professional training, and ubiquitous guidelines. "Either you do the job or you don't do the job," a logistics coordinator told me emphatically in 2004, referring to a temporary structure he had once had to construct with logs and mud in the absence of recommended materials.[43] Once in the field, kits can be pulled apart, partially used, and reordered. But even when emptied and enjoying an afterlife—such as the bench I once sat on in northern Uganda—their battered modules serve as a material reminder of a larger, mobile network.

Partly derived from the artifacts of other organizations (if not a master plan), MSF's logistics system in turn influenced the larger humanitarian enterprise. In 1988 the World Health Organization endorsed MSF's classic kit by adopting it as the "new emergency health kit," and the International Committee of the Red Cross (ICRC), where several former MSF figures migrated, purchased many of its corresponding guidelines.[44] Unlike the WHO's more scientific guides, MSF's were emphatically operational in outlook and directed to teams in the field. Early editions invited users to copy them, and consequently borrowed elements circulated widely, some even reappearing in the WHO's growing repertoire.[45] However unorthodox, MSF's journey to professional respectability was now complete.

Technopolitics of Emergency

The first point I wish to stress analytically is that MSF's kit system represents a self-consciously *global* system, mobile and adaptable to "limited resource environments" worldwide. Though parts of it may be flexible in application, the result is not at all fluid in the sense of flowing around community involvement.[46] Indeed, the kit system is the exact opposite of local knowledge in the traditional sense of geographic and cultural specificity in place. Rather, it represents a mobile, transitional variety of limited intervention, modifying and partially reconstructing a local environment around specific artifacts and a set script. Though in practice it may require considerable negotiation to enact (in keeping with actor network theory), its very concept strives to streamline that potential negotiation through provisions that reconstitute a minimal operating environment. The cold chain system used in vaccine distribution serves as a useful general analog in this regard. Just as a cold chain extends the essential environment of a vaccine alongside the vaccine itself with different forms of refrigeration, the kit system extends the essential environment for biomedicine into the landscape of a disaster. To ensure reliability and quality, MSF is willing to ship almost anything anywhere during an emergency.

Deeply invested in a practical logic of standards, the kit system reflects something of Bruno Latour's analysis of circulating inscriptions as "immutable mobiles."[47] MSF's guidelines and toolkits collect and distill local clinical knowledge into a portable map of frontline medicine. Developed and refined through practice, they connect one outbreak or crisis to another. In this sense the cholera epidemic in Thailand travels to stabilize the cholera epidemic in the Congo. Together, as a vast chain, the kit assemblage standardizes disaster through responding to it worldwide. Such a characterization reveals the degree to which biomedical knowledge and practice depends on infrastructure, and the background work necessary to translate it into a new setting. MSF's classic emergency formation generated a "culture of standardization" (as one logistician proudly put it to me) in which speed and control were paramount values. Beyond obvious incompatibilities, local concerns could emerge later.

Second, I wish to emphasize that the kit system is not the product of either corporate or state need, economic goals, or defined political strategy. Rather, it stems from a humanitarian focus on the moral imperative of responding to immediate human suffering. To be sure, the greater logic of standardization has a long history in both military and business settings. Moreover, MSF's tool chests draw from commercial commodities, and the group's administration maintains plenty of balance sheets. However, the central motivation for its decisions derives from valuing human life rather than profit.[48] And although MSF may often find itself in a position of temporary governance relative to a population in crisis, that governance remains ever partial and impermanent as it refuses the responsibility of rule. Thus any analysis should never lose sight of the fact that the kits were designed to respond to emergency settings in which the instrumental goal is temporary stabilization. Standardization was never an end to itself, nor was it part of an effort to reshape or capture economic terrain.

The defining role of crisis has grown all the more clear as MSF has extended its activities beyond emergency interventions into an array of other projects: targeting specific diseases over a longer term, advocating policy positions, and even facilitating pharmaceutical research and production. In these contexts, the logic of the kit no longer holds sway. Instead MSF missions purchase a greater variety of materials from local sources and place orders for items in bulk rather than in prepackaged assemblies. Even the kits themselves have experienced alterations, with outsourcing and flexibility playing an increased role in their production.[49] Once beyond emergency settings, MSF missions re-enter a larger world of exchange and circulation, and standardization melts away.

To illustrate this last point, let us return again to Uganda in the post-Cold War period. Two decades after the initial forays there, several sections of MSF ran a variety of programs in the country. Among these were a garage to maintain and repair vehicles and for a project providing antiretroviral medications to an increasing number of AIDS patients.[50] Located in Kampala, the garage was the domain of a veteran French logistician, a taciturn but dedicated man who nursed it as a longer-term venture amid MSF's many short-lived interventions. In addition to servicing the vehicles of MSF-France and MSF-Switzerland based in Uganda, it also cared for some in Sudan and some in the Democratic Republic of Congo, volatile settings where parts were not available. To further augment its bottom line, it undertook contract work for other NGOs. Well equipped with standards, catalogs, and a computerized ordering system connecting it to MSF's depot, the garage exemplified stabilized humanitarian infrastructure. At the same time, however, its continued existence was under continual threat, not only from the turnover rate of MSF's fluid administration and their varying visions but also from the pressures of competing interests on the part of the local mechanics who labored there. Once trained, they would often leave for a better-paying position, and even when on the job they did not always work with the fervor the director expected. As the director noted wryly, they were, after all, driven less by humanitarian ideals than by a search for their livelihoods. The garage also faced potential competition from commercial rivals that threatened to undercut it, and the impatience of field personnel in project sites who wanted to circumvent central control and make purchases directly. "It's a constant battle," the director acknowledged, especially since some parts could be found more cheaply in local markets, and their quality was improving. Though he was a firm believer in the value of the kit system and in the advantages of using standard, well-selected materials, the director emphasized that MSF's logistics network was really designed for emergency missions. A stable entity such as the garage regularly interacted with the local economy, each small transaction pulling it away from the institutional orbit.

Similarly, efforts to address specific diseases and broader health inequities altered MSF's technical circulatory system, exposing its limits. The project in the northern town of Arua was part of an ambitious, worldwide foray into HIV/AIDS medicine. After years of resisting extensive involvement with the disease, the organization threw itself into the movement to demonstrate the feasibility of treating poor people in "resource limited settings," rolling out a wave of anti-retroviral projects in 2001. MSF added Uganda to the list a year later, locating the project in a region in which it had extensive experience. By 2004, the Arua clinic served more than 1,000 patients, and it was set to expand further. In one sense the AIDS clinic constituted a meta-kit. By combining experience from multiple locations, MSF could create a mobile set of treatment protocols, less dependent on full-scale laboratory support and adapted to shifting personnel. In this way no project would be open to the charge of representing only an anomaly, since the larger chain was clearly replicable. In another sense, however, the AIDS clinic exposed the limits of the kit approach. MSF's initial commitment was to five years of treatment. The therapy provided, however, would have to last a lifetime, since the drugs produced temporary remission rather than a cure. MSF's approach depended on imported materials, personnel, and funding, none easily substitutable in a provincial town. Members of the team worried about these issues, even as they worked frenetically to expand patient rolls in the face of tremendous demand. "It's not an emergency project, but most days we work at this speed," the mission head told me, wondering how it would all keep going. At the same time, as patients improved they began to refocus on the hardships of their everyday life, and to seek support and counsel well beyond medical therapy. Although sympathetic, MSF was poorly equipped to respond to matters of poverty, unemployment, and family expectations. The translation of treatment from rich countries to poor ones could not alter the structural imbalance between contexts in economic terms. That particular crisis exceeded the boundaries of a shipping container.

For some members of MSF, the kit system can now appear a constraint, a self-created frontier limiting the organization's creativity and the larger humanitarian project. A humanitarian affairs officer with MSF-Holland described the general dilemma to me in 2006:

> The kit made us good specialists in a closed camp setting. . . . We just don't seem to know what to do with open settings. Whether urban Rio or Chad, when we have low density and widely dispersed populations we have more problems. What I see is an institutionalization of closed settings. . . . We're used to thinking we have to have kit to act.[51]

A pointed blog posting written in 2009 by an individual in MSF-France's research arm related a caustic anecdote of field experience: What a triumphal headline hailed as the donation of "60 metric tons of essential drugs" from a UN agency appeared on the ground as one hastily delivered box, its instructions in a language that mystified the Congolese aid workers who unpacked it and its contents conforming to outdated protocols. The author of the posting queried the bureaucratizing force of what he termed "kit culture," noting that "in Congo as well as many other places, the kit has become more than a tool, it is increasingly the embodiment of THE humanitarian gesture itself, as if dropping a kit constitutes the raison d'être of humanitarian interventions." UN agencies, he added in conclusion, were hardly alone in this regard.[52] If MSF was also guilty of lapsing into kit culture, as he implied, the rise of military humanitarianism further unsettled the organization by combining life and death into a single logistical form. When the US Air Force appeared over Afghanistan after the attacks of September 2001, its planes dropped aid packets as well as bombs. The different strands in Lindqvist's story had woven back together.

The end of the Cold War featured high-profile episodes of humanitarian crisis, along with a proliferation of NGOs responding to them. By then a template for global humanitarian action was well established, with the means to extend a minimal medical apparatus worldwide. The general form of this apparatus and its components were less surprising or innovative than their combination, orientation, and effects. In response to perceived conditions of emergency, efforts at standardization focused on basic provisions for life and medical care. The result was a limited and temporary infrastructure for interruption, highly mobile and concentrated on immediate needs. Amid the debris of decolonization and superpower struggles, humanitarians devised means for crisis response that were simultaneously effective and ephemeral.

Discussing MSF's kit system with me in 2003, Rony Brauman described it as "an apparatus [*dispositif*] in Foucault's sense" as well as "a logic for action." Widely read and resolutely critical, the former president of MSF-France attributed a degree of strategic coherence to the organization's material engagement, however contingent its origins and practice.[53] A faith in action has defined MSF's ethics, and a sense of emergency has delineated its ethos. Its goal, as Brauman reiterated at the end of the 2006 documentary film *L'aventure MSF*, is not to transform the world, but simply to assist people in moments of distress so that they might be alive for any future reconstruction. To this end the kit system has proved an admirable asset, at least for certain conditions. Yet how does this ethical action fit into a larger political calculus, and into the broader "kit culture" of humanitarianism?

Recall that the technical safety net produced in the name of medical humanitarianism has few claims to long-term "development" or "capacity building." It promises no utopian liberation, and offers few political guidelines. Rather, it remains relentlessly focused on present suffering and basic health, deferring future states in favor of assuring survival. The politics of life and death here is a minimalist one, concentrated on the maintenance of survival rather than on the extension of a more complex political regime.[54] This point is most graphically embodied by one of MSF's simplest tools: a thin strip used to measure the middle upper arm circumference (MUAC) of children below the age of 5. When tightened around the arm, the MUAC bracelet provides an indication of child's nutritional state, providing one means for rapid assessment of malnutrition in a population. Cheap to produce, easy to comprehend, and brightly colored (with a warning scale running from safe green into dangerous red), it is a compelling object in technical and visual terms. MSF-USA has featured this tool in a publicity campaign aimed at children, calling it "The Bracelet of Life." The sense of life involved, however, remains as thin as the means used to measure it. Painfully lean arms may result in access to a therapeutic feeding center, or at least a package of the high-protein peanut mixture known as Plumpy'nut—a potential means to continued survival. They may also contribute to the statistical representation of population's distress. But the safe measure of green only indicates reasonable nutrition at the moment of measurement. It can promise nothing more about overall health, let alone a full or happy childhood. That would take a far more expansive set of tools, and a less attenuated sense of politics.

Rather than nuclear annihilation, the global Cold War ultimately bequeathed a new era of lower-intensity suffering along the edges of political power and economic exchange.[55] The establishment of a fast and efficient logistics system for humanitarian action changed everything and nothing, altering the landscape of civilian anguish without resolving it. Particular lives could now be spared, at least in the short run, from certain forms of distress. The situations that imperiled them, however, too often found eternal return, as emergencies grew "chronic" or re-emerged. Since the demographics of suffering usually far outweigh any response, it would be a gross misnomer to call the greater humanitarian apparatus anything like a solution to global states of disaster. Rather, the humanitarian kit culture represents nothing more—and also nothing less—than the means to effective measurement of the human agony of political failure through its temporary alleviation: a "bracelet of life" for a suffering planet.

Notes

1. See Lindqvist 2001.

2. Lindqvist 2001: 72–73. On the significance of visual media to conceptions of suffering, see Boltanski 1999; Sontag 2003.

3. On the AK-47, see Clapperton Mavhunga's essay in this volume. For more on proxy wars and their significance during the post-Vietnam period, see Westad 2005; Mamdani 2004. I follow MSF's historiographic perspective.

4. For more on the concept of "technopolitics," see Hecht 1998.

5. See Barnett and Weiss 2008; Fassin 2007; Feher et al. 2007; Fisher 1997; Fox 1995.

6. See Druett 2000; Lynn 1993.

7. JRCIRC 1944: i.

8. JRCIRC 1944.

9. JRCIRC 1944: i.

10. JRCIRC 1944: ii.

11. JRCIRC 1948: 245.

12. The 2004 edition of the Red Cross Logistics Field Manual does not mention this historical antecedent, though it does note the dispatching of family parcels during the war, and the gradual evolution of purchasing and transport between the late 1970s and the early 1980s (ICRC 2004: 24). For other discussions of humanitarian logistics see Cock 2003; Kaatrud et al. 2003; Payet 1996.

13. Brauman (1996: 76) and Rieff (2002: 75, 86, 166) caustically suggest how little protection this would offer. The apotheosis of the Holocaust as the extreme of evil may well have occurred a generation later, in the 1960s. See Rabinow 2003: 22; Novick 1999.

14. See de Waal 1997; Rieff 2002. The case of Biafra is both paradigmatic and complex; although remembered as a televised war, Benthall (1993: 102) suggests that newspaper and magazine photos may actually have been more influential. Like Benthall, Waters (2004) emphasizes the significance of religious actors in influencing European and American perceptions of the conflict. Here I merely follow the mythic version of MSF's origin. For a more complete history of the group's origins and early squabbles, see Vallaeys 2004.

15. See Calhoun 2008.

16. Lindqvist 2003. For more on colonial medicine in Africa, see Headrick 1994 and Vaughn 1991. On the complex colonial history surrounding the term "civil society," see Comaroff and Comaroff 1999. While the Red Cross focused on civilizing conflicts between "civilized" countries, it helped transform both military medicine and international law. See Forsythe 2005; Hutchinson 1996.

17. The SAMU system emphasized mobile intensive care units that could offer rapid treatment by doctors, a step closer to Dominique-Jean Larrey's *hôpital ambulant* of the Napoleonic Wars than to the American model of paramedic transport to the emergency room. See Haller 1992; Zink 2006. When recounting MSF's history, as in my own interview with him, MSF-France's former leader and *philosophe* Rony Brauman often emphasizes the significance of SAMU, travel, and media technologies. See, e.g., Brauman 2006: 58; Tanguy 1999: 226–244; Emmanuelli 1991: 21–25. Bertrand Taithe further emphasizes the significance of SAMU in shaping MSF's particular sense of emergency (2004: 150–151).

18. "Dans leur sale d'attente 2 milliards d'hommes." The publicity campaign is outlined in *Bulletin Médecins Sans Frontières* 6 (1977).

19. Kouchner 1991: 327. Brauman 2006: 39–70 provides a more detailed and critical self-reflection. Although often opposed on such matters as a "right to interference" and state humanitarianism as described below, Kouchner and Brauman have both defined their ethics around a response to suffering, understood in medical terms.

20. Westad 2005; Mamdani 2004.

21. The Biafrans planned to sail a hospital ship to the rescue— something they subsequently did amid considerable media coverage. Their critics within MSF, more focused on improving field capacity, denounced this as an overly symbolic and superficial gesture. See Vallaeys 2004: 275–306.

22. Although consistently popular with the French public, Kouchner has long been a controversial figure. The French journalist Pierre Péan (2009) has accused him of naiveté, corruption, and neo-conservatism. For another caustic assessment of Kouchner as a generational icon, see Ross 2002: 147–169. See also Allen and Styan 2000; Taithe 2004: 147–158. Within MSF, Kouchner's public legacy has proved a source of continued frustration; for example, when the organization won the Nobel Peace Prize in 1999, it found itself misidentified in the press with his positions. (See Vallaeys 2004.: 749–50.) Nonetheless, Kouchner has undeniably presented a coherent vision of moral intervention, one implicit in the phrase "sans frontières" if no longer embraced by MSF.

23. For a portrayal of the ethos and images of the time, see Guibert et al. 2009.

24. Vallaeys 2004; de Waal 1997.

25. Though there are now 19 national sections of the larger movement, the central five remain European: MSF-France (founded in 1971), MSF-Belgium (1980), MSF-Switzerland (1980), MSF-Holland (1984), and MSF-Spain (1986). For the purposes of this essay I am treating MSF as a single entity, since the national sections share a general logistical approach. However, the sections remain effectively autonomous, even if linked by flows of funds and personnel, by a charter, and by a loose international association. There have been moments of extreme acrimony and near civil war, particularly among the largest three national sections (France, Belgium, and Holland).

26. Leopold 2005.

27. See the papers collected in Dodge and Wiebe 1985. Karl-Eric Knutsson uses the evocative phrases "at best chaotic" and "disaster within a disaster" in his chapter, "Preparedness for Disaster Operations." As a number of the contributors note, the Karamoja famine could be traced not only to drought, but also to a background of social factors, including colonial land management policies in the region and increased availability of automatic weapons that altered the balance of cattle raids. For an incisive analysis of the principle of preparedness in an American context, see Lakoff 2007.

28. Well 1985: 177–182. The model for Well's strike force was a Swedish government team known as the Swedish Special Unit, whose efficient work in the West Nile region received accolades from several contributors to the volume.

29. Knutsson 1985: 187–188.

30. Brauman 1980: 9–12. Unlike Oxfam or Action Contre la Faim (Action Against Hunger, or ACF), MSF focuses primarily on therapeutic aspects of nutrition. Nonetheless, famine relief has played a significant role in its history, notably in Ethiopia in 1985.

31. This and all other descriptions refer to the 2003 English edition of the MSF Catalogue.

32. Though MSF may remain an association of doctors in nominal terms, in 2001–02 only 25% of its expatriate volunteers fit that category, another 32% being nurses or paramedics. In addition to the 1,605 field posts that cycle, the organization counted 13,320 "national" staff hired locally. See *MSF Activity Report 2001–2*: 97.

33. Pinel and Nzakou 2006; Vallaeys 2004: 374–384.

34. This account of the origins of MSF's kit system draws from an interview with Jacques Pinel conducted in French by Johanna Rankin on December 21, 2004. Ms. Rankin, then working as an intern with MSF, kindly included questions about the kit system on my behalf. The translation is mine, from the transcript of the exchange included as an appendix in Ms. Rankin's undergraduate honors thesis (2005: 152–170). In this interview, Pinel— like every MSF logistician I have ever queried—presents the kit system as an inspired but ultimately matter-of-fact response to an inherent technical problem.

35. Vidal and Pinel 2009: 30–34.

36. Vidal and Pinel 2009: 32.

37. Rankin 2005: 166–67.

38. MSF Catalogue, 2003 edition.

39. Vidal and Pinel 2009: 34.

40. Pinel 2006.

41. MSF Field Library List, as recorded by the author in Brussels, July 2003.

42. Innovations include insect netting on vehicle grilles to simplify maintenance and experiments to improve a portable system for mixing food used in nutritional therapy. In 2003, the logistics director of MSF-Belgium told me: "One of MSF's luxuries is that we have the means to do R&D. Many others don't, but we have both will and resources. . . .The market usually favors things that are expensive and use a lot of energy. We want to try and find things that are less so, for example solar panels or a bike as an energy source."

43. Field notes, Kampala, 2004. The point extends to medical action as well. Describing a 1983 mission in Sudan for a special issue of MSF-France's house journal on surgery, an MSF veteran outlined the challenge of adapting surgical discipline to

field conditions, including harmonizing surgeons from different backgrounds around a common standard of suture thread. See Falhun 2007.

44. The ICRC also began significant logistics developments in the late 1970s, and established a unit to centralize vehicle purchase and management in 1984. See ICRC 2004: 24. After the 1996 crisis in eastern Zaire, the UN established a Joint Logistics Center to better coordinate between agencies. See Kaatrud 2003.

45. Vidal and Pinel 2009: 33. As an example of its field focus, MSF's cholera kit diverged from WHO's model, in part because the former derived from the group's experiences in Malawi, as opposed to the latter's reliance on the findings of a research institute in Bangladesh. Since MSF had encountered higher incidence rates than expected in its own epidemiological studies and was running emergency programs, it emphasized treatment of acute cases and included more intravenous treatments relative to oral ones (Corty 2009: 84–85).

46. de Laet and Mol 2000.

47. Latour 1987: 226–227; Latour 1999.

48. By the end of the century, MSF had emerged as a relatively wealthy and financially independent NGO, with over 340 million euros in annual income, 80% of it derived from private sources (*MSF Activity Report 2001–2002*: 96). It could thus maintain a high measure of independent capacity rather than relying on donor agencies or foundations for the bulk of its operations. Although private fundraising through public appeals entails image management and thus responds to market logic in the broader sense suggested by Pierre Bourdieu, it is not reducible to economic profit.

49. For example, now that the kit concept has spread and the humanitarian market has expanded, many kits are no longer manufactured in-house at either MSF-Logistique in Bordeaux (the primary logistics depot for MSF-France, MSF-Switzerland and MSF-Spain) or MSF Supply in Brussels (a similar unit for MSF-Belgium, formerly named Transfer). Instead of maintaining a proprietary logistics center, MSF Holland largely relies on agreements with established suppliers to provide it with materials on a flexible, rapid-response basis.

50. Observations and quotations are drawn from author's field notes, Kampala, July 2003 and May 2004.

51. Field notes, Amsterdam, July 2006.

52. See Jean-Hervé Jézéquel, "Kit Culture," at http://www.msf-crash.org.

53. Foucault's use of the term "dispositif" is notoriously slippery. In a 1977 interview he described it as a "heterogeneous ensemble" involving a range of discursive and non-discursive elements, and as a formation that "has as its major function at a given historical moment that of responding to an *urgent need*." See Foucault 1980: 194–195; Rabinow 2003: 49–55; Cock 2005. Here I take Brauman's observation seri-

ously to situate MSF's moral reasoning about emergency—its "regime of living" (Collier and Lakoff 2005)—amid the larger humanitarian enterprise it both enacts and criticizes.

54. For an elaboration of the theme of life in crisis, see Redfield 2005. Humanitarians, it is important to note, approach survival from the opposite direction of Agamben's deadly sovereign (1998), refusing to justify killing. At the same time, as Fassin suggests (2007), the politics of human inequality can reveal an uncomfortable limit in their moral reasoning.

55. On landscapes of conflict, extra-formal networks, and civilian suffering, see Nordstrom 2004.

Bibliography

Abele, Johannes. 2000. *Kernkraft in der DDR: zwischen nationaler Industriepolitik und sozialistischer Zusammenarbeit 1963–1990*. Hannah-Arendt-Institut für Totalitarismusforschung.

Abraham, Itty. 1998. *The Making of the Indian Atomic Bomb: Science, Secrecy and the Postcolonial State*. St. Martin's Press and Zed Books.

Abraham, Itty. 2006. The Ambivalence of Nuclear Histories. *Osiris* 21: 49–65.

Abraham, Itty, ed. 2009. *South Asian Cultures of the Bomb: Atomic Publics and the State in India and Pakistan*. Indiana University Press.

Adas, Michael. 1989. *Machines as the Measure of Men: Science, Technology, and Ideologies of Western Dominance*. Cornell University Press.

Adas, Michael. 1997. A Field Matures: Technology, Science, and Western Colonialism. *Technology and Culture* 38: 478–487.

Adas, Michael. 2005. *Dominance by Design: Technological Imperatives and America's Civilizing Mission*. Harvard University Press.

Ademjumobi, Saheed A. 2007. *The History of Ethiopia*. Greenwood.

African National Congress. 1975. The Nuclear Conspiracy: FRG Collaborates to Strengthen Apartheid. PDW.

Agamben, Giorgio. 1998. *Homo Sacer: Sovereign Power and Bare Life*. Stanford University Press.

Albrecht, Ulrich, Andreas Heinemann-Grueder, and Arend Wellmann. 1992. *Die Spezialisten: Deutsche Naturwissenschaftler und Techniker in der Sowjetunion nach 1945*. Dietz.

Alden, Chris, Daniel Large, and Ricardo Soares de Oliveira, eds. 2008. *China Returns to Africa: A Rising Power and a Continent Embrace*. Columbia University Press.

Al-Elawy, Ibrahim S. al-Abdullah. 1976. The Influence of Oil Upon Settlement in al-Hasa Oasis, Saudi Arabia. PhD dissertation, University of Durham.

Alexander, Yonah. 1966. *International Technical Assistance Experts: A Case Study of the U.N. Experience*. Praeger.

Al-'Idrīs, Muhammad Hassan. 1992. *Al-haya al-idāriyya fī sanjaq al-ihsa al-'uthmānī*. Dar al-Mutanabi.

Allen, G. C., and A. G. Donnithorne. 1957. *Western Enterprise in Indonesia and Malaya*. Allen & Unwin.

Allen, Tim, and David Styan. 2000. A Right to Interfere? Bernard Kouchner and the New Humanitarianism. *Journal of International Development* 12: 825–842.

Al-Subay'ī, Abdallah bin Nasir. 1989. *Iktishāf al-naft wa ātharihā 'ala al-haya al-iqtisādiyya fī al-mantiqa al-sharqiyya, 1933–1960*, second edition. Sharif.

Al-Subay'ī, Abdallah bin Nasir. 1999. *Al-hakim wa al-idāra fī al-ahsa' wa al-qatif wa Qatar ithna' al-hakim al-'uthmānī al-thānī, 1871–1913*. King Fahd National Publishers.

Al-Subay'ī, Abdallah bin Nasir. 1999. *Iqtisād al-ahsa' wa al-qatif wa qatar ithna' al-hakim al-'uthmānī al-thānī, 1871–1913*. King Fahd National Publishers.

Amsden, Alice H. 1989. *Asia's Next Giant: South Korea and Late Industrialization*. Oxford University Press.

Appadurai, Arjun. 1996. *Modernity at Large: Cultural Dimensions of Globalization*. University of Minnesota Press.

Appiah, Kwame Anthony. 1992. *In My Father's House: Africa in the Philosophy of Culture*. Oxford University Press.

Arnold, David. 2005. Europe, Technology, and Colonialism in the 20th Century. *History and Technology* 21 (1): 85–106.

Asmolov, Vladimir G., et al. 2004. *Atomnaia energetika: otsenki proshlogo, realii nastoiashchego, ozhidaniia budushchego*. IzdAt.

Avirgan, Tony. 1975. Rhodesia's ANC to Use Tanzanian Camps. *Guardian*, April 21.

Baberowski, Jörg. 2003. *Der Feind ist überall: Stalinismus im Kaukasus*. Deutsche Verlags-Anstalt.

Bacevich, Andrew J. 2002. *American Empire: The Realities and Consequences of U.S. Diplomacy*. Harvard University Press.

Badics, József. 1934. Irányított mezőgazdaság. *Közgazdasági Szemle* 58 (77): 153–167.

Balás, Károly. 1939. Az állami beavatkozás mértéke a gazdasági életbe. *Közgazdasági Szemle* 58 (82): 915–920.

Bandeira, Luiz de Vianna Moniz. 2000. *Brazil in Latin America*. WHU.

Bárányos, Károly. 1944. A magyar földmivelésügy átmenetgazdasági problémái. In *A mez⊠gazdaság és az átmenet a békegazdaságra*. Dárányi Ignác Agrártudományos Társaság Kiadványai.

Barnett, Michael, and Thomas G. Weiss. 2008. *Humanitarianism in Question: Politics, Power, Ethics*. Cornell University Press.

Barron, John. 1974. *KGB*. Hodder and Stoughton.

Bayart, Jean François. 2000. Africa in the World: A History of Extraversion. *African Affairs* 99 (395): 217–267.

Beach, David N. 1971. The Rising in South-Western Mashonaland, 1896–7. PhD Thesis, University of London.

Beach, David N. 1993. *The Shona and Their Neighbours*. Blackwell.

Bell, J. B. 1971. *The Myth of the Guerrilla*. Knopf.

Bello, Walden. 1991. Moment of Decision: The Philippines, the Pacific, and the U.S. Bases. In *The Sun Never Sets*, ed. J. Gerson and B. Birchard. South End.

Bender, Thomas. 2002. *Rethinking American History in a Global Age*. University of California Press.

Bender, Thomas. 2006. *A Nation among Nations: America's Place in World History*. Hill and Wang.

Bendix, Reinhard. 1956. *Work and Authority in Industry: Ideologies of Management in the Course of Industrialization*. Harper and Row.

Benthall, Jonathan. 1993. *Disasters, Relief and the Media*. I.B. Tauris.

Berend, T. Iván, and György Ránki. 1958. *Magyarország gyáripara. A második világháború el⊠tt és a háború id⊠szakában (1933–1944)*. Akadémiai Kiadó.

Beresford, David. 2008. Chinese Ship Carries Arms Cargo to Mugabe Regime. *Guardian*, April 18.

Bereznai, Aurél. 1943. *Munkaer⊠ tervgazdálkodás a mez⊠gazdaságban*. Magyar Közigazgatástudományi Intézet.

Berkers, Eric, et al. 2004. *Geodesie: De aarde verdeeld en verbeeld, berekend en getekend*. Walburgpers.

Bethkenhagen, Jochen. 1986. Nuclear Energy Policy in the USSR and Eastern Europe. *Economic Bulletin* 23 (9): 5–11.

Bhila, Hoyini K. 1982. *Trade and Politics in a Shona Kingdom*. Longman.

Bijker, Wiebe E., Thomas P. Hughes, and Trevor Pinch, eds. 1987. *The Social Construction of Technological Systems*. MIT Press.

Bill, James A. 1984. Islam, Politics, and Shi'ism in the Gulf. *Middle East Insight* 3 (3): 3–12.

Blaker, James R. 1990. *United States Overseas Basing: Anatomy of the Dilemna*. Praeger.

Blelloch, David. 1957. Bold New Programme: A Review of United Nations Technical Assistance. *International Affairs* 33 (1): 36–50.

Boltanski, Luc. [1993] 1999. *Distant Suffering: Morality, Media and Politics*. Cambridge University Press.

Bopela, Thula, and Daluxolo Luthuli. 2005. *Umkhonto weSizwe: Fighting for a Divided People*. Galago.

Borstelmann, Thomas. 1993. *Apartheid's Reluctant Uncle: The United States and Southern Africa in the Early Cold War*. Oxford University Press.

Bockman, Johanna, and Gil Eyal. 2002. Eastern Europe as a Laboratory for Economic Knowledge: The Transnational Roots of Neo-Liberalism. *American Journal of Sociology* 108 (2): 310–352.

Bojkó, Béla. 1997. *Államkapitalizmus Magyarországon, 1919–1945*. Püski.

Borhi, László. 2004. *Hungary in the Cold War 1945–1956: Between the United States and the Soviet Union*. Central European University Press.

Borstelmann, Thomas. 2001. *The Cold War and the Color Line: American Race Relations in the Global Arena*. Harvard University Press.

Branco, Lúcio Castelo. 1983. *Staat, Raum und Macht in Brasilien*. Fink.

Brand, W. 1979. *1879 HVA 1979: Honderd Jaar Geschiedenis der Verenigde HVA Maatschappijen NV*. HVA.

Brauman, Rony. 1996. *Humanitaire, le dilemme*. Editions Textuel.

Brauman, Rony. 2006. *Penser dans l'urgence: parcours critique d'un humanitaire*. Seuil.

Brickhill, Jeremy. Daring to Storm the Heavens: The Military Strategy of ZAPU 1976–79. In *Soldiers in Zimbabwe's Liberation War*. Heinemann.

Bright, Charles, and Michael Geyer. 2002. Where in the World Is America? The History of the United States in the Global Age. In *Rethinking American History in a Global Age*, ed. T. Bender. Cornell University Press.

Broderman, Arvid. 1948. Scientists Study International Tensions. *Unesco Courier* I (1): 7.

Broscious, S. David. 1999. Longing for International Control, Banking on America's Superiority: Harry S. Truman's Approach to Nuclear Weapons. In *Cold War Statesmen Confront the Bomb*, ed. J. Gaddis et al. Oxford University Press.

Brown, J. F. 1966. *The New Eastern Europe: The Khrushchev Era and After*. Praeger.

Brzezinski, Zbigniew K. 1967. *The Soviet Bloc: Unity and Conflict*. Harvard University Press.

Buchmann, Armando. 2004. *Construção de Brasília. Uma 'Mensagem a Garcia.' Documentário*. Thesaurus.

Buhl, Lance C. 1974. Mariners and Machines: Resistance to Technological Change in the American Navy, 1865–1869. *Journal of American History* 61 (3): 703–727.

Bunce, Valerie. 1985. The Empire Strikes Back: The Evolution of the Eastern Bloc from a Soviet Asset to a Soviet Liability. *International Organization* 39 (1): 1–46.

Bunce, Valerie. 1999. *Subversive Institutions: The Design and the Destruction of Socialism and the State*. Cambridge University Press.

Bunn, George. 1992. *Arms Control by Committee: Managing Negotiations with the Russians*. Stanford University Press.

Burnett, Christina Duffy. 2005. The Edges of Empire and the Limits of Sovereignty: American Guano Islands. *American Quarterly* 57 (3): 779–803.

Calhoun, Craig. 2008. The Imperative to Reduce Suffering: Charity, Progress and Emergencies in the Field of Humanitarian Action. In *Humanitarianism in Question*, ed. M. Barnett and T. Weiss. Cornell University Press.

Carter, John D. 1958. Airway to the Middle East. In *Army Air Forces in World War II*, volume VII, ed. W. Craven and J. Cate. University of Chicago Press.

Castells, Manuel. [1996] 2000. *The Rise of the Network Society*. Blackwell.

Cate, James Lea, and E. Kathleen Williams. 1948. The Air Corps Prepares for War, 1939–41. In *Army Air Forces in World War II*, volume 1, ed. W. Craven and J. Cate. University of Chicago Press.

Catholic Commission for Justice and Peace and Legal Resources Foundation. 1999. *Breaking the Silence, Building True Peace: A Report on the Disturbances in Matabeleland and the Midlands, 1980–1988: A Summary*. CCJP and LRF.

Caustin, H. E. 1967. United Nations Technical Assistance in an African Setting. *African Affairs* 66: 113–126.

Cawte, Alice. 1992. *Atomic Australia, 1944–1990*. New South Wales Press.

Cervenka, Zdenek, and Barbara Rogers. 1978. *The Nuclear Axis: Secret collaboration between West Germany and South Africa*. Julian Friedmann Books.

Chambers, David, and Richard Gillespie. 2000. Locality in the History of Science: Colonial Science, Technoscience, and Indigenous Knowledge. *Osiris* 15: 221–240.

Child, John. 1979. Geopolitical Thinking in Latin America. *Latin American Research Review* 14 (2): 89–111.

Citino, Nathan. 2002. *From Arab Nationalism to OPEC: Eisenhower, King Sa'ud, and the Making of US-Saudi Relations*. Indiana University Press.

Clarren, Rebecca. 2006. Paradise Lost: Greed, Sex Slavery, Forced Abortions and Right-Wing Moralists. *Ms.*, spring.

Cleveland, Reginald M. 1946. *Air Transport at War*. Harper.

Cloud, John. 2002. American Cartographic Transformations during the Cold War. *Cartography and Geographic Information Science* 29 (3): 261–282.

Cock, Emile. 2005. *Le dispositf humanitaire: Géopolitique de la générosité*. Harmattan.

Collier, Stephen J., and Andrew Lakoff. 2005. On Regimes of Living. In *Global Assemblages*, ed. A. Ong and S. Collier. Blackwell.

Collins, H. M. 1974. The T.E.A. Set: Tacit Knowledge and Scientific Networks. *Science Studies* 4: 165–186.

Collins, Martin J. 2002. *Cold War Laboratory: RAND, the Air Force, and the American State, 1945–1950*. Smithsonian Institution Press.

Comaroff, John, and Jean Comaroff, eds. 1999. *Civil Society and the Political Imagination in Africa: Critical Perspectives*. University of Chicago Press.

Connelly, John. 2000. *Captive University: the Sovietization of East German, Czech and Polish Higher Education, 1945–1956*. University of North Carolina Press.

Connelly, Matthew. 2002. *A Diplomatic Revolution: Algeria's Fight for Independence and the Origins of the Post-Cold War Era*. Oxford University Press.

Connelly, Matthew. 2008. *Fatal Misconception: The Struggle to Control World Population*. Harvard University Press.

Converse, Elliot V. [1984] 2005. *Circling the Earth: United States Plans for a Postwar Overseas Military Base System, 1942–48*. Air University Press.

Cooke, Morris L., and João Alberto. 1942. Copy of a Release made in Brazil To FDR and Vargas, November 30th 1942, Roosevelt Presidential Library, Morris L. Cooke Papers (1936–1945), Box 283: M. L. Cooke as Chief. U.S. Technical Mission to Brazil.

Cooke, Morris L. 1944. *Brazil on the March: A Study in International Cooperation*. Whittlesey House.

Cooper, Frederick. 2004. Empire Multiplied: A Review Essay. *Comparative Studies in Society and History* 46 (2): 247–272.

Cooper, Frederick. 2005. *Colonialism in Question: Theory, Knowledge, History*. University of California Press.

Cooper, Frederick, and Randall M. Packard, eds. 1997. *International Development and the Social Sciences: Essays on the History and Politics of Knowledge*. University of California Press.

Cooper, Frederick, and Ann Laura Stoler, eds. 1997. *Tensions of Empire: Colonial Cultures in a Bourgeois World*. University of California Press.

Corbisier, Roland. 1960. Brasilia and National Development. *Módulo* 3 (18).

Corn, Joseph J. [1983] 2002. *The Winged Gospel: America's Romance with Aviation*. Johns Hopkins University Press.

Corty, Jean-François. 2009. Choléra: depister et traiter hors des murs de l'hôpital. In *Innovations médicales en situations humanitaire*, ed. J.-H. Bradol and C. Vidal. Harmattan.

Costa, Lúcio. 1959. Architectural Development in Brazil. *Brazilian-American Survey* 9: 63–71.

Costa, Lúcio. 1960. Pilot Plan for Brasilia. *Módulo* 3 (18).

Couto e Silva, Golbery do. 1967. *Geopolítica do Brasil*. José Olympio.

Crampton, Richard J. 1997. *Eastern Europe in the Twentieth Century—and After*, second edition. Routledge.

Craven, Wesley Frank, and James Lea Cate, eds. 1958. *Army Air Forces in World War II*, volume VII. University of Chicago Press.

Cronin, Marionne. 2007. Northern Visions: Aerial Surveying and the Canadian Mining Industry, 1918–1928. *Technology and Culture* 48 (2): 303–330.

Cullather, Nick. 2004. Miracles of Modernization: The Green Revolution and the Apotheosis of Technology. *Diplomatic History* 28 (2): 227–254.

Dabengwa, Dumiso. 1995. ZIPRA in the Zimbabwe War of National Liberation. In *Soldiers in Zimbabwe's Liberation War*. Heinemann.

Daneel, Martinus L. 1970. *Guerrilla Snuff*. Baobab Books.

Davies, R. E. G. 1964. *A History of the World's Airlines*. Putnam.

Dawson, Jane I. 1996. *Eco-Nationalism, Anti-Nuclear Activism, and National Identity in Russia, Lithuania, and Ukraine*. Duke University Press.

de Grazia, Victoria. 2005. *Irresistible Empire*. Harvard University Press.

de Laet, Marianne, and Annemarie Mol. 2000. The Zimbabwe Bush Pump: Mechanics of a Fluid Technology. *Social Studies of Science* 30 (2): 225–263.

Deletant, Dennis, and Mihail E. Ionescu, eds. 2004. *Romania and the Warsaw Pact, 1955–1989: selected documents*. Politeia-SNSPA.

Deleuze, Gilles, and Félix Guattari. [1980] 1987. *A Thousand Plateaus: Capitalism and Schizophrenia*. University of Minnesota Press.

Denoon, Donald, et al., eds. 2003. *The Cambridge History of Pacific Islanders*. Cambridge University Press.

DeNovo, John A. 1955. Petroleum and the United States Navy before World War I. *Mississippi Valley Historical Review* 44 (4): 641–656.

de Waal, Alex. 1997. *Famine Crimes: Politics and the Disaster Relief Industry in Africa.* Currey.

Diouf, Mamadou. 1997. Senegalese Development: From Mass Mobilization to Technocratic Elitism. In *International Development and the Social Sciences*, ed. F. Cooper and R. Packard. University of California Press.

Diouf, Mamadou. 2000. The Senegalese Murid Trade Diaspora and the Making of a Vernacular Cosmopolitanism. *Public Culture* 12: 679–702.

Dodge, Cole P. 1985. The West Nile Emergency. In *Crisis in Uganda*, ed. C. Dodge and P. Wiebe. Pergamon.

Dodge, Cole P., and Paul D. Wiebe, eds. 1985. *Crisis in Uganda: The Breakdown of Health Services.* Pergamon.

Dodge, D., W. Gauden, B. Kipila, I. Kost, G. Menas, and M. Zelina. 1983. *Warsaw Pact Economic Integration.* Defense Intelligence Agency.

Dower, John. 1987. *War Without Mercy: Race and Power in the Pacific War.* Pantheon.

Doyon, Denis F. 1991. Middle East Bases. Model for the Future. In *The Sun Never Sets*, ed. J. Gerson and B. Birchard. South End.

Druett, Joan. 2000. *Rough Medicine: Surgeons at Sea in the Age of Sail.* Routledge.

Edis, Richard. 1993. *Peak of Limuria: The Story of Diego Garcia.* Bellew.

Edwards, Paul N. 1996. *The Closed World: Computers and the Politics of Discourse in Cold War America.* MIT Press.

Eisenman, Joshua, Eric Heginbotham, and Derek Mitchell, eds. 2007. *China and the Developing World: Beijing's Strategy for the Twenty-First Century.* Sharpe.

Ellert, Henrik. 1989. *Rhodesian Front War, Counterinsurgency and Guerrilla War in Rhodesia, 1962–1980.* Mambo.

Emelyanov, V. S. 1958. The Development of International Cooperation by the USSR in the Peaceful Uses of Atomic Energy, Session 23b (Development of international collaboration in the field of atomic energy), P/2415 USSR. In Proceedings of the Second United Nations International Conference on the Peaceful Uses of Atomic Energy.

Emmanuelli, Xavier. 1991. *Les prédateurs de l'action humanitaire.* Albin Michel.

Engel, Jeffrey A., ed. 2007. *Local Consequences of the Global Cold War.* Stanford University Press.

Engeln, Ralf. 2001. *Uransklaven Oder Sonnensucher? Die Sowjetische AG Wismut in der SBZ/DDR, 1946–1953.* Klartext.

Engerman, David C., Nils Gilman, Mark H. Haefele, and Michael E. Latham, eds. 2003. *Staging Growth: Modernization, Development, and the Global Cold War*. University of Massachusetts Press.

Enloe, Cynthia. [1989] 2000. *Bananas, Beaches, and Bases: Making Feminist Sense of International Politics*. University of California Press.

Eriksen, Tore Linné, ed. 2000. *Norway and National Liberation in Southern Africa*. Nordic Africa Institute.

Escobar, Arturo. 1995. *Encountering Development: The Making and Unmaking of the Third World*. Princeton University Press.

Evans, Michael. 1981. Fighting against Chimurenga: An Analysis of Counterinsurgency in Rhodesia, 1972–'79. *Historical Association of Zimbabwe Local Series* 37: 21–25.

Evinger, William R. 1998. *Directory of U.S. Military Bases Worldwide*. Oryx.

Fairchild, Johnson E. 1941. Alaska in Relation to National Defense: The Value of the Territory. *Annals of the Association of American Geographers*. Association of American Geographers 31 (2): 105–112.

Falhun, Olivier. 2007. Probing Surgery. *Messages*: 1–3.

Fassin, Didier. 2007. Humanitarianism as a Politics of Life. *Public Culture* 19 (3): 499–520.

Feher, Michel, Gaëlle Krikorian, and Yates McKee, eds. 2007. *Nongovernmental Politics*. Zone Books.

Ferguson, James. 1990. *The Anti-Politics Machine: "Development," Depoliticization, and Bureaucratic Power in Lesotho*. Cambridge University Press.

Ferguson, James. 2006. *Global Shadows: Essays on Africa in the Neoliberal World Order*. Duke University Press.

Ferguson, Niall. 2004. *Colossus: The Price of America's Empire*. Penguin.

Fermi, Laura. 1957. *Atoms for the World: United States Participation in the Conference on the Peaceful Uses of Atomic Energy*. University of Chicago Press.

Ferraz, Francisco César. 2005. *Os brasileiros e a Segunda Guerra Mundial*. Jorge Zahar.

Fiennes, Ranulph. 1975. *Where Soldiers Fear to Tread*. Hodder and Stoughton.

Finch, Ron. 1986. *Exporting Danger*. Black Rose Books.

Firth, Stewart, and Karin von Strokirch. 2003. A Nuclear Pacific. In *The Cambridge History of Pacific Islanders*, ed. D. Denoon et al. Cambridge University Press.

Fischer, David. 1997. *History of the International Atomic Energy Agency: The First Forty Years*. IAEA.

Fisher, William F. 1997. Doing Good? The Politics and Antipolitics of NGO Practices. *Annual Review of Anthropology* 26: 439–464.

Fishkin, Shelley Fisher. 2005. Crossroads of Cultures: The Transnational Turn in American Studies. *American Quarterly* 57 (1): 17–57.

Fitzgerald, Donald T. 2000. The Machine in the Pacific. In *Science and the Pacific War*, ed. R. Macleod. Kluwer.

Fleron, Frederic J., Erik P. Hoffman, and Robin F. Laird, eds. 1991. *Soviet Foreign Policy: Classic and Contemporary Issues*. Aldine de Gruyter.

Forland, Astrid. 1997. Negotiating Supranational Rules: The Genesis of the International Atomic Energy Agency Safeguards System. Dr. Art. Thesis, University of Bergen.

Forsythe, David P. 2005. *The Humanitarians: The International Committee of the Red Cross*. Cambridge University Press.

Foucault, Michel. 1980. The Confession of the Flesh. In *Power/Knowledge*, ed. C. Gordon. Pantheon.

Fox, Renée C. 1995. Medical Humanitarianism and Human Rights: Reflections on Doctors Without Borders and Doctors of the World. *Social Science & Medicine* 41 (12): 1607–1626.

Franck, Peter G., and Dorothea Seelye Franck. 1951. Implementation of United Nations Technical Assistance Programs. *International Conciliation* 468: 59–80.

Franck, Peter G. 1955. Technical Assistance through the United Nations—The U.N. Mission to Afghanistan, 1950–1953. In *Hands Across Frontiers*, ed. H. Teaf, Jr., and P. Franck. Cornell University Press.

Frankel, S. Herbert. 1952. United Nations Primer for Development. *Quarterly Journal of Economics* 66 (3): 301–326.

Fraser, Cary. 1994. *Ambivalent Anti-Colonialism: The United States and the Genesis of West Indian Independence, 1940–1964*. Greenwood.

Friedmann, John R. P. 1955. Developmental Planning in Haiti—A Critique of the U.N. Report. *Economic Development and Cultural Change* 4 (1): 39–54.

FRUS (Foreign Relations of the United States). 1958–1960. Volume V: American Republics.

FRUS. 1964–1968. Volume XXXI: American Republics.

Fuller, R. Buckminster. 1983. A Compendium of Certain Engineering Principles Pertinent to Brazil's Control of Impending Acceleration in Its Industrialization, Mechanical Engineering Section. U.S. Board of Economic Warfare, August 13th 1943. In *Critical Path*. St. Martin's Griffin.

Furtado, Celso. 1971. *Von der Republik der Oligarchen zum Militärstaat (1967), Brasilien Heute: Beiträge zur politischen, wirtschaftlichen und sozio-kulturellen Situation Brasiliens.* Athenäum.

Gaddis, John Lewis. 1999. *Cold War Statesmen Confront the Bomb: Nuclear Diplomacy since 1945.* Oxford University Press.

Gaddis, John Lewis. 2005. *The Cold War: A New History.* Penguin.

Garon, Sheldon. 1987. *The State and Labor in Modern Japan.* University of California Press.

Gerson, Joseph. 1991. The Sun Never Sets. In *The Sun Never Sets*, ed. J. Gerson and B. Birchard. South End.

Geschiere, Peter. 1997. *The Modernity of Witchcraft.* University of Wisconsin Press.

Gilberto, Astrud. 1965. "Non-Stop to Brazil." On *The Shadow of Your Smile.* Verve Records.

Gilchrist, Huntington. 1959. Technical Aid from the United Nations—as Seen in Pakistan. *International Organization* 13 (4): 505–519.

Gilinsky, Victor, and Bruce L. R. Smith. 1968. Civilian Nuclear Power and Foreign Policy. *Orbis* 12 (3): 816–830.

Gilman, Nils. 2003. *Mandarins of the Future: Modernization Theory in Cold War America.* Johns Hopkins University Press.

Ginsburgs, George. 1960. The Soviet Union and International Co-Operation in the Peaceful Use of Atomic Energy: Bilateral Agreements. *American Journal of International Law* 54 (3): 605–614.

Gittings, John. 1964. Co-Operation and Conflict in Sino-Soviet Relations. *International Affairs* 40 (1): 60–75.

Gleijeses, Piero. 2002. *Conflicting Missions: Havana, Washington, and Africa, 1959–1976.* University of North Carolina Press.

Goedhart, Adriaan. 1999. *Eerherstel voor de Plantage: Uit de Geschiedenis van de Handelsvereniging 'Amsterdam' (HVA) 1879–1983.* Albini.

Goedkoop, J. A. M. 1990. Handelsvereeniging 'Amsterdam' 1945–1958, Herstel en Heroriëntatie. *Jaarboek voor de Geschiedenis van Bedrijf en Techniek* 7: 219–240.

Goldberg, Jacob. 1986. The Shi'i Minority in Saudi Arabia. In *Shi'ism and Social Protest*, ed. J. Cole and N. Keddie. Yale University Press.

Goldschmitt, Bertrand. 1980. The Negotiation of the Non-Proliferation Treaty (NPT). *International Atomic Energy Agency Bulletin* 22 (3–4): 73–80.

Goncharenko, Sergei. 1998. Sino-Soviet Military Cooperation. In *Brothers in Arms*, ed. O. Westad. Woodrow Wilson Center Press and Stanford University Press.

Goncharov, V. V. 2001. Pervyi period razvitiia atomnoi energetiki v SSSR. In *Istoriia atomnoi energetiki Sovetskogo Soiuza i Rossii*, ed. V. Sidorenko. IzdAt.

González, Roberto J. 2001. *Zapotec Science: Farming and Food in the Northern Sierra of Oaxaca*. University of Texas Press.

Gordon, Andrew. 1985. *The Evolution of Labor Relations in Japan: Heavy Industry, 1853–1955*. Harvard University Press.

Gourevitch, Peter. 1986. *Politics in Hard Times: Comparative Responses to International Crises*. Cornell University Press.

Government of Zimbabwe. 1989. *A Guide to the Heroes Acre*.

Government of Zimbabwe. 1999. *Obituary: The Late Brigadier-General Charles Ruocha Gumbo*.

Gowing, Margaret. 1974. *Independence and Deterrence: Britain and Atomic Energy, 1945–1952*, volume 1. St. Martin's Press.

Gregory, Paul R., and Robert C. Stuart. 1994. *Soviet and Post-Soviet Economic Structure and Performance*. HarperCollins.

Greig, Ian. 1977. *The Communist Challenge to Africa*. Foreign Affairs.

Guibert, Emmanuel, Didier Lefèvre, and Frédéric Lemercier. 2009. *The Photographer*. First Second.

Guillén, Mauro. 1994. *Models of Models of Management: Work, Authority, and Organization in a Comparative Perspective*. University of Chicago Press.

Gupta, Akhil. 1998. *Postcolonial Developments: Agriculture in the Making of Modern India*. Duke University Press.

Haenel, Michael. 1998. *Das Ende vor dem Ende: Zur Rolle der DDR-Energiewirtschaft beim Systemwechsel 1980–1990*. University of Alberta.

Hall, Melvin, and Walter Peck. 1941. Wings for the Trojan Horse. *Foreign Affairs* 19 (2): 349–369.

Haller, John. 1992. *Farmcarts to Fords: A History of the Military Ambulance, 1790–1925*. Southern Illinois University Press.

Halliday, Fred. 1986. *The Making of the Second Cold War*. Verso.

Hanlon, David. 1998. *Remaking Micronesia: Discourses over Development in a Pacific Territory, 1944–1982*. University of Hawaii Press.

Harding, Sandra. 1998. *Is Science Multicultural? Poscolonialisms, Feminisms, and Empistemologies*. Indiana University Press.

Hardt, Michael, and Antonio Negri. 2000. *Empire*. Cambridge University Press.

Harrison, Hope. 2003. *Driving the Soviets Up the Wall: Soviet-East German Relations, 1953–1961*. Princeton University Press.

Headrick, Daniel R. 1981. *The Tools of Empire: Technology and European Imperialism in the Nineteenth Century*. Oxford University Press.

Headrick, Daniel R. 1988. *The Tentacles of Progress: Technology Transfer in the Age of Imperialism*. Oxford University Press.

Headrick, Daniel R. 1991. *Invisible Weapon: Telecommunications and International Politics, 1851–1945*. Oxford University Press.

Headrick, Daniel R. 1997. Radio Versus Cable: International Telecommunications Before Satellites. In *Beyond the Ionosphere*, ed. A. Butrica. NASA.

Headrick, Rita. 1994. *Colonialism, Health and Illness in French Equatorial Africa, 1885–1935*. African Studies Association Press.

Hecht, Gabrielle. 1998/2009. *The Radiance of France: Nuclear Power and National Identity after World War II*. MIT Press.

Hecht, Gabrielle. 2003. Globalization Meets Frankenstein? Reflections on Terorrism, Nuclearity, and Global Technopolitical Discourse. *History and Technology* 19 (1): 1–8.

Hecht, Gabrielle. 2006a. Negotiating Global Nuclearities: Apartheid, Decolonization, and the Cold War in the Making of the IAEA. *Osiris* 21: 25–48.

Hecht, Gabrielle. 2006b. Nuclear Ontologies. *Constellations* 13 (3): 320–331.

Hecht, Gabrielle. 2007. A Cosmogram for Nuclear Things. *Isis* 98, March: 100–108.

Hecht, Gabrielle, and Paul N. Edwards. 2008. *The Technopolitics of Cold War: Toward a Transregional Perspective*. American Historical Association.

Hecht, Gabrielle. 2009. Africa and the Nuclear World: Labor, Occupational Health, and the Transnational Production of Uranium. *Comparative Studies in Society and History* 51 (4): 896–926.

Hecht, Gabrielle. 2010. The Power of Nuclear Things. *Technology and Culture* 51 (1): 1–30.

Helmreich, Jonathan E. 1986. *Gathering Rare Ores: The Diplomacy of Uranium Acquisition, 1943–1954*. Princeton University Press.

Henrikson, Alan K. 1975. The Map as an 'Idea': The Role of Cartographic Imagery During the Second World War. *American Cartographer* 2 (1): 19–53.

Hevi, Emmanuel John. 1963. *An African Student in China*. Pall Mall.

Hewett, Ed A., et al. 1986. 1986 Panel on the Soviet Economic Outlook. *Soviet Economy* 2 (1): 3–18.

Hewlett, Richard G., and Jack M. Holl. 1989. *Atoms for Peace and War, 1953–1961: Eisenhower and the Atomic Energy Commission*. University of California Press.

Hezoucky, Frantisek. 2000. Temelin NPP Status: The Challenge of Safety Improvements. Paper presented at The Uranium Institute—25th Annual International Symposium, London.

Hietala, Thomas R. [1985] 2003. *Manifest Design: American Exceptionalism and Empire*. Cornell University Press.

Hilton, Stanley E. 1975. Brazilian Diplomacy and the Washington-Rio de Janeiro 'Axis' during the World War II Era. *Hispanic American Historical Review* 59 (2): 201–231.

Hirsch, Francine. 2000. Toward an Empire of Nations: Border-Making and the Formation of Soviet National Identities. *Russian Review* 59 (2): 201–226.

Hirsch, Francine. 2005. *Empire of Nations: Ethnographic Knowledge and the Making of the Soviet Union*. Cornell University Press.

Hoag, Heather, and May-Britt Öhman. 2008. Turning Water into Power: Debates over the Development of the Tanzania's Rufiji River Basin, 1945–1985. *Technology and Culture* 49 (3): 624–651.

Hodges, D. C., and R. E. Abu Shanab, eds. 1972. *NLF National Liberation Fronts 1960/1970*. Morrow.

Hoebink, Paul. 1988. Geven is nemen: De Nederlandse Ontwikkelingshulp aan Tanzania en Sri Lanka. PhD dissertation, Nijmegen University.

Holland, Heidi. 2008. *Dinner with Mugabe*. Penguin.

Holloway, David, and Jane M. O. Sharp, eds. 1984. *The Warsaw Pact: Alliance in Transition?* Cornell University Press.

Holzmann, Philip. 1971. Engineering and Technical Reports.

Howe, K. R., Robert C. Kiste, and Brij Lal, eds. 1994. *Tides of History*. University of Hawaii Press.

Hsieh, Alice Langley. 1964. The Sino-Soviet Nuclear Dialogue, 1963. *Journal of Conflict Resolution* 8 (2): 99–115.

Hughes, Thomas P. 1983. *Networks of Power: Electrification in Western Society, 1880–1930*. Johns Hopkins University Press.

Hugill, Peter J. 1999. *Global Communications since 1844: Geopolitics and Technology*. Johns Hopkins University Press.

Hutchison, A. 1975. *China's African Revolution*. Hutchison.

Hutchinson, John F. 1996. *Champions of Charity: War and the Rise of the Red Cross*. Westview.

Ibhawoh, Bonny, and J. I. Dibua. 2003. Deconstructing Ujamaa: The Legacy of Julius Nyerere in the Quest for Social and Economic Development. *African Journal of Political Science* 8 (1): 59–83.

Ihrig, Károly. 1935. A mezőgazdaság irányítása. *Magyar Szemle* 23: 123–131.

Institute for Inter-American Affairs. 1954. *The Development of Brazil*.

Institute for Inter-American Affairs. 1955. *Brazilian Technical Studies*.

International Committee of the Red Cross. 2004. *Logistics Field Manual*.

Ivanov, Ivan D. 1987. Restructuring the Mechanism of Foreign Economic Relations in the USSR. *Soviet Economy* 3 (3): 192–218.

Izsák, Lajos, and Miklós Kun, eds. 1996. *Moszkvának jelentjük. Titkos documentumok 1944–1948*. Századvég Kiadó.

Jaarverslag Verenigde, H. V. A. Maatschappijen NV, 1975 (21 April 1976). Verslag van de Directie: 14.

Jalal, Ayesha. 1985. *The Sole Spokesman: Jinnah, the Muslim League and the demand for Pakistan*. Cambridge University Press.

Janos, Andrew C. 2000. *East Central Europe in the Modern World: The Politics of the Borderlands from Pre- to Postcommunism*. Stanford University Press.

Johnson, Chalmers. 1982. *MITI and the Japanese Miracle: The Growth of Industrial Policy, 1925–1975*. Stanford University Press.

Johnson, Chalmers. 1982/1999. The Developmental State: Odyssey of a Concept. In *The Developmental State*, ed. M. Woo-Cummings. Cornell University Press.

Johnson, Chalmers. 2000. *Blowback: The Costs and Consequences of American Empire*. Holt.

Johnson, Chalmers. 2004. *The Sorrows of Empire: Militarism, Secrecy, and the End of the Republic*. Metropolitan Books.

Johnson, Chalmers. 2006. *Nemesis: The Last Days of the American Republic*. Metropolitan Books.

Johnson, M. Dujon. 2006. *Race and Racism in the Chinas: Chinese Attitudes Towards Africans and African-Americans*. AuthorHouse.

Jonasson, Jonas A. 1958. The Army Airways Communications System. In *Army Air Forces in World War II*, volume VII, ed. W. Craven and J. Cate. University of Chicago Press.

Jones, Christopher D. 1981. *Soviet Influence in Eastern Europe*. Praeger.

JRCIRC (Joint Relief Commission of the International Red Cross). 1944. *Materia Medica Minimalis*. ICRC and LRCS.

JRCIRC. 1948. *Report of the Joint Relief Commission of the International Red Cross*. ICRC and LRCS.

Jones, Toby C. 2006. Rebellion on the Saudi Periphery: Modernity, Marginalization, and the Shi'a Uprising of 1979. *International Journal of Middle East Studies* 38 (2): 213–233.

Josephson, Paul R. 1996. Atomic-Powered Communism: Nuclear Culture in the Postwar USSR. *Slavic Review* 55 (2): 297–324.

Kaatrud, David, Ramina Samii, and Luk N. Van Wassenhove. 2003. UN Joint Logistics Center: A Coordinated Response to Common Humanitarian Logistics Concerns. *Forced Migration Review* 18: 11–14.

Kaplan, Amy. 2005. Where is Guantanamo? *American Quarterly* 57 (3): 831–857.

Karlsch, Rainer. 2007. Uran für Moskau: Die Wismut—Eine populäre Geschichte. Ch. Links.

Karumbidza, John Blessing. 2006. Can China Save Zimbabwe's Economy? *Pambazuka News*.

Kelly, John Dunham. 2003. U.S. Power, After 9/11 and Before It: If Not an Empire, Then What? *Public Culture* 15 (2): 347–369.

Kennedy, Michael. 2004. Ireland's Role in Post-War Transatlantic Aviation and Its Implication for the Defence of the North Atlantic Area. Presented at annual meeting of Society for the History of Technology.

Kennedy, Paul M. 1971. Imperial Cable Communications and Strategy, 1870–1912. *English Historical Review* 86, October: 728–775.

Kent, Mary Day. 1991. Panama: Protecting the United States' Backyard. In *The Sun Never Sets*, ed. J. Gerson and B. Birchard. South End.

Kingdom of Saudi Arabia, Ministry of Agriculture. 1964. Al-Hassa Irrigation and Drainage Project, Part I: Civil Work.

Kneese, Eduardo de Mello. 1960. Why Brasília? *Brasília, Acropole—Revista Mensal*, 5–16.

Knell, Mark, and Christine Rider, eds. 1992. *Socialist Economies in Transition: Appraisals of the Market Mechanism*. Elgar.

Knutsson, Karl-Eric. 1985. Preparedness for Disaster Operations. In *Crisis in Uganda*, ed. C. Dodge and P. Wiebe. Pergamon.

Koriakin, Iurii I. 2002. *Okrestnosti iadernoi energetiki Rossii: novye vyzovy*. GUP NIKIET.

Kornai, János. 1992. *The Socialist System: The Political Economy of Communism*. Princeton University Press.

Kornbluh, Peter. 2004. Brazil Marks 40th Anniversary of Military Coup. Declassified Documents Shed Light on U.S. Role. National Security Archive, George Washington University.

Kouchner, Bernard. 1991. *Le Malheur des autres*. Odile Jacob.

Kovács, Imre. 1940. *Szovjet-Oroszország agrárpolitikája*. Cserépfalvi Kiadó.

Kovrig, Bennett. 1979. *Communism in Hungary: From Kun to Kádár*. Hoover Institution Press.

Kramer, Mark. 2003. The Collapse of East European Communism and the Repercussions within the Soviet Union (Part 1). *Journal of Cold War Studies* 5 (4): 178–256.

Kramer, Mark. 2004. The Collapse of East European Communism and the Repercussions within the Soviet Union (Part 2). *Journal of Cold War Studies* 6 (4): 3–64.

Kramer, Mark. 2005. The Collapse of East European Communism and the Repercussions within the Soviet Union (Part 3). *Journal of Cold War Studies* 7 (1): 3–96.

Kramish, Arnold. 1959. *Atomic Energy in the Soviet Union*. Stanford University Press.

Kramish, Arnold. 1963. *The Peaceful Atom in Foreign Policy*. Harper & Row.

Kranakis, Eda. 1997. *Constructing a Bridge: An Exploration of Engineering Culture, Design, and Research in Nineteenth-Century France and America*. MIT Press.

Kraus, Theresa L. 1986. The Establishment of United States Army Air Corps Bases in Brazil, 1938–1945. PhD dissertation, University of Maryland.

Krige, John. 2006a. *American Hegemony and the Postwar Reconstruction of Science in Europe*. MIT Press.

Krige, John. 2006b. Atoms for Peace, Scientific Internationalism, and Scientific Intelligence. *Osiris* 21: 161–181.

Krige, John. 2008. The Peaceful Atom as Political Weapon: Euratom and American Foreign Policy in the Late 1950s. *Historical Studies in the Natural Sciences* 38 (1): 9–48.

Kriger, Norma. *Zimbabwe's Guerrilla War: Peasant Voices*. Cambridge University Press.

Kubitschek, Juscelino de Oliveira. 1955. *Diretrizes gerais do plano nacional de desenvolvimento*. Estabelecimentos Gráf. Santa Maria.

Kubitschek, Juscelino de Oliveira. 1975. *Por que Construí Brasília*. Bloch.

Kubitschek, Juscelino de Oliveira. 1978. *Meo Caminho para Brasília*, volume 3: "Cinqüenta Anos em Cinco." Bloch.

Kuisel, Richard. 1993. *Seducing the French: The Dilemma of Americanization.* University of California Press.

LaFeber, Walter. 1963. *The New Empire; An Interpretation of American Expansion, 1860–1998.* Cornell University Press.

LaFeber, Walter. 2003. The Unites States and Europe in an Age of American Unilateralism. In *The American Century in Europe*, ed. R. Moore and M. Vaudagna. Cornell University Press.

Lakoff, Andrew. 2007. Preparing for the Next Emergency. *Public Culture* 19 (2): 247–271.

Lan, David. 1985. *Guns and Rain: Guerrillas and Spirit Mediums in Zimbabwe.* Currey.

Lapidus, Gail W., Victor Zaslavsky, and Philip Goldman, eds. 1992. *From Union to Commonwealth: Nationalism and Separatism in the Soviet Republics.* Cambridge University Press.

Latham, Michael E. 2000. *Modernization as Ideology: American Social Science and "Nation Building" in the Kennedy Era.* University of North Carolina Press.

Latour, Bruno. 1987. *Science in Action: How to Follow Scientists and Engineers through Society.* Harvard University Press.

Latour, Bruno. 1993/1991. *We Have Never Been Modern.* Harvard University Press.

Latour, Bruno. 1999. *Pandora's Hope: Essays on the Reality of Science Studies.* Harvard University Press.

Leacock, Ruth. 1990. *Requiem for Revolution. The United States and Brazil*: Kent State University Press, 1961–1969.

Le Corbusier. 1998. *The City of To-Morrow and Its Planning (1929), Essential Le Corbusier: L'Esprit Nouveau Articles.* Architectural Press.

Le Corbusier. 1935. *Aircraft.* The Studio.

Leffler, Melvyn P. 1992. *A Preponderance of Power: National Securiy, the Truman Administration, and the Cold War.* Stanford University Press.

Legge, J. D. 1972. *Sukarno: A Political Biography.* Allen Lane/Penguin.

Legvold, Robert. 1970. *Soviet Policy in West Africa.* Harvard University Press.

Lencsés, Ferenc. 1982. *Mezőgazdasági idénymunkások a negyvenes években.* Akadémiai Kiadó.

Lenin, Vladimir I. 1950. *Sochineniia*, fourth edition. Gospolitizdat.

Leopold, Mark. 2005. *Inside West Nile: Violence, History and Representation on an African Frontier.* School of American Research Press.

Lepawsky, Albert. 1952. The Bolivian Operation: New Trends in Technical Assistance. *International Conciliation* 479: 103–140.

Lepgold, Joseph. 1997. Azores. In *Encyclopedia of U.S. Foreign Relations*, ed. O. Holsti et al. Oxford University Press.

Leslie, Stuart W. 1993. *The Cold War and American Science*. Columbia University Press.

Leslie, Stuart W., and Robert Kargon. 2006. Exporting MIT: Science, Technology, and Nation-Building in India and Iran. *Osiris* 21: 110–130.

Lessing, Pieter. 1962. *Africa's Red Harvest*. Michael Joseph.

Levcik, Friedrich, and Jiri Skolka. 1984. *East-West Technology Transfer: Study of Czechoslovakia: The Place of Technology Transfer in the Economic Relations between Czechoslovakia and the OECD Countries*. Organisation for Economic Co-Operation and Development.

Lévi-Strauss, Claude. 1973/1955. *Tristes tropiques*. Cape.

Lewis, John Wilson, and Xue Litai. 1988. *China Builds the Bomb*. Stanford University Press.

Lieven, Dominic. 2000. *Empire: The Russian Empire and Its Rivals*. Murray.

Lindee, Susan. 1994. *Suffering Made Real: American Science and the Survivors at Hiroshima*. University of Chicago Press.

Lindqvist, Sven. 2001. *A History of Bombing*. New Press.

Lockwood, Agnes Nelms. 1956. Indians of the Andes. *International Conciliation* 508: 355–431.

Lonsdale, John M. 1968. Some Origins of Nationalism in East Africa. *Journal of African History* 9 (1): 119–146.

Lorwin, Lewis L. 1940. Public Works and Employment Planning in Germany, 1933–1939. Prepared for the National Resources Planning Board. Washington, D.C., November 1, 1940.

Louis, William Roger. 1978. *Imperialism at Bay: The United States and the Decolonization of the British Empire, 1941–1945*. Oxford University Press.

Lovász, János. 1942. A gazdasági élet hullámmozgásai és az új gazdasági rend. *Új Európa* 1 (2): 41–44.

Lowen, Rebecca S. 1997. *Creating the Cold War University: The Transformation of Stanford*. University of California Press.

Ludwig, Armin K. 1980. *Brasília's First Decade: A Study of Its Urban Morphology and Urban Support Systems*. University of Massachusetts.

Lundestad, Geir. 1986. Empire by Invitation? The United States and Western Europe, 1945–1952. *Journal of Peace Research* 23 (3): 263–277.

Lutfī, Tala't Ibrahīm. 1986. *Athar mashr☒' al-rai wa al-sarraf 'ala mantiqat al-Ihsa': dir☒sa f☒ al-tughayr al-ijtim☒'i al-quraw☒ bil-Mamlika al-'Arabiyya Al-Sa'udiyya*. King Saud University Press.

Lynn, John A., ed. 1993. *Feeding Mars: Logistics in Western Warfare from the Middle Ages to the Present*. Westview.

Maat, Harro. 2001. *Science Cultivating Practice: A History of Agricultural Science in the Netherlands and Its Colohies, 1803–1986*. Kluwer.

MacKenzie, Donald. 1990. *Inventing Accuracy: A Historical Sociology of Nuclear Missile Guidance*. MIT Press.

MacLeod, Roy M., ed. 2000. *Science and the Pacific War: Science and Survival in the Pacific, 1939–1945*. Kluwer.

MacLeod, Roy, ed. 2000. Nature and Empire: Science and the Colonial Enterprise. *Osiris* 15.

Magyary, Zoltán. 1930. A magyar közigazgatás racionalizálása. A Debreceni Tisza István Tudományos Társaság I. Osztályának Kiadványai III. kötet, 3. szám.

Mahan, Alfred Thayer. 1890. *The Influence of Sea Power upon History: 1660–1783*. Historical Association.

Maier, Charles. 1975. *Recasting Bourgeois Europe: Stabilization in France, Germany, and Italy in the Decade after World War I*. Princeton University Press.

Maier, Charles. 2006. *Among Empires: American Ascendancy and Its Predecessors*. Harvard University Press.

Malraux, André. 1959. *La Capitale de l'Espoir*. Brasília: Presidência da República, Serviço de Documentação.

Mamdani, Mahmood. 2001. *When Victims Become Killers: Colonialism, Nativism, and the Genocide in Rwanda*. Princeton University Press.

Mamdani, Mahmood. 2004. *Good Muslim, Bad Muslim: America, the Cold War, and the Roots of Terror*. Pantheon.

Mancke, Elizabeth. 1999. Early Modern Expansion and the Politicization of Oceanic Space. *Geographical Review* 89 (2): 225–236.

Mandaville, Jon E. 1970. The Ottoman Province of al-Hasa in the Sixteenth and Seventeenth Centuries. *Journal of the American Oriental Society*. American Oriental Society 90 (3): 504–506.

Marcus, Harold G. 2002. *A History of Ethiopia*, updated edition. University of California Press.

Mastny, Vojtech, and Malcolm Byrne, eds. 2005. *A Cardboard Castle? An Inside History of the Warsaw Pact, 1955–1991*. Central European University Press.

Matolcsy, Mátyás. 1943. Terménybeszolgáltatási módszerek. *Uj Európa* 1 (1): 18–23.

Maurer, Bill. 2001. Islands in the Net: Re-wiring Technological and Financial Circuits in the Offshore Caribbean. *Comparative Studies in Society and History* 43: 467–501.

Mavhunga, Clapperton. 2003. Firearms Diffusion, Exotic and Indigenous Knowledge Systems in the Lowveld Frontier, South Eastern Zimbabwe, 1870–1920. *Comparative Technology Transfer and Society* 1 (2): 201–231.

Maxey, Rees. 1975. *The Fight for Zimbabwe*. Rex Collings.

Mbembe, Achille. 2000. At the Edge of the World: Boundaries, Territoriality and Sovereignty in Africa. *Public Culture* 12 (1): 259–284.

Mbembe, Achille. 2001. *On the Postcolony*. University of California Press.

Mbembe, Achille, and Sarah Nuttall eds. 2004. Johannesburg, the Elusive Metropolis. *Public Culture* 16: 3.

McNay, John T. 2001. *Acheson and Empire: The British Accent in American Foreign Policy*. University of Missouri Press.

Mead, Margaret, ed. 1953. *Cultural Patterns and Technical Change*. UNESCO.

Mehos, Donna and Suzanne M. Moon n.d. Constituting and Reconstituting Economies of Expertise Decolonization, Cold War and the Global Circulation of Technical Experts. Unpublished manuscript.

Melo, Afrânio. 1962. Road of the Century: The Belém-Brasilia Highway. *Américas* 14 (11): 1–6.

Memon, A. Sreedhara. 2001. *Triumph and Tragedy in Travancore: Annals of Sir CP's Sixteen Years*. Current Books.

Menon, Vapal Pangunni. 1956. *The Story of the Integration of the Indian States*. Orient Longman.

Merkle, Judith. 1980. *Management and Ideology: The Legacy of the International Scientific Movement*. University of California Press.

Mesquita, Raul A. de A. 1958. Aeroporto de Brasília. *Epuc. Engenharia Arquitetura* 2 (8): 198–199.

Miller, Francis Pickens. 1941. The Atlantic Area. *Foreign Affairs* 20 (4): 727–728.

Mindlin, Henrique E. 1965. *Neues Bauen in Brasilien*. Callwey.

Mitchell, Timothy. 2002. *Rule of Experts: Egypt, Techno-Politics, Modernity*. University of California Press.

Modelski, George A. 1959. *Atomic Energy in the Communist Bloc*. Cambridge University Press.

Moody, Walton S. 1995. *Building a Strategic Air Force*. Air Force History and Museums Program.

Moon, Katharine H.S. 1997. *Sex among Allies: Military Prostitution in U.S.-Korea Relations*. Columbia University Press.

Moon, Suzanne. 1998. Takeoff or Self-Sufficiency? Ideologies of Development in Indonesia, 1957–1961. *Technology and Culture* 39: 187–212.

Moon, Suzanne. 2007. *Technology and Ethical Idealism: A History of Development in the Netherlands East Indies*. CNWS.

Moorcraft, Paul L. 1985. *Chimurenga—the War in Rhodesia, 1965–80*. Galago.

Moorcraft, Paul. 1990. *African Nemesis: War and Revolutions in Southern Africa, 1945–2010*. Brassey.

Moore, David Chioni. 2001. Is the Post- in Postcolonial the Post- in Post-Soviet? Toward a Global Postcolonial Critique. *PMLA* 116 (1): 111–128.

Moore, Donald. 1995. The Zimbabwe People's Army: Strategic Innovation or More of the Same? In *Soldiers in Zimbabwe's Liberation War*, ed. N. Bhebe and T. Ranger. Heinemann.

Morgan, Theodore. 1953. The Underdeveloped Area Expert: South Asia Model. *Economic Development and Cultural Change* 2 (1): 27–31.

Morgenstienne, Munthe Christopher. 2003. *Denmark and National Liberation in Southern Africa: A Flexible Response*. Nordic Africa Institute.

Morrison, David. 1964. *The USSR and Africa*. Oxford University Press.

Mubako, Simbi. 2007. Tongo: Legend, Role Model. *Herald* (Zimbabwe), August 14.

Mudenge, Stan. 1986. *A Political History of Munhumutapa*. Zimbabwe Publishing Company.

Mudimbe, Valentin Y. 1988. *The Invention of Africa: Gnosis, Philosophy, and the Order of Knowledge*. Indiana University Press.

Müller, Wolfgang D. 2001. *Geschichte der Kernenergie in der DDR: Kernforschung und Kerntechnik im Schatten des Sozialismus*. Schäffer-Poeschel.

Munochiveyi, Munyaradzi Bryn. 2008. It Was Difficult in Zimbabwe: A History of Imprisonment, Detention and Confinement during Zimbabwe's Liberation Struggle, 1960–1980. PhD thesis, University of Minnesota.

Murray, Robin. 1981. From Colony to Contract: HVA and the Retreat from Land, unpublished manuscript, Institute of Development Studies, University of Sussex. Available in KITLV Library, Leiden, The Netherlands.

Nagy, Iván (Edgár). 1941. *Szovjet-Oroszország kollektív mezögazdasági termelése*. Cserépfalvi Kiadó.

Narkiewicz, Olga A. 1986. *Eastern Europe 1968–1984*. Croom Helm.

Nash, June C. 1979. *We Eat the Mines and the Mines Eat Us: Dependency and Exploitation in Bolivian Tin Mines*. Columbia University Press.

Neal, Marian. 1951. United Nations Technical Assistance Programs in Haiti. *International Conciliation* 468: 81–118.

Nero, Karen. 2003. The End of Insularity. In *The Cambridge History of Pacific Islanders*, ed. D. Denoon et al. Cambridge University Press.

Newby-Fraser, A. R. 1979. *Chain Reaction: Twenty Years of Nuclear Research and Development in South Africa*. Atomic Energy Board.

Newman, Wendy. 1978. *The Politics of Energy in the Soviet Bloc*. Center for International Studies, Massachusetts Institute of Technology.

Nichols, Roy. 1933. Navassa: A Forgotten Acquisition. *American Historical Review* 38 (3): 505–510.

Nkomo, Joshua. 1984. *Nkomo: The Story of My Life*. Methuen.

Nolan, Mary. 1994. *Visions of Modernity: American Business and the Modernization of Germany*. Oxford University Press.

Noorani, A. G. 2003. C.P. and Independent Travancore. *Frontline* (www.hinduonnet.com/fline) 20 (13).

Nordstrom, Carolyn. 2004. *Shadows of War: Violence, Power, and International Profiteering in the Twenty-First Century*. University of California Press.

Norman, Andrew. 2004. *Robert Mugabe and the Betrayal of Zimbabwe*. MacFarland.

Northrop, Douglas. 2004. *Veiled Empire: Gender and Power in Stalinist Central Asia*. Cornell University Press.

Novick, Peter. 1999. *The Holocaust in American Life*. Houghton Mifflin.

Nyarota, Geoff. 2006. *Against the Grain: Memoirs of a Zimbabwean Newsman*. Zebra.

Nye, Joseph S. 1990. *Bound to Lead: The Changing Nature of American Power*. Basic Books.

Nye, Joseph S. 2002. *The American Paradox of Power: Why the World's Only Superpower Can't Go It Alone*. Oxford University Press.

Tuathail, Ó. Gearóid. 1996. *Critical Geopolitics: The Politics of Writing Global Space*. Routledge.

Odom, William E. 2006. The Cold War Origins of the U.S. Central Command. *Journal of Cold War Studies* 8 (2): 52–82.

Office of Technology Assessment, U.S. Congress. 1981. *Technology and Soviet Energy Availability.*

Office of Technology Assessment, U.S. Congress. 2005. *Nuclear Safeguards and the International Atomic Energy Agency.*

Oldberg, Ingmar ed. 1984. *Proceedings of a Symposium on Unity and Conflict in the Warsaw Pact: Stockholm, Nov. 18–19, 1982.* Swedish National Defence Research Institute.

de Oliveira, Márcio. 1998. Gaston Bachelard e o imaginário das cidades: imagens da construção de Brasília. *Revista Sociedade e Estado* 13 (1): 225–240.

Orent, Beatrice, and Pauline Reinsch. 1941. The Sovereignty over the Islands in the Pacific. *American Journal of International Law* 35 (3): 443–461.

Organization for the Islamic Revolution in Arabia. 1979. *Ahdāth nufimbir (al-muharram) 1979 fī sa'udiyya.*

Orwell, George. 1945. You and the Atomic Bomb. *Tribune*, October 19.

Oudshoorn, Nelly, and Trevor Pinch, eds. 2003. *How Users Matter: The Co-Construction of Users and Technology.* MIT Press.

Oudshoorn, Nelly. 2003. *The Male Pill: A Biography of a Technology in the Making.* Duke University Press.

Ouwerkerk, Louise. 1994. *No Elephants for the Maharaja: Social and Political Change in the Princely State of Travancore, 1921–1947.* Manohar.

Ong, Aihwa, and Stephen J. Collier, eds. 2005. *Global Assemblages: Technology, Politics, and Ethics as Anthropological Problems.* Blackwell.

Owen, David. 1950. The United Nations Program of Technical Assistance. *Annals of the American Academy of Political and Social Science* 270: 109–117.

Owen, David. 1959. The United Nations Expanded Programme of Technical Assistance-A Multilateral Approach. *Annals of the American Academy of Political and Social Science* 323: 25–32.

Palan, Ronen. 1998. Trying to Have Your Cake and Eating It: How and Why the State System has Created Off-Shore. *International Studies Quarterly* 42 (4): 625–648.

Palasik, Mária. 2000. *A jogállamiság megteremtésének kísérlete és kudarca Magyarországon 1944–1949.* Napvilág Kiadó.

Palladino, Paolo, and Michael Worboys. 1993. Science and Imperialism. *Isis* 84: 91–102.

Pascoe, David. 2001. *Airspaces.* Reaktion Books.

Payet, Marc. 1996. *Logs: Les hommes-orchestres de l'humanitaire*. Éditions Alternatives.

Péan, Pierre. 2009. *Le Monde selon K.* Fayard.

Pearson, Raymond. 2002. *The Rise and Fall of the Soviet Empire*, second edition. Palgrave.

Pells, Richard. 1997. *Not Like Us: How Europeans Have Loved, Hated, and Transformed American Culture Since World War II*. Basic Books.

Perkins, Whitney T. 1962. *Denial of Empire: The United States and Its Dependencies*. Sythoff.

Perkovich, George. 1999. *India's Nuclear Bomb: The Impact on Global Proliferation*. University of California Press.

Perras, Galen Roger. 2003. *Stepping Stones to Nowhere: The Aleutian Islands, Alaska, and American Military Strategy, 1867–1945*. University of British Columbia Press.

Peta, Basildon. 2004. US Seeks 'Coalition' to Force Zimbabwe Change. *Independent* (UK), August 25.

Péteri, György. 1991. Academic Elite into Scientific Cadres: A Statistical Contribution to the History of the Hungarian Academy of Sciences 1945–49. *Soviet Studies* 43 (2): 281–299.

Péteri, György. 1998. A fordulat a magyar közgazdaság-tudományban. A Magyar Gazdaságkutató Intézettől a Közgazdaság-Tudományi Intézetig. In *A fordulat évei. 1947–1949*, ed. É. Standeisky et al. 1956-os Intézet.

Petö Iván and Sándor Szakács. 1985. A hazai gazdaság négy évtizedének története, 1945–1985. Közgazdasági és Jogi Könyvkiadó.

Petros'iants, Andranik M. 1993. *Dorogi zhizni, kotorye vybirali nas*. Energoatomizdat.

Pfaffenberger, Bryan. 1992. Technological Dramas. *Science, Technology & Human Values* 17 (3): 282–312.

Pfaffenberger, Bryan. 1993. The Factory as Artefact. In *Technological Choices: Transformation in Material Cultures since the Neolithic*, ed. P. Lemonnier. Routledge.

Pilger, John. 2004a. Stealing a Nation. Granada Television.

Pilger, John. 2004b. Paradise Cleansed. *Guardian* (London), October 11.

Pinel, Jacques, and Irène Nzakou. 2006. *Log Story*. Messages. MSF-France.

Pinto, Rui. 1973. *The Making of a Middle Cadre*. Liberation Support Movement Information.

Polach, J. G. 1968. Nuclear Energy in Czechoslovakia: A Study in Frustration. *Orbis* 12 (fall): 831–851.

Prashad, Vijay. 2007. *The Darker Nations: A People's History of the Third World.* New Press.

Raath, Jan, and Catherine Philp. 2008. Robert Mugabe Warns Zimbabwe's Voters: "How Can a Pen Fight a Gun?" *Times* (London), June 17.

Rabinbach, Anson. 1990. *The Human Motor: Energy, Fatigue, and the Origins of Modernity.* University of California Press.

Rabinow, Paul. 2003. *Anthropos Today: Reflections on Modern Equipment.* Princeton University Press.

Rainer, János. 1996. *Nagy Imre: Politikai életrajz, Els⊠ kötet, 1896–1953.* 1956-os Intézet.

Rajab, Zayn al-Abadīn al-Rahman. 1980. Dirāsat fī mawāradihā al-ma'iyya wa ta'thīrihī fī al-istikhdām al-rīfi al-'ard. *ad-Darah* 16 (3): 122–124.

Rajan, S. Ravi. 2006. *Modernizing Nature: Forestry and Imperial Eco-development.* Oxford University Press.

Rangaswami, Vanaja. 1981. *The Story of Integration: A New Interpretation.* Manohar.

Ranger, Terence. 1999. *Voices from the Rock: Nature Culture and History in the Matopo Hills of Zimbabwe.* Currey.

Ranger, Terence. 1967. *Revolt in Southern Rhodesia, 1896–97: A Study in African Resistance.* Heinemann.

Ranger, Terence. 1976. Towards a Usable African Past. In *African Studies Since 1945*, ed. C. Fyfe. Longman.

Ranger, Terence. 1995. *Are We Not Also Men? Samkange Family and African Politics in Zimbabwe, 1920–64.* Currey.

Rankin, Johanna. 2005. A New Frontier for Humanitarianism? Médecins Sans Frontières Responds to Neglected Diseases. Undergraduate honors thesis, University of North Carolina, Chapel Hill.

Ray, Deborah W. 1973. Pan American Airways and the Trans-African Air Base Program of World War II. PhD dissertation, New York University.

Redfield, Peter. 2000. *Space in the Tropics: From Convicts to Rockets in French Guiana.* University of California Press.

Redfield, Peter. 2005. Doctors, Borders and Life in Crisis. *Cultural Anthropology* 20 (3): 328–361.

Reilly, James P., et al. 1973. Free geometric adjustment of the DOC/DOD Cooperative Worldwide Geodetic Satellite (BC-4) Network. Reports of the Department of Geodetic Science, Ohio State University Research Foundation.

Reisinger, William M. 1983. East European Military Expenditures in the 1970s: Collective Good or Bargaining Offer? *International Organization* 37 (1): 143–155.

Reisinger, William M. 1992. *Energy and the Soviet Bloc: Alliance Politics after Stalin.* Cornell University Press.

Reitzer, Béla. 1941. A munkapiac helyzete a munkaközvetítés reformja előtt. *Közgazdasági Szemle* 65 (84): 996–1013.

Rézler, Gyula. 1940. *A paraszt Szovjet-Oroszországban.* Stádium Sajtóvállalat.

Rieff, David. 2002. *A Bed for the Night: Humanitarianism in Crisis.* Simon and Schuster.

Ritschel, Daniel. 1997. *The Politics of Planning: The Debate on Economic Planning in Britain in the 1930s.* Clarendon.

Robinson, Thomas W., and David L. Shambaugh. 1994. *Chinese Foreign Policy: Theory and Practice.* Clarendon.

Roman, Eric. 1996. *Hungary and the Victor Powers, 1945–1950.* St. Martin's Press.

Roosevelt, Franklin D. [1946] 1970. February 23, 1942. In *Nothing to Fear.* Books for Libraries Press.

Roskin, Michael. 2002. *The Rebirth of East Europe*, fourth edition. Prentice-Hall.

Ross, Kristin. 2002. *May '68 and Its Afterlives.* University of Chicago Press.

Rostow, Walt W. 1960. *The Stages of Economic Growth: A Non-Communist Manifesto.* Cambridge University Press.

Rotberg, Robert I, ed. 2008. *China into Africa: Trade, Aid, and Influence.* World Peace Foundation and Brookings Institution Press.

Rowen, Henry S., and Charles Wolf, eds. 1987. *The Future of the Soviet Empire.* St. Martin's Press.

Rubinstein, Alvin Z. 1992. *Soviet Foreign Policy since World War II: Imperial and Global.* HarperCollins.

Ruskola, Teemu. 2005. Canton Is Not Boston: The Invention of American Imperial Sovereignty. *American Quarterly* 57 (3): 859–884.

Sanders, Christopher T. 2000. *Leasehold Empire: Overseas Garrisons.* Oxford University Press.

Scarnecchia, Timothy, Jocelyn Alexander, et al. 2009. Lessons of Zimbabwe. *London Review of Books* 31 (1).

Schaaf, C. Hart. 1960. The Role of Resident Representative of the U.N. Technical Assistance Board. *International Organization* 14 (4): 548–562.

Schake, Kurt Wayne. 1998. *Strategic Frontier: American Bomber Bases Overseas 1950–1960*. University of Trondheim.

Scheinman, Lawrence. 1987. *The International Atomic Energy Agency and World Nuclear Order*. Resources for the Future.

Schiavone, Giuseppe. 1981. *The Institutions of COMECON*. Holmes & Meier.

Schlesinger, Arthur, Jr. 1986. *The Cycles of American History*. Houghton Mifflin.

Schmid, Hellmut H. 1974. *Three-Dimensional Triangulation with Satellites*. National Oceanic and Atmospheric Administration.

Schmid, Sonja. 2005. Envisioning a Technological State: Reactor Design Choices and Political Legitimacy in the Soviet Union and Russia. PhD dissertation, Cornell University.

Schmitt, Carl. [1950] 2003. *The Nomos of the Earth in the International Law of the Jus Publicum Europaeum*. Telos.

Schulten, Susan. 1998. Richard Edes Harrison and the Challenge to American Cartography. *Imago Mundi* 50: 174–188.

Scott, Harriet F., and William F. Scott. 1979. *The Armed Forces of the USSR*. Westview.

Scott, James C. 1998. *Seeing Like a State: How Certain Schemes to Improve the Human Condition Have Failed*. Yale University Press.

Seager, Robert, II. 1953. Years Before Mahan: The Unofficial Case for the New Navy, 1880–1890. *Mississippi Valley Historical Review* 40 (3): 491–512.

Sebestyen, Victor. 2009. *Revolution 1989: The Fall of the Soviet Empire*. Weidenfeld and Nicholson.

Selcher, Wayne A. 1976. Brazilian Relations with Portuguese Africa in the Context of the Elusive "Luso-Brazilian Community." *Journal of Inter-American Studies and World Affairs* 18 (1): 25–58.

Sellström, Tor. 2002. *Sweden and National Liberation in Southern Africa*, volumes I and II. Nordic Africa Institute.

Selvage, Douglas. 2001. *The Warsaw Pact and Nuclear Nonproliferation: 1936–1965*. Woodrow Wilson International Center for Scholars.

Seton-Watson, Hugh. 1962. *The New Imperialism: A Background Book*. Bodley Head.

Sharp, Walter. 1948. The Specialized Agencies and the United Nations: Progress Report II. *International Organization* 2 (2): 247–267.

Sharp, Walter R. 1953. The Institutional Framework for Technical Assistance. *International Organization* 7 (3): 342–379.

Sharp, Walter R. 1956. The United Nations System in Egypt: A Country Survey of Field Operations. *International Organization* 10 (2): 235–260.

Shay, Reg, and Chris Vermaak. 1971. *The Silent War*. Galaxie.

Shearer, David R. 1996. *Industry, State, and Society in Stalin's Russia, 1926–1934*. Cornell University Press.

Shofield, Andrew. 1965. *Modern Capitalism: The Changing Balance of Public and Private Power*. Oxford University Press.

Shoko, Tabona. 2006. My Bones Shall Rise Again: War Veterans, Spirits and Land Reform in Zimbabwe. Working Paper 68, African Studies Centre, Leiden.

Shubin, Vladimir. 2008. *The Hot 'Cold War': The USSR in Southern Africa*. University of KwaZulu-Natal Press.

Sidorenko, Viktor A. 1997. Nuclear Power in the Soviet Union and in Russia. *Nuclear Engineering and Design* 173: 3–20.

Sidorenko, Viktor A. 2001. Upravlenie atomnoi energetikoi. In *Istoriia atomnoi energetiki Sovetskogo Soiuza i Rossii*, volume 1, ed. V. Sidorenko. IzdAt.

Siklos, Pierre. 1991. *War Finance, Reconstruction, Hyperinflation and Stabilization in Hungary, 1938–1948*. Macmillan.

Silva, Ana Paula, and Maria Paula Diogo. 2006. From Host to Hostage: Portugal, Britain, and the Atlantic Telegraph Networks. In *Networking Europe*, ed. E. van der Vleuten and A. Kaijser. Science History Publications.

Simons, Thomas W. 1993. *Eastern Europe in the Postwar World*, second edition. St. Martin's Press.

Sithole, Ndabaningi. 1959. *African Nationalism*. Oxford University Press.

Sithole, Vesta. 2006. *My Life with an Unsung Hero*. AuthorHouse.

Skidmore, Thomas E. 1967. *Politics in Brazil, 1930–1964: An Experiment in Democracy*. Oxford University Press.

Skidmore, Thomas E. 1988. *The Politics of Military Rule in Brazil, 1964–1985*. Oxford University Press.

Slezkine, Yuri. 1994. The USSR as a Communal Apartment, or How a Socialist State Promoted Ethnic Particularism. *Slavic Review* 53 (2): 414–452.

Slezkine, Yuri. 2000. Imperialism as the Highest Stage of Socialism. *Russian Review* 59 (2): 227–234.

Slotten, Hugh R. 2002. Satellite Communications, Globalization, and the Cold War. *Technology and Culture* 43 (2): 315–350.

S. M. 1955. Atoms for Peace in the U.N. *Bulletin of the Atomic Scientists* 11 (1): 24–27.

Smith, Alan H. 1983. *The Planned Economies of Eastern Europe.* Holmes and Meier.

Smith, Dan. 1980. *South Africa's Nuclear Capability.* World Campaign Against Military and Nuclear Collaboration with South Africa.

Smith, Neil. 1996. Spaces of Vulnerability: The Space of Flows and the Politics of Scale. *Critique of Anthropology* 16 (1): 63–77.

Smith, Neil. 2003. *American Empire: Roosevelt's Geographer and the Prelude to Globalization.* University of California Press.

Sobell, Vladimir. 1984. *The Red Market: Industrial Co-operation and Specialisation in Comecon.* Gower.

Sobell, Vladimir. 1990. *The CMEA in Crisis: Toward a New European Order?* Center for Strategic and International Studies.

Sontag, Susan. 2003. *Regarding the Pain of Others.* Picador.

Sparrow, Bartolomew H. 2006. *The Insular Cases and the Emergence of American Empire.* University Press of Kansas.

Spykman, Nicholas J. 1944. *The Geography of Peace.* Harcourt, Brace.

Standeisky, Éva, Gyula Kozák, Gábor Pataki, and János Rainer. 1998. *A fordulat évei. Politika, képzömüvészet, épitészet. 1947–1949.* 1956-os Intézet.

Star, Susan Leigh. 1989. The Structure of Ill-Structured Solutions: Boundary Objects and Heterogeneous Distributed Problem Solving. In *Distributed Artificial Intelligence*, volume 2, ed. L. Gasser and M. Huhns. Pitman.

Steuer, György. 1938. A legkisebb földmunkabérek megállapítása. *Katolikus Szemle* 3 (2): 601–611.

Stokes, Raymond G. 2003. Book Review: *Uransklaven oder Sonnensucher? Die Sowjetische AG Wismut in der SBZ/DDR 1946–1953* by Ralf Engeln. *Slavic Review* 62 (4): 827–828.

Stone, David R. 2008. CMEA's International Investment Bank and the Crisis of Developed Socialism. *Journal of Cold War Studies* 10 (3): 48–77.

Stone, Randall. 1996. *Satellites and Commissars: Strategy and Conflict in the Politics of Soviet-Bloc Trade.* Princeton University Press.

Stuart, Peter C. 1999. *Isles of Empire: The United States and Its Overseas Possessions.* University Press of America.

Sufrin, Sydney. 1966. *Technical Assistance—Theory and Guidelines.* Syracuse University Press.

Sundararajan, Saroja. 2003. *Sir C.P. Ramaswamy Aiyar: A Biography*. Allied.

Sutter, John O. 1959a. Indonesianisasi: A Historical Survey of the Role of Politics of a Changing Economy from the Second World War to the Eve of the General Elections (1940–1955). PhD dissertation, Cornell University.

Sutter, John O. 1959b. *Indonesianisasi: Politics in a Changing Economy, 1940–1955*. Cornell University Southeast Asia Program.

Swift, Richard. 1957. Personnel Problems and the United Nations Secretariat. *International Organization* 11 (2): 228–247.

Szabó, Zoltán. 1937. *A tardi helyzet*. Cserépfalvi Kiadása.

Szabó, Zoltán. 1947a. A hároméves gazdasági terv. In *Ipari ujjáépítésünk*, ed. S. Tonelli. Forum Hungaricum Kiadás.

Szabó, Zoltán. 1947b. Merre Haladunk? In *Ipari ujjáépítésünk*, ed. S. Tonelli. Forum Hungaricum Kiadás.

Taithe, Bernard. 2004. Reinventing (French) Universalism: Religion, Humanitarianism and the 'French Doctors'. *Modern and Contemporary France* 12 (2): 147–158.

Talbot, Phillips. 1949. Kashmir and Hyderabad. *World Politics* 1 (3): 321–332.

Tanguy, Joelle. 1999. The Médecins Sans Frontières Experience. In *Framework for Survival*, ed. K. Cahill. Routledge.

Tatarnikov, Viktor P. 2002. Atomnaia elektroenergetika (s VVER i drugimi reaktorami). In *Istoriia atomnoi energetiki Sovetskogo Soiuza i Rossii. Istoriia VVER*, volume 2, ed. V. Sidorenko. IzdAt.

Tekere, Edgar. 2007. *A Lifetime of Struggle*. Sapes.

Tongogara, Josiah. 1978. The War Is Here, Everywhere. *Zimbabwe News*.

Tongogara, Josiah. 1978. Umtali Attack: The Beginning of the End of the War. *Zimbabwe News*.

Tracy, William. 1965. The Restless Sands. *Aramco World* 16 (2).

Trager, F. N., and Helen G. Trager. 1962. Exporting and Training Experts. *Review of Politics* 24 (1): 88–108.

Trend, Harry G. 1976. Soviet Crude Oil Price to Comecon May Be a Third Higher in 1977. Open Society Archives, box 127, folder 1, report 269.

Trotsky, Leon. 1960. *The History of the Russian Revolution*. University of Michigan Press.

Tungamirai, Josiah. 1995. Recruitment to ZANLA: Building Up a War Machine. In *Soldiers in Zimbabwe's Liberation War*, volume 1, ed. N. Bhebe and T. Ranger. University of Zimbabwe Publications.

Turnbull, David. 2000. *Masons, Tricksters, and Cartographers: Comparative Studies in the Sociology of Scientific and Indigenous Knowledge*. Harwood.

Ungváry, Krisztián. 1998. *Budapest ostroma*. Corvina Kiadó.

United Nations. 1949. *Mission to Haiti: Report of the United Nations Mission of Technical Assistance to the Republic of Haiti.*

United Nations. 1953. *World Against Want: An Account of the U.N. Technical Assistance Programme for Economic Development.*

United Nations. 2008. *Report of the Fact-Finding Mission to Zimbabwe to Assess the Scope and Impact of Operation Murambatsvina by the UN Special Envoy on Human Settlements Issues in Zimbabwe Mrs. Anna Kajumulo Tobaijuka.*

U.S. Department of Commerce. 1966. The Coast and Geodetic Survey. Its Products and Service.

U.S. Department of State. 1949. Report of Joint Brazil-United States Technical Commission. *Federal Reserve Bulletin*, April: 361–373.

Valenta, Jiri. 1975. The Soviet-Cuban Intervention in Angola. *Studies in Comparative Communism* 11 (1–2): 3–33.

Vallaeys, Anne. 2004. *Médecins sans frontières: La biographie*. Fayard.

Van de Kerkhof, J. P. 'Defeatism is our worst enemy': Rehabilitation, Reorientation and *Indonesianisasi* at Internatio and HVA, 1945–1958. http://www.indie-indonesie.nl.

Van der Eng, Pierre. 2003. Marshall Aid as a Catalyst in the Decolonization of Indonesia, 1947–1949. In *The Decolonization Reader*, ed. J. Le Sueur. Routledge.

van Elteren, Mel. 2006. *Americanism and Americanization: A Critical History of Domestic and Global Influence*. McFarland.

van Oosterhout, Dianne. 2008. From Colonial to Postcolonial Irrigation Technology: Technological Romanticism and the Revival of Colonial Water Tanks in Java, Indonesia. *Technology and Culture* 49 (3): 701–726.

van Oosterhout, Dianne. n.d. Technopolitical Mapping and Contested Geographies of Nationalism: Constructing and deconstructing Cold War spaces and structure in Indonesia. Manuscript.

van Vleck, Jenifer L. 2007. The "Logic of the Air": Aviation and the Globalism of the "American Century." *New Global Studies* 1 (1). http://www.bepress.com.

van Zandt, J. Parker. 1944. *The Geography of World Air Transport*. Brookings Institution.

Vaughn, Megan. 1991. *Curing Their Ills: Colonial Power and African Illness*. Stanford University Press.

Venter, Al J. 1974. *The Zambesi Salient: Conflict in Southern Africa*. Timmins.

Vidal, Claudine, and Jacques Pinel. 2009. Les 'satellites' de MSF: Une strátegie à l'origine de pratiques médicales différentes. In *Innovations médicales en situations humanitaires*, ed. J.-H. Bradol and C. Vidal. Harmattan.

Vidal, F. S. 1954. Date Culture in the Oasis of Al-Hasa. *Middle East Journal* 8 (4): 417–428.

Vidal, F. S. 1955. *The Oasis of al-Hasa*. Arabian American Oil Company.

Vine, David. 2009. *Island of Shame: The Secret History of the U.S. Military Base on Diego Garcia*. Princeton University Press.

Virilio, Paul. [1975] 1978. *Fahren, fahren, fahren . . . .* Merve.

Vitalis, Robert. 2006. *America's Kingdom: Mythmaking on the Saudi Oil Frontier*. Stanford University Press.

Von Eschen, Penny. 1997. *Race against Empire: Black Americans and Anticolonialism, 1937–1957*. Cornell University Press.

Wade, Robert. 1990. *Governing the Market: Economic theory and the Role of Government in East Asian Industrialization*. Princeton University Press.

Walker, J. Samuel. 1997. *Prompt and Utter Destruction: Truman and the Use of Atomic Bombs against Japan*. University of North Carolina Press.

Walker, Paul. 1991. U.S. Military Power Projection Abroad. In *The Sun Never Sets*, ed. J. Gerson and B. Birchard. South End.

Wall, Patrick. 1975. *Prelude to Détente: An In-depth Report on Southern Africa*. Stacey.

Wallace, Henry A. 1944. America Can Get It. In *Democracy Reborn*. Reynal & Hitchcock.

Wallace, William V. 1976. *Czechoslovakia*. Westview.

Wallace, William V., and Roger A. Clarke. 1986. *Comecon, Trade and the West*. Pinter.

Wang, Jessica. 1999. *American Science in an Age of Anxiety: Scientists, Anti-Communism, and the Cold War*. University of North Carolina Press.

Waters, Ken. 2004. Influencing the Message: The Role of Catholic Missionaries in Media Coverage of the Nigerian Civil War. *Catholic Historical Review* 90 (4): 697–718.

Weigert, Hans W. 1946. U.S. Strategic Bases and Collective Security. *Foreign Affairs* 25 (1): 250–262.

Weigert, Hans W., and Vilhjalmur Stefansson, eds. 1944. *Compass of the World: A Symposium on Political Geography*. Macmillan.

Weiss, Leonard. 2003. Atoms for Peace. *Bulletin of the Atomic Scientists* 59 (6): 34–44.

Well, Melissa. 1985. The Relief Operation in Karamoja: What Was Learned and What Needs Improvement. In *Crisis in Uganda*, ed. C. Dodge and P. Wiebe. Pergamon.

Westad, Odd Arne, ed. 1998. *Brothers in Arms: The Rise and Fall of the Sino-Soviet Alliance, 1945–1963*. Woodrow Wilson Center Press and Stanford University Press.

Westad, Odd Arne. 2005. *The Global Cold War: Third World Interventions and the Making of Our Times*. Cambridge University Press.

Whitman, Edward C. 2005. SOSUS the 'Secret Weapon'. *Undersea Warfare*. http://www.navy.mil/.

Wilczynski, J. 1974. *Technology in Comecon: Acceleration of Technological Progress through Economic Planning and the Market*. Praeger.

Wilheim, Jorge. 1960. *Brasilia 1970. Revisited*. Acropole—Revista Mensal.

Williams, William Appleman. 1959. *Tragedy of American Diplomacy*. Norton.

Williams, William Appleman. 1969. *The Roots of the Modern Empire: A Study of the Growth and Shaping of Social Consciousness in a Marketplace Society*. Random House.

Williams, William Appleman. 1980. *Empire as a Way of Life*. Oxford University Press.

Willkie, Wendell. [1943] 1966. *One World*. University of Illinois Press.

Willner, A. R. 1953. The Foreign Expert in Indonesia: Problems of Adjustability and Contribution. *Economic Development and Cultural Change* 2 (1): 71–80.

Winchester, Simon. 2001. Diego Garcia. *Granta* 23, March.

Winner, Langdon. 1986. Do Artifacts Have Politics? In *The Whale and the Reactor: A Search for Limits in an Age of High Technology*. University of Chicago Press.

Wisnik, Guilherme. 2004. Doomed to Modernity. In *Brazil's Modern Architecture*, ed. E. Andreoli and A. Forty. Phaidon.

Woo-Cumings, Meredith, ed. 1999. *The Developmental State: Odyssey of a Concept in the Developmental State*. Cornell University Press.

Wright, Allan. 1972. *Valley of the Ironwoods: A Personal Record of Ten Years Served as District Commissioner in Rhodesia's Largest Administrative Area, Nuanetsi, in the South-Eastern Lowveld*. T.V. Bulpin.

Wright, Richard. 1956. *The Color Curtain: A Report on the Bandung Conference*. World.

Yaqub, Salim. 2006. *Containing Arab Nationalism: The Eisenhower Doctrine and the Middle East*. University of North Carolina Press.

Young, John W. 1996. *Cold War Europe, 1945–1991: A Political History*. Arnold and St. Martin's Press.

Zewde, Bahru. 2001. *A History of Modern Ethiopia 1855–1991*, second edition. Currey.

Zink, Brian. 2006. *Anyone, Anywhere, Anytime: A History of Emergency Medicine*. Mosby Elsevier.

About the Authors

Itty Abraham is an associate professor of Government and Asian Studies at the University of Texas at Austin, where he holds the Marlene and Morton Meyerson Centennial Chair. He is the editor of *South Asian Cultures of the Bomb: Atomic Publics and the State in India and Pakistan* (2009).

Lars Denicke wrote his Ph.D. thesis, titled Global/Airport: The Geopolitics of Air Transportation, at the Humboldt-Universität zu Berlin. His research focuses on media theory, sovereignty, and visual culture. He works as a curator. Recently he edited *Prepare for Pictopia* (Pictoplasma, 2009), a volume of essays on figuration in contemporary art and design.

Gabrielle Hecht is an associate professor of History at the University of Michigan. Her first book, *The Radiance of France: Nuclear Power and National Identity after World War II* (MIT Press 1998, new edition 2009), received awards from the American Historical Association and the Society for the History of Technology. She is completing a manuscript on uranium from Africa and the power of nuclear things, drawing on research conducted in Gabon, Madagascar, Namibia, and South Africa.

Toby C. Jones is an assistant professor of history at Rutgers University. His research focuses on the history of oil, development, state building, and environmental politics in the Middle East. His book *Desert Kingdom: How Oil and Water Forged Modern Saudi Arabia* was published by Harvard University Press in 2010. He is working on a book for Harvard University Press titled *America's Oil Wars*.

Martha Lampland, an associate professor of Sociology and Science Studies at the University of, California at San Diego, is the author of *The Object of Labor: Commodification in Socialist Hungary* (University of Chicago Press, 1995) and a co-editor of *Altering States: Ethnographies of Transition in Eastern Europe and the Soviet Union* (University of Michigan Press, 2000) and of

Standards and their Stories: How Quantifying, Classifying, and Formalizing Practices Shape Everyday Life (Cornell University Press, 2009).

Clapperton Chakanetsa Mavhunga is an assistant professor in the Program in Science, Technology, and Society at the Massachusetts Institute of Technology. His publications include "The Glass Fortress: Zimbabwe's Cyber-Guerrilla Warfare" (*Journal of International Affairs*, 2009). He is completing two monographs on mobility and human-animal interactions in Africa.

Donna C. Mehos, a senior researcher at the Technische Universiteit Delft, has written on various topics in nineteenth- and twentieth-century science, technology, and culture, and, with Tinecke M. Egyedi, has edited the volume *Inverse Infrastructures* (Edward Elgar, expected 2010). She coordinates a project to synthesize the results of the multidisciplinary academic research program Next Generation Infrastructures for non-academic stakeholders.

Suzanne Moon is an assistant professor in the Department of the History of Science at the University of Oklahoma. Her recent publications include "Justice, Geography, and Steel: Technology and National Identity in Indonesian Industrialization" (*Osiris*, 2009) and *Technology and Ethical Idealism: A History of Development in the Netherlands East Indies* (Leiden University Press, 2007). She is editor-in-chief of *Technology and Culture*.

Ruth Oldenziel is a professor at the Technical University Eindhoven. Her publications include *Cold War Kitchen: Americanization, Technology, and European Users*, edited with Karin Zachmann (MIT Press, 2009), *Gender and Technology: A Reader*, edited with Nina Lerman and Arwen Mohun (Johns Hopkins University Press, 2003), and *Making Technology Masculine* (Amsterdam University Press, 1999). She is working with Mikael Hård on a monograph tentatively titled *Appropriating America, Making Europe*.

Peter Redfield is an associate professor of Anthropology at the University of North Carolina at Chapel Hill. The author of *Space in the Tropics: From Convicts to Rockets in French Guiana* (University of California Press, 2000), he is completing an ethnographic study of the organization Médecins sans Frontières. A related essay ("Doctors, Borders and Life in Crisis"), published in the journal *Cultural Anthropology*, received the 2006 Cultural Horizons Prize from the Society for Cultural Anthropology.

Sonja D. Schmid is an assistant professor in the Department of Science and Technology in Society at Virginia Polytechnic Institute and State University, where she teaches history and sociology of technology. She has held postdoctoral positions at Stanford University and at the James Martin Center for Nonproliferation Studies. She is revising a manuscript on reactor-design choices for the Soviet civilian nuclear industry.

Index

African National Congress, 237
Asian Development Bank, 56
Atomic Bomb Casualty Commission, 4, 5
Atomic Energy Board (South Africa), 82–90
Atomic Energy Commission (India), 113
Atomic Energy Commission (US), 81, 82
Atoms for Peace, 77, 107, 131, 133, 145
Authoritarianism, 209–226
Aviation, 191–202
Azores, 10, 13, 20–24, 190, 199

Baker Island, 18
Bandung Conference, 241
Baruch Plan, 107
Bedouins, 216
Ben Bella, Ahmed, 243
Berlin Airlift, 21
Bermuda, 27
Bhabha, Homi J., 111–116
Bhatnagar, Shanti Swarup, 111–113
Biafra, 271– 273
Bikini Atoll, 22, 23
Bilateral technical assistance, 126
Board for Atomic Energy Research (India), 110–112
Bohunice nuclear reactor, 137, 142
Botswana, 245
Brasilia, 185, 193–197, 202–204
Brauman, Rony, 273

Brazil, 54, 185–205
British atomic arsenal, 76, 77, 91, 93
British Empire, 15, 16
British Indian Ocean Territory, 24, 25
British island possessions, 15, 16
Bulgaria, 142
Bureaucratic reorganization, 167–174

California Arabian Standard Oil Company, 212, 213
Cambodia, 273, 277
Capitalism, 5, 166–179
Cargo planes, 192
Cartography of air travel, 189, 190, 198, 204
Chagos Archipelago, 24
Chamber of Mines (South Africa), 83
Chevchenko, Alexei, 240
Chikerema, James, 239
Chimurenga, 247–250, 257
China, 135, 139, 143, 231– 237, 241–243
Christmas Island, 27
City planning, 186–189
Coast and Geodetic Survey, 26–30
Cold War, 1–12, 177, 178, 209–213, 226, 258–260, 270, 273, 274, 286
  First, 21–24
  Global, 6–9, 231–235
  nostalgia for, 75
  Second, 24–30

Cold War islands, 33–37
Combined Development Trust, 107, 108
Commisariat à l'Ènergie Atomique, 82–86
Communications networks, 17–21, 27, 28, 36, 37
Communism, 5, 201, 202, 213, 235, 239, 240, 246
Communist Party, 165, 176, 177, 239, 240
Communist Youth League (Zimbabwe), 242
Containment policy, 20, 198, 199
Cooke Mission, 192–197, 203
Cooke, Morris L., 192
Corry, Andrew, 108, 109
Costa, Lúcio, 187, 188, 194, 195
Council for Mutual Economic Assistance, 125, 128–130, 133–135, 140–146
Council of Scientific and Industrial Research (India), 110–112
Crisis response, 268–286
Cross-fertilization, 58–61
Cuba, 16, 17, 201, 231, 241, 245
Curie, Marie, 102
Czechoslovakia, 132–143
Czech Republic, 145

Date groves, 214
Date market, 215–217
Defense networks, 19, 20
Democratic Republic of Congo, 282, 284
Department of Defense (US), 26–30
Department of the Interior (US), 23
Deprovincialization, 157, 158
Developmentalism, 5, 6
Diego Garcia, 20, 21, 26–32
Diplomatic history, 1–3
Displaced populations, 5, 21–24, 34, 35
Doctors Without Borders, 268–286
Dulles, John Foster, 201
Dutch East Indies, 46

Earth mapping, 32
Eastern Europe, 125–146
East Germany, 132, 133, 137–142
Ebeye Island, 22–24
Echelon spy network, 28, 36
Ecology, 211–226
Economic and Social Council, 57
Economic institutionalization, 166–174
Economics Research Institute (Hungary), 173
Egypt, 237–243
Eisenhower, Dwight D., 77, 78, 86, 102, 107, 131, 201
Emergency, 281–286
Enewatak Atoll, 22, 23
Espionage training, 240
Ethiopia, 45–56, 274
Euratom, 91, 134
Exclusive Economic Zone, 31, 32, 36
Expanded Program of Technical Assistance for the Economic Development of Less-Developed Countries, 45, 58–67
Expertise, 43–69, 166–177
Export Processing Zone, 31, 32
Extraterritoriality, 20, 21, 31–37

Famine, 274
First World, 5, 6
Fissionable materials, 78
Flexibility, 43–45, 64–67, 283–286
Flight metaphor, 185–187
Food and Agricultural Organization, 56, 57
Foreign Economic Administration (India), 109
France, 75, 76, 82, 86, 91
Free Trade Zone, 31, 32
Frente de Libertação de Moçambique, 242–245, 249
Fuller, Richard Buckminster, 203
Furtado, Celso, 202, 203

Index

Gabon, 82, 85
Geisel, Ernesto, 204
Geneva Peace Conference, 255
Genocide, 272
Geographic intelligence, 15–17
German Democratic Republic, 132, 133, 137–142
Germany, 87, 102, 125, 161, 162, 174, 175
Ghana, 243
Global crisis response, 267–286
Global humanitarianism, 267–286
Globalism, 185, 186
Global North, 231–234, 241, 260
Global Positioning System, 23, 28, 29, 33
Global South, 231–235, 260
Goedhart, Adriaan, 56
Golbery do Couto e Silva, 197, 199, 204, 205
Goldschmidt, Bertrand, 84, 85
Goulart, João, 202
Great Depression, 157, 159
Greenland, 20, 22, 190
Groves, Leslie, 108, 132
Guam, 14, 16, 17, 27, 30, 33
Guano Island Act, 18
Guerilla training, 238–250
Guns, 235–238, 256–258, 267, 268

Haile Selassie I, 49, 50, 52
Haiti, 57
Handelsvereeniging "Amsterdam," 44–56, 67, 68
Hawaiian Islands, 16, 18, 27
HIV/AIDS, 282, 283
Holocaust, 271
Hopkins and Williams Limited, 102, 108, 109
Houphouët-Boigny, 234
Howland Island, 18
Humanitarian infrastructure, 283–286
Humanitarianism, 267–286

Humanitarian kit, 276–281
Humanitarian logistics, 276–286
Humanitarian non-governmental organizations, 284–286
Hungarian-Soviet joint ventures, 163
Hungary, 132, 133, 142, 155–179
Husted, Hugh, 83, 84

Iceland, 20, 190
Idi Amin Dada, 243
India, 5, 6, 101–117
Indonesia, 45–56
Industrial Cooperative Program, 56
Industrial rationalization, 5, 157, 158
Institute for Economic Science (Hungary), 173, 174
Institutionalization, 166–174
Insular Cases, 16, 17
International Atomic Energy Agency, 8, 76–81, 91, 139
International Bank for Reconstruction and Development, 56, 57
International Commission of the Red Cross, 270, 271
International Committee of the Red Cross, 280, 281
International Labor Organization, 57
International Monetary Fund, 57
International Red Cross, 270, 271, 280, 281
Iran, 93, 94
Iron Curtain, 236
Irrigation, 211–226
Island possessions, 7, 13–37
Italconsult, 221, 222

Jarvis Island, 18, 32
Java, 45, 46
Jet age, 200, 201
Jet-nozzle uranium enrichment process, 87, 88
Johnston Atoll, 14, 18, 22, 33

Joint Institute on Nuclear Research, 135
Joint Relief Commission, 270, 271

Karamoja, 275, 276
Kaunda, Kenneth, 232–234
Kennan, George, 20
Kerala, 101–110, 113–117
Keynes, John Maynard, 157, 158
Khmer Rouge, 273
Khruschev, Nikita, 129, 137–139, 143
Kilombero Sugar Company, 53, 54
Király, Béla, 177
Kirimati, 27
Kongwa, 244
Kortleve, A., 49, 50
Kouchner, Bernard, 272–274
Krishnan, K. S., 113
Kubitschek, Juscelino, 186, 194, 195, 200–202
Kwajalein, 21–24

Labor policy, 160, 161, 177
Latour, Bruno, 186
Le Corbusier, 186–188
Lend-Lease agreements, 19, 20
Liberation Committee of the Organization of African Unity, 243
Lindsay Lighting and Chemical Company, 108
Lodge, Henry Cabot, 16
Long-Range Objectives Group, 20
Luena, 245

Machel, Samora, 244
Mackinder, Halford John, 197
Madagascar, 82
Mahan, Alfred Thayer, 15–17
Malenkov, Georgii, 137
Malraux, André, 188
Managerialism, 157–160
Maps, 189, 190, 198, 204
Maralinga, 5
Marine Amphibious Brigade, 30

Marshall Islands, 5, 14, 17, 22, 23
Marshall Plan, 47, 143, 165
Maruroa, 5
Marxism, 242, 246, 247
Marxist-Leninist state building, 166–174
Materia Medica Minimalis, 268–271, 276
Matsudov, Latyp, 239
Mbeki, Thebo, 94
Mbita, Hashim, 244
Mead, Margaret, 63
Médecins du Monde, 274
Médecins Sans Frontières, 268–270, 272–286
Mengistu Haile Mariam, 53
Menon, V. P., 104, 105, 113, 114
Mercury Project, 28, 33, 36
Micronesia, 14
Midway Island, 16, 18, 30, 32
Military deployments, 24–30
Ministry of Agriculture and Water (Saudi Arabia), 216–223
Ministry of Agriculture (Hungary), 168, 177, 178
Ministry of State Farms (Hungary), 178
Minty, Abdul, 93, 94
Mir system, 142
Missile testing, 23, 24
MK, 237
Mobile medical supplies, 268–286
Mobutu Sese Seko, 233
Modernization, 185, 188, 204, 205
Monazite, 102, 108
Monopolies, 159, 160
Morogoro, 244
Mountbatten, Louis, 105
Movement for Democratic Change (Zimbabwe), 257
Mozambique, 244, 245, 273
Mpoko, Pelekezela, 240
Mugabe, Robert, 234, 236, 248, 252–260
Multilateral collaboration, 126

Muslim minorities, 209, 210, 214, 215, 219–226
Muzorewa, Abel, 255

Nachingwea, 244
Namibia, 92
Nanjing Military Academy, 242
Nasser, Gamal Abdel, 243
Nasserism, 210
National Bank, 161, 162
National Democratic Party (Zimbabwe), 235, 237
National Imagery and Mapping Agency, 15
Nationalism, 7, 235, 246–250
Nationalization, 53, 157, 158
NATO, 130, 133
Navajo Nation, 5
Naval-to-air-power transition, 17, 18
Naval warfare, 29
Navassa Island, 14–17, 32
NAVSTAR, 23
Ndlovu, Akim, 240
Negritude, 246
Nehru, Jawaharlal, 111–113, 116
Netherlands, 45, 46, 53–55
Netherlands Development Finance Company, 54
Neto, Agostino, 245
Niemeyer, Oscar, 187, 188
Niger, 82
Nigeria, 243, 271, 272
Nkomo, Joshua, 236, 237, 241, 248, 251, 252, 255–257
Nkrumah, Kwame, 243, 255
Non-Aligned Movement, 94, 241
Non-governmental organizations, 284–286
North Atlantic Treaty Organization, 130, 133
Northern Hemisphere, 198, 199, 204
Northern Marianas, 14, 17, 31
Nuclear Fuel Corporation, 86–89

Nuclear non-proliferation, 90–94
Nuclear Non-Proliferation Treaty, 76, 80, 81, 88, 91, 92
Nuclear states, 7–9
Nuclear technology transfer, 7–9, 125–146
Nuclear testing, 4, 5, 23, 24
Nuclear weapons, 3–5
Nuclear weapons inspectors, 79, 80
Nxele, Albert, 244
Nyerere, Julius, 54, 243, 244, 255

OAO-2, 33
Oasis, 211–226
The Oasis of Al-Hasa (Vidal), 214, 215
Obasanjo, Olusegun, 243
Oil, 209–211, 224
Oil crisis of 1973, 141, 142
Okinawa, 22
One World (Wilkie), 189
Operation Pan-America, 201
Organization for the Islamic Revolution in the Arabian Peninsula, 210, 211
Organization of African Unity, 243
Orwell, George, 232
Owen, David, 58–65

Pacific Islands, 14–18, 21–24, 28
Pakistan, 104, 113
Palau, 14, 17
Palmyra Island, 18
Pan-Africanism, 243–250
Pan American Airways, 190, 191
Patel, Sardar Vallabhai, 104, 105
People's Caretaker Council (Zimbabwe), 236, 237
People's Economic Council (Hungary), 177, 178
Philipp Holzmann A.G., 220
The Philippines, 17
Pinel, Jacques, 277–279
Place-based knowledge, 44, 46
Point Four Program, 5, 58, 66

Political indoctrination, 235, 246–250
Population displacement, 5, 21–24, 34, 35
Portability, 43–69
Portable personalities, 64–67
Puerto Rico, 14, 16, 17, 22, 27, 199

Racism, 213, 214, 239, 242
Radioactivity, 102, 103
Red Cross, 270, 271, 280, 281
Resident Representatives, 63, 64
Reversed development, 185, 186, 193–197, 204, 205
Rhodesia, 233, 234, 246, 248–260
Rhodesian Security Forces, 237
Rogue nuclear states, 75
Romania, 144
Roosevelt, Franklin D., 17–20
Roosevelt, Theodore, 16–18
Rostow, W. W., 185
Roux, A. J. A., 89, 90

Sahara, 5
Saipan, 31
Sanitization, 269
Satellite systems, 26–30, 33, 36
Saudi Arabia, 209–226
Savimbi, Jonas, 233, 234
Schlesinger, Arthur Jr., 13
Schmidt, Gerhard Carl, 102
Schumann, A. W. S., 89, 90
Science and technology studies, 1–3
Seabees, 18
Service d'aide médicale urgente, 272
Seychelles Islands, 27
Shi'is, 209–215, 219–226
Silundika, Tarcissius George, 237
Sino-Soviet bloc, 233
Sino-Soviet split, 143
Sithole, Ndabaningi, 236, 248, 255
Smith, Ian, 246, 251, 253, 255
Socialism, 166–179
Sole, Donald, 79, 82

Sound Surveillance System, 32
South Africa, 75–94
Soviet influence, 174, 175
Soviet Union, 125–146, 163, 234, 236–241, 245
Spanish-American-Cuban War, 15
Special Forces, 24
Stalinism, 155, 156, 165
Stalin, Joseph, 127–130
Standard Oil Company, 212, 213
State planning, 158–166
St. Helena, 27
Stork, 51, 54
Strategic Defense Initiative, 33
Strategic Island Concept, 20, 21, 24
Sudan, 282
Sugar agro-business, 45–56
Sukarno, 45, 46
Sunnis, 214, 215, 221, 222, 225
Supply networks, 17–21

Taiwan, 241
Tanganyika, 53, 54
Tanzania, 243, 244
Tanzania-Zambia Railway, 244
Technical Assistance Board, 58–67
Technical Cooperation Administration, 58
Technological diversity, 60–67
Technological thickness, 33, 34
Tembwe, 244, 245
Territorial leasing, 19–21
Third World, 6, 241
Thorium, 102, 103, 107
Thorium Limited, 108, 109
Three Year Plan (Hungary), 163–165
Tinian Island, 27
Tongogara, Josiah, 242, 243, 249
Transportation, 188
Transport networks, 17–21, 36, 37
Travancore, 101–110, 113–117
Travancore-India Joint Committee on Atomic Energy, 112
Travancore Minerals Company, 102

Traveling consultants, 43–69
Triangulation satellite program, 26–30, 33, 36
Tristan da Cunha, 27
Truman, Harry, 5, 58
Turning Point Strategy, 251, 252

Uganda, 243, 275, 276, 282, 283
Umkonto weSizwe, 237
United Nations, 107
  Conference on the Peaceful Uses of Atomic Energy, 131
  Rehabilitation and Relief Administration, 57
  Resident Representatives, 63
  Scientific and Cultural Organization, 57
  Secretariat, 57
  technical aid programs, 43–45, 56–68
Uranium, 76–94, 102, 103, 132, 133
Urban planning, 186–189
US Agency for International Development, 58–60
US atomic arsenals, 91
US logistical network, 189–197
US Virgin Islands, 14, 16, 31

Vidal, Federico S., 214, 215, 217, 220
Vietnam, 273
Virilio, Paul, 194, 195
Vorster, B. J., 87, 88, 93

Wahhabism, 209
Wake Island, 18, 27
Wakuti A.G., 220, 221, 224
Warsaw Pact, 125–146
Weaponization, 235–239, 244, 245, 248–250, 256, 258, 267, 268
Weigert, Hans W., 199
Westlands Farm, 246
West Nile, 275, 276
Willkie, Wendell, 189
Wonji project, 51

World Bank, 67
World Campaign against Military and Nuclear Collaboration with South Africa, 93
World Health Organization, 57, 280, 281
World War II, 4, 5, 47, 106, 162–167, 186, 188–197, 201–204, 267, 269, 270
Wright, Richard, 236

Youth Pioneers, 242

Zambia, 245, 246
Zimbabwe, 233–260